STATISTICAL METHODS

An Introductory Text

REVISED SECOND EDITION

J Medhi
Professor Emeritus
Gauhati University
Guwahati, India

NEW AGE INTERNATIONAL (P) LIMITED, PUBLISHERS
LONDON • NEW DELHI • NAIROBI
Bangalore • Chennai • Cochin • Guwahati • Hyderabad • Kolkata • Lucknow • Mumbai
Visit us at **www.newagepublishers.com**

Published by New Age International (P) Ltd., Publishers
First Edition: 1992
Second Edition: 2006
Reprint: 2018

GLOBAL OFFICES

- **New Delhi** — **NEW AGE INTERNATIONAL (P) LIMITED, PUBLISHERS**
  7/30 A, Daryaganj, New Delhi-110002, (INDIA)
  Tel.: (011) 23253771, 23253472, **Telefax:** 23267437, 43551305
  E-mail: contactus@newagepublishers.com • Visit us at www.newagepublishers.com
- **London** — **NEW AGE INTERNATIONAL (UK) LTD.**
  27 Old Gloucester Street, London, WC1N 3AX, UK
  E-mail: info@newacademicscience.co.uk • Visit us at www.newacademicscience.co.uk
- **Nairobi** — **NEW AGE GOLDEN (EAST AFRICA) LTD.**
  Ground Floor, Westlands Arcade, Chiromo Road (Next to Naivas Supermarket)
  Westlands, Nairobi, KENYA, **Tel.:** 00-254-713848772, 00-254-725700286
  E-mail: kenya@newagepublishers.com

BRANCHES

- **Bangalore** — 37/10, 8th Cross (Near Hanuman Temple), Azad Nagar, Chamarajpet, Bangalore- 560 018
  Tel.: (080) 26756823, **Telefax:** 26756820, **E-mail: bangalore@newagepublishers.com**
- **Chennai** — 26, Damodaran Street, T. Nagar, Chennai-600 017, **Tel.:** (044) 24353401
  Telefax: 24351463, **E-mail: chennai@newagepublishers.com**
- **Cochin** — CC-39/1016, Carrier Station Road, Ernakulam South, Cochin-682 016
  Tel.: (0484) 2377303, **Telefax:** 4051304, **E-mail: cochin@newagepublishers.com**
- **Guwahati** — Hemsen Complex, Mohd. Shah Road, Paltan Bazar, Near Starline Hotel
  Guwahati-781 008, **Tel.:** (0361) 2513881, **Telefax:** 2543669
  E-mail: guwahati@newagepublishers.com
- **Hyderabad** — 105, 1st Floor, Madhiray Kaveri Tower, 3-2-19, Azam Jahi Road, Near Kumar Theater
  Nimboliadda Kachiguda, Hyderabad-500 027, **Tel.:** (040) 24652456, **Telefax:** 24652457
  E-mail: hyderabad@newagepublishers.com
- **Kolkata** — RDB Chambers (Formerly Lotus Cinema) 106A, 1st Floor, S N Banerjee Road
  Kolkata-700 014, **Tel.:** (033) 22273773, **Telefax:** 22275247
  E-mail: kolkata@newagepublishers.com
- **Lucknow** — 16-A, Jopling Road, Lucknow-226 001, **Tel.:** (0522) 2209578, 4045297, **Telefax:** 2204098
  E-mail: lucknow@newagepublishers.com
- **Mumbai** — 142C, Victor House, Ground Floor, N.M. Joshi Marg, Lower Parel, Mumbai-400 013
  Tel.: (022) 24927869, **Telefax:** 24915415, **E-mail: mumbai@newagepublishers.com**
- **New Delhi** — 22, Golden House, Daryaganj, New Delhi-110 002, **Tel.:** (011) 23262368, 23262370
  Telefax: 43551305, **E-mail: sales@newagepublishers.com**

ISBN: 978-81-224-1957-3
C-18-05-11402

Printed in India at Tanya Printers, Delhi.

NEW AGE INTERNATIONAL (P) LIMITED, PUBLISHERS
7/30 A, Daryaganj, New Delhi - 110002
Visit us at **www.newagepublishers.com**
(CIN: U74899DL1966PTC004618)

To the endearing memory of
my sister NILIMA

na jāyate mriyate vā kadācan

nā 'yam bhūtva bhavitā na bhūyah

ajo nityah sasvato' yaṁ purāno

na hanyate hanyamāne sarīre

The soul is never born nor dies; nor does it exist on coming into being.
For it is unborn, eternal, everlasting and primeval;
even though the body is slain, the soul is not.

Bhagavadgitā, II, 20

Preface to Second Edition

Statistical methodology has been recognised as an important tool for analysis and interpretation of data in natural, biological, agricultural and engineering sciences as well as in social sciences. The subject has great relevance in several disciplines like economics, commerce, psychology, geography, geology, forestry, agriculture, veterinary, biochemistry, pharmacology etc. in addition to mathematics, statistics, engineering, and business management. Thus there is a very large number of students of several disciplines who need to pursue a course of statistics and statistical methodology at the introductory and intermediate or higher level. The book has been prepared keeping in view the diverse audience with varying orientation.

It is true that there are a number of books at this level. Considering the large, varied and growing audience, it is felt that there is scope for more text books. Texts by different authors have appeal for different sections of this large readership.

The book is divided into 3 parts. Part I deals with Descriptive Statistics, and Part II with Probability and Mathematical Statistics. Part III covers some special topics in Applied Statistics; discussion of some matters of general interest, not usually included in other books at this level, is an added feature of Part III. Mathematics used have been mostly of high school level.

The book, designed for degree/PG level courses in India for economics and other social sciences (as well as for MBA, MCA, CA etc. courses), would also be suitable for use for a one/two semester course in Introductory Statistics in UK, US and Canadian Universities. Part II alone could be used as a text for a one-semester beginners' level course in Probability and Statistics. This part would also be useful for research workers requiring knowledge of statistical methodology for their research work.

This book has been written with a view to present a qualitative understanding of the subject rather than present a surfeit of formulas and of manipulative examples/exercises only. It is hoped that this approach and objective would be appreciated by a large section of the teachers and students. The theory has been explained in some detail, supplemented with notes, where considered necessary. Several worked examples have been provided to bring home how the theory works and how the methodology could be applied. A large number of exercises are given, answers to most of which (for all Chapters of Part II and Chapter 12 of Part III) are provided. Emphasis has been on the understanding of the basic theoretical content and methodology.

A number of computer programs are given. The computer programs have been presented in some detail keeping the beginners in view. Students who may have access to some computer would find them very interesting. Research workers and analysts who would have to undertake statistical analysis of data would find the programs useful.

The text has grown from the author's long 40 years' experience of teaching and research, in India and abroad in Canada and USA. He has learnt a lot from the students; this has been useful in presenting the material.

The author's motivation was further heightened by the very complimentary and encouraging review (in the *American Mathematical Monthly,* December 1982) that recommended his earlier text *Stochastic Processes* as *the clear choice* as a text book. This prompted him further to prepare a book in statistics at the introductory level in a similar attractive and readable manner.

I have received encouragement and assistance from a large number of friends and colleagues in this endeavour. The work was undertaken at the request of a friend, Mr. A. Machwe of Wiley Eastern Ltd., My one-time students and colleagues, Profs. S. K. Dutta of Arya Vidyapith College, D. Chakravarty of Handique Girls' College and D. Nath of Gauhati University have rendered enormous help, by going through the manuscript as well as by assisting in the preparation of the solutions of the exercises. S. K. Dutta also read the proofs along with me with meticulous care and offered valuable suggestions which led to improvement of the text at several places. The Computer programs have been prepared with the expertise of and help rendered by S. K. Sinha, of Department of Computer Science, Gauhati University. L. Choudhury, Dept of Statistics, Gauhati University and P.C. Jha, Dept of O.R'., Delhi University have also been of help.

The Times of India have kindly accorded permission to reproduce some figures appearing in the esteemed daily. The Biometrika Trust, London have been kind enough to permit reproduction of the statistical tables. I am indeed thankful to them all.

My elder son Deepankar Medhi, of Computer Telecommunications Program, University of Missouri-Kansas and my elder daughter Shakuntala Choudhury of AT&T Technology Systems, Bedminster, NJ have rendered immense help.

Lastly, it is my wife Prity, who bore the tedium of a 'Professor's wife' with great patience and understanding and whose constant support and encouragement made the writing possible.

I hope teachers and students of different disciplines would consider the book and give it a trial. Suggestions and comments received toward improvement of the book in later editions will be highly appreciated.

U.N. Bezbarua Road.
Silpukhuri West,
Guwahati-781 003, Assam, INDIA

This is a corrected new Edition with updated data.

JYOTIPRASAD MEDHI

Contents

Part III. Special Topics : Applied Statistics

PART I
Descriptive Statistics

"Measure, measure, measure;
measure again and again;
measure the difference,
and the difference of the difference."

Dictum of Galileo Galilei
(1564-1642)

Chapter 1

Introduction

1.1. THE MEANING OF STATISTICS

"The age of chivalry is gone: that of sophisters, economists, and calculators has succeeded", thus said Burke.[1] Calculations and quantifications have pervaded all spheres. This is the age of facts, figures and statistics. One is faced with statistics everywhere: in newspapers and other media, home and in connection with one's at work in office. A citizen has to understand the implication and significance of statistics he comes across. "Statistical thinking will be one day as necessary for efficient citizenship as the ability to read and write" observed Wells[2]. That day seems to have arrived. For effective participation in a democracy and for performing one's task efficiently, one has to understand the underlying meaning of the statistical data that he frequently encountered.

Apart from this, statistics and statistical methodology have increasingly been used in all disciplines of social sciences, pure and applied sciences, as well as in several areas of humanities. It covers all fields of study wherein quantitative data appear and need analysis.

The term *statistics* is used in two senses. As a plural noun it is used for denoting and referring to numerical and quantitive information e.g., labour statistics, vital statistics, and so on. As a singular noun, the term *statistics* is used to denote the science of collecting, analysing and interpreting numerical data relating to an aggregate of individuals. We thus have the following definitions of statistics:

Encyclopaedia Britanica (1969): As is commonly understood now a days, statistics is a mathematical discipline concerned with the study of masses of quantitative data of any kind.

Encyclopaedia Americana (1968): As a name of a field of study, statistics refers to the science and art of obtaining and analyzing quantitative data with a view to make sound inferences in the face of uncertainty.

International Encyclopaedia of Social Sciences (1968): Statistics deals with the inferential process, in particular, with the planning and analysis of experiments and surveys, with the nature of observational errors and sources of variability that obscure underlying patterns with the efficient summarizing of sets of data.

It is to be noted that statistics primarily deals with phenomena in which the occurrences of the event under study cannot be predicted with certainty.

---

1 Edmund Burke (1729-1797) was a great British writer and orator. who is well known for his attack on Pitt and Governor General Warren Hastings for their policies and actions in India.

2 Herbert George Wells (1866-1946) was a celebrated British thinker and writer. author of novels and science fiction.

Numerical data obtained in the face of uncertainty constitute statistical data. Suppose we study the heights of students aged 10 years in a large city. We cannot predict with certainty the height of an individual student: the data of heights of students constitute statistical data. Further we note that there will be variation in heights of students, though they are of the same age. *Uncertainty and variability* are two characteristics of statistical data. Not all quantitive data are statistics. Suppose we prepare a multiplication table by writing the product of 5 and 8 with the integers 1, 2, 3,... in a tabular form. The quantitive data so obtained are not statistical as there is no uncertainty involved in obtaining such data. Further, individual measurements are, not statistical. Statistics is concerned with masses of numerical data. We may summarise as follows:

Statistics is a collection of concepts and methods used for collection, analysis and interpretation of masses of numerical data relating to a certain area of investigation as well as for drawing valid conclusions in situations in the face of uncertainty and variability. Statistics in this sense denotes statistical methodology.

1.2. HISTORY OF STATISTICS

Historically speaking, collection of data began as early as recorded history. It is stated that even the ancient Babylonians collected data on population. Such data were considered important to the state. The word *statistics* was derived from the latin word '*status*' meaning state. Statistics was thought of as mass of data relating to state. The word census is of Roman origin and dates back to the first or second century A.D.: the magistrates in Rome were required to prepare registers of the population by which the state could determine the liability of adult males for military service as well as for imposition of taxes. Apart from census of population and wealth, census of land was also taken. As reported by the Greek historian Herodotus, a census of all lands in Egypt was taken as early as in 1400 B.C. There are reports of collection of data on population, wealth and land also in other countries, such as Greece and China. In Kautilya's *Artha Shastra*, there is reference to collection of such data in ancient India. Gradually the sphere of data collection and census spread in several other directions; and now it encompasses a variety of areas. The word statistics is thus used today in a much broader context.

"Observations on the London bill of mortality" of 1662 marked the start of tabulation and publication of data. Along with progress in collection of data, progress was being achieved in the development of a framework for analysis of data through mathematics. These two aspects developed independently as two parallel streams. The theory of probability was emerging as a potentially rich field of mathematical investigations. The organisations and workers dedicated to collection and tabulation of data created an increasing need for scientific analysis of data. One of the pioneers who recognised the need for analysis of collected data and the capability provided by mathematical framework in such analysis was a Belgian astronomer Quetelet[3], by name. He studied mathematics and probability under the renowned French mathematicians Fourier[4] and

---

3 Adolphe Quetelet (1796-1874), was a Belgian mathematician, astronomer and statistician.

4 Joseph Fourier (1768-1830) was a French mathematician.

Laplace[5] and his work comprised of collection and tabulation of data in the Royal Belgium Observatory. He visualised the possibility of fusing the two streams, the collection of data and the analysis of data through mathematical concepts, into a general body of techniques, now known as statistical methodology or statistics, for short. Incidentally Quetelet was the first to formulate the concept of the '*average man*' (*l 'homme moyen'*).

1.3. THE ROLE, SCOPE AND LIMITATIONS OF STATISTICS

The three main components of statistics are the collection of data, analysis of data and interpretation of facts therefrom; statistical methodology is our guide in all these spheres. At the stage of collection of data, statistics indicates what statistical methodology is to be used for the collection of data, so that logical conclusions can be drawn therefrom later. The branch is known as *sampling* or *experimental design*. After this first stage, statistics is concerned with methodologies to be adopted for summarizing the data and to obtain its salient features from the vast mass of original data. This branch of statistics is termed as *descriptive statistics*. The final stage is the one in which inferences are drawn and logical conclusions arrived at. The branch of statistics containing the relevant statistical methodology for this is termed *statistical inference*.

The scope of statistics is indeed very vast. Apart from helping elicit an intelligent assessment from a body of figures and facts, statistics is an indispensible tool for any scientific enquiry – right from the stage of planning the enquiry to the stage of conclusion. It applies to almost all sciences: pure and applied, physical, natural, biological, medical, agricultural and engineering. It also finds applications in social and management sciences, in commerce, business and industry.

Of the social sciences, economics leans most heavily on statistical methods for analysis of data relating to micro as well as to macro economics, from demand analysis to National Income analysis. The impact of mathematics and statistics has led to the development of new disciplines like Econometrics and Economic Statistics. There is wide scope of statistics in handling and analysing data relating to socio-economic, demographic and political processes.

Statistics is now considered an indispensible tool in analysis of activities relating to business, commerce and industry. For example, statistical methods are useful in determining trends of growth in business, in making proper production plans after consumer and market surveys and in the process of production itself, while statistical concepts and methods of quality control and reliability prove invaluable in industrial engineering.

Statistics figures prominently in agricultural and biological sciences, in determining the precise role of various factors in the growth and development of the organism or plant under study. In medical sciences, the effect of drugs on individuals is one area where statistical methods are widely used. Applications of statistical methods to education and psychology have led to development of a new discipline 'Psychometry'. In fact in almost all investigations wherefrom quantitive data is obtained, whatever the subject, statistical analysis is applicable.

---

5 Pierre-Simon Laplace (1749-1827) was a French mathematician, astronomer and physicist.

While the basic concept and methodology are the same, the emphasis varies from subject to subject. Certain techniques are used more often in one field than another.

There are, however, certain *limitations of statistics*.

For instance, *statistics deals with aggregates of items and not with individual items*. In fact data on an item, considered individually and separately does not constitute statistical data. Statistics studies characteristics relating only to aggregates or groups.

Statistics deals only with quantitative data. If the situation or phenomenon under study yields only qualitative data, which cannot be meaningfully converted to quantitative data, statistical methods cannot be applied to draw valid conclusions. For example, a phenomenon like poverty is qualitative and as such is not amenable to statistical analysis. It is possible to do so only by assigning suitable quantitative measures to such a phenomenon.

Statistical laws hold good only for the averages (or the average individual). It may not be true for a particular individual. When we find addition of a certain nutrient to food to chicken results in increase in weight – it is for the average and may not be true for an individual.

Statistics does not and cannot study a phenomenon in its entirety. It may not be able to reveal and pinpoint anything about the underlying factors responsible for the behaviour of individuals under study.

Last, but not the least, *statistics are liable to be misused and misrepresented*. Inadequate and faulty procedure of data collection as well as inappropriate comparisons may lead to disastrous results. As it is well known "there are three kinds of lies-lies, white lies and statistics".

1.4. SOME BASIC CONCEPTS

1.4.1. Population and Sample

A set or group of observations relating to a phenomenon under statistical investigation is known as statistical population, or simply population. The population comprises of all the *potential* observations that can be conceived with reference to the phenomenon under study. The population is finite or infinite according to whether the set contains a finite or infinite number of observations. Observations on population may relate to any source, animate or inanimate: human beings, families, commodities, business establishments and so on. For example, the phenomenon under enquiry may be the height of all children between the ages 10 and 11 in a certain district. Here the source, children, are human beings, the observations or measurements of heights constitute the population. If the phenomenon of enquiry is the expenditure incurred on food items only by families in a state, then the observations (which are the amounts spent in food items) will constitute the population, the sources being the families. Here the observations for all families may not be actually available – we refer to potential observations; implying that it should be possible to get an observation for each family in the state.

Now the process of collection of observations from all the elements of the source is, in general, expensive, time consuming and difficult, if not impossible. Some items from the source are selected through a valid statistical procedure, and observations are taken from these items. This set of observations from the selected items is a subset of the set of observations for source. The set of data actually collected through a process of observation or experimentation from selected items of the source is called a *sample.*

A sample is taken in order to gather information about the true state and characteristics of the population, A sample has thus to be truly representative of the population.

1.4.2. Parameter and Statistic

A parameter is a descriptive measure of some characteristic of the population. It is often possible to have a good idea of the population from a few descriptive measures. A measure which is calculated from all the observations on the population is a parameter of the population. For example, if an average is computed from the set of all observations on the population, the average will be a *parameter* of the population: the average is *population average.*

A corresponding descriptive measure can be obtained from the observations contained in a sample of the population. A descriptive measure computed from the observations in the sample is called a *statistic.* An average computed from observation in a sample is a *statistic*; it is called the *sample average.* Statistic is used here in the singular and its plural is statistics. When more than one descriptive measure are obtained from a sample, we get as many *statistics.*

It is the usual convention to denote the parameters (or population parameters) by Greek letters and the statistics by roman letters. For example, the population mean is denoted by μ whereas the sample mean is denoted by $\overline{x}$; the population variance is denoted by σ^2, the sample variance by s^2.

It may be noted that while a parameter is constant for a population, the corresponding statistic may vary from sample to sample. For example, if an investigator considers sample of heights of 50 students and another investigator considers a sample of 50 (or 100) students the sample mean will, in general, be different.

It is said : 'The quiet statisticians have changed our world–not by discovering new facts or technical developments but by changing the ways, we reason, experiment and form our opinions about it'.

EXERCISES-I

1. Explain the meaning and scope of statistics.
2. What are the basic elements or statistical data? Do the numerical figures representing the dates on a calendar constitute statistical data? If not, why not?
3. Explain what is meant by statistical population. Suppose that the age on 1st January of each member of the population of a state is available. What will constitute statistical population?
4. Explain the difference between a parameter and a statistic,
5. What are the limitations of statistics? Enumerate them.

Chapter 2

Collection of Data

2.1. INTRODUCTION

We have already explained what is meant by statistical data. The numerical facts or measurements obtained in the course of an enquiry into a phenomenon, marked by uncertainty, constitute *statistical data*. The statistical data may be already available or may have to be collected by an investigator or an agency. Data is termed *primary* when the reference is to data collected for the first time by the investigator (or on his behalf) and is termed *secondary* when the data are taken from records or data already available. The Meteorological Department regularly collects data on different aspects of the weather and climate such as amount of rainfall, humidity, maximum and minimum temperature of a certain place. These constitute *primary data*. To someone using them for a certain investigation the data will be *secondary*.

There are two principal methods of data collection – through a census or through a sample survey. *Census* implies complete enumeration of each and every element of the source. Data obtained by taking relevant measurement or observation of each and every element of the source constitute *census data*. When only some selected elements of the source (selected according to some valid procedure) are taken and measurements or observations of these selected elements are recorded – the data is said to be collected through a sample enquiry and is said to be *sample data*.

The advantages of sample method over the census method of enquiry are the following:

(1) *Reduced cost*: The sample method is more economical.

(2) *Greater scope*: Complete enumeration is sometimes neither desirable nor feasible. In such cases only the sample method is to be adopted.

Moreover it is possible to collect more information in a sample enquiry than in a complete count.

(3) *Greater speed*: Data can be collected more quickly and summarised with a sample than with a complete count or census.

(4) *Greater accuracy*: It is possible to engage better trained personnel for collection of data in the case of a sample enquiry than in a complete count. Processing of data is also much easier with sample data. All these factors lead to greater accuracy in data collected.

There are various techniques of sample enquiry suitable to particular situations. The methodologies and techniques constitute a specific area of statistics known as Sample Survey.

2.2. METHODS OF COLLECTION OF PRIMARY DATA

Primary data is data collected for the first time through census or sample. There are several ways of collecting such data. These are:

(1) direct personal interview or observation,
(2) indirect personal interview or observation,
(3) mailed questionnaire, and
(4) schedules through enumerators.

The usual procedure of collecting and recording information is through a questionnaire or a schedule. The relevant matters of the phenomenon under enquiry are put in the form of questions in a questionnaire or schedule. A distinction is made between a questionnaire and a schedule in that a questionnaire contains a set of questions, the answers to which are recorded by the informant himself, whereas in a schedule answers are recorded by the investigator or an enumerator on his behalf.

In the direct personal method, the investigator collects the information directly from the sources concerned.

The merits of such a method are:

(1) Information so collected is more accurate, reliable and useful. The investigator can check and countercheck the information and get in the form in which he desires.
(2) The investigator can put alternative questions suited to the educational and cultural level of the persons concerned.
(3) In such cases, information can be collected by eliminating the bias and prejudices of the persons concerned.

The demerits are:

(1) Such a method can be adopted only when the enquiry is intensive and localised to a locality or a group. This cannot be used when the enquiry is extensive or is to be done in large/areas.
(2) Such an enquiry is subjective in the sense that the intelligence, tact, skill as well as personal bias of the investigator are all reflected in the process.

The indirect personal investigation is through some agencies who have intimate knowledge of the phenomenon under enquiry. For example, an investigator may collect information about cost of cultivation through perosnal observation directly from the cultivators or indirectly from village headmen who have intimate knowledge of such matters.

The merits of this method are:

(1) It is less time consuming and expensive.
(2) As information can be collected from more knowledgeable persons, these are expected to be more useful and reliable.
(3) As fewer persons need be contacted, the enquiry could be more extensive than in case of direct personal enquiry.

The demerits are:

(1) The information collected is subjective and is subject to the personal bias of the persons from whom it is collected.
(2) One has to be very careful about the selection of such persons – not only their knowledge but their personal attributes affect the quality of data. Great caution is called for in dealing with such a situation.

In the mailed questionnaire method, a questionnaire in the form of a set of questions is sent by mail to the persons from whom information is to be collected. They, in their turn, are expected to answer the questions and also to supply additional information and comments, where, necessary and mail them back to the investigator.

The merits of such a method are:

(1) It is much less time consuming and is economical.
(2) A much larger coverage can be made as, people in distant places can be reached without much difficulty.
(3) It is advantageous in a situation where the persons concerned move to far away places. For example, in an enquiry relating to old students of a college, such a method may be useful as students move out and away after leaving the institution.

The demerits are:

(1) The method can be adopted only in case of enlightened and educated people.
(2) As persons are approached directly, the proportion of *non-response* is usually much larger. People do not have the time to spare nor are they willing to take the trouble of writing the answers themselves and of returning the questionnaire. Sometimes people also do not like to record information in their own handwriting and very often avoid answering delicate questions.

In censuses and large scale surveys, enumerators are engaged to collect information from the persons concerned. They gather information in schedules specially prepared for the purpose in the form of answers given by the respondents to specific questions. In the case of a census, enumerators visit every member of the source in the zones or areas specifically allotted to them and in the case of a sample survey, they visit those members who come under their sample procedure. This method is applied in censuses and in most other extensive enquiries designed to cover larger areas or population.

The merits of this method are:

(1) This method is the only one possible in case of extensive censuses as well as sample enquiries.
(2) There is a much lesser degree of subjectivity on the part of interviewer in this method.
(3) This method is useful where the scope and coverage are large enough.

The demerits are:

(1) It is expensive and time consuming
(2) Thorough-training of the enumerators is needed before they are sent to the field. It also needs an organisation to handle the whole process of appointment, training and supervision of enumeration work.

2.3. FRAMING OF QUESTIONNAIRE OR SCHEDULE

Great care is to be taken in drafting a questionnaire or schedule, as this is the medium through which information is collected. Further it is also to be seen that the information collected is usable.

Apart from care, expertise such as skill, wisdom, experience of the phenomenon under enquiry are needed in drafting a questionnaire or schedule. Though there are no hard and fast rules for designing a questionnaire, there are a few general points which should be borne in mind.

These are:

(1) The questions put should be clear, concise and unambiguous.
(2) Delicate questions are to be put with great care – often indirect questions should be put to get answers to some pertinent point. It is sometimes desirable to avoid very delicate questions.
(3) The size of the questionnaire/schedule should be small. It saves time, both for the enumerator and the respondent. A large questionnaire is likely to exhaust the patience of the respondent.
(4) There should be a natural, logical order in which questions are put.
(5) It should be noted that the information collected through questions should be such that it is usable.

2.4. 1981 POPULATION CENSUS SCHEDULES

The main schedules used in the 1981 Population Census of India were as follows:

(1) *Houselist*: This was used for identifying houses and recording house members. This is done six to nine months before the census. Through this list, data about physically handicapped persons residing in a household were also collected.
(2) *Household schedule*: It was used for recording particulars relating to members of the household such as religion, language mainly spoken in the household as well as for recording details of the construction of the house occupied by the household.
(3) *Individual slip*: This slip contained questions in three parts; demographic, social and economic/particulars of each and every individual.
(4) *Degree holders and technical personnel cards*: These cards which were canvassed on a sample basis covered details regarding academic

and professional qualifications and training, etc., of the persons covered.

(5) *Enterprise list*: Such a list was canvassed along with the houselist, listing the enterprises for obtaining information regarding the enterprise as well as of the number of persons employed therein.

Some of the salient features of 1981 census are noted below:

(1) Information relating to physically handicapped persons.

(2) Information on fertility aspects not only on currently married but on all ever married women. Age at marriage, number of children surviving by age, number of children even-born by sex, etc., were ascertained.

(3) Information regarding attendance at school/college.

(4) Information on migration, including reasons for migration from place of last residence.

(5) A profile of the working characteristics of the population including information relating to main workers, marginal workers and non-workers.

For Census of India 1991, 2001, see Section 13.7.

EXERCISES-2

1. Explain the census and sample methods of data collection. Enumerate the advantages of sample method over complete enumeration.
2. Distinguish between primary data and secondary data. Mention some sources of secondary data.
3. Give an account of the principal methods of collection of primary data. Discuss their merits and demerits.
4. Explain the difference between a questionnaire and a schedule. Enumerate the requirements of a good questionnaire.
5. You are required to conduct a survey of educated unemployment in your district. Outline a draft questionnaire to be used for this purpose.
6. A survey is to be conducted of the household (cottage) industries in your district Draft a questionnaire to be canvassed for the survey.

Chapter 3

Presentation and Classification of Data

3.1. INTRODUCTION

The methods of collection of statistical data were described in the last chapter. The data so collected is known as raw data. The raw data which is, in general, huge and unwieldy, needs be organised and presented in meaningful and readily comprehensible form in order to facilitate further statistical analysis.

There are three broad ways of presenting data. These are:

(1) Textual presentation;

(2) Tabular presentation;

(3) Graphic or diagrammatic presentation.

In textual presentation, data is presented along with the text. For comprehension, one has to read through and scan the whole text in order to grasp its meaning and implications. This is a serious disadvantage, though certain points can be emphasised in such a presentation. It is often necessary and advantageous to represent such data in either of the two other forms of representation.

In a tabular representation, data are arranged in a systematic way in rows and columns. Huge and unwieldy raw data can be neatly condensed in a table, by classifying data according to suitable groups or classes.

Graphic and diagrammatic representations are useful ways or devices of presenting data for quick and ready comprehension.

An example of textual representation is given below: The enumerated population of India as revealed by the censuses were 251 321 213 in 1921, 278 977238 in 1931, 318 660 580 in 1941, 361 088 090 in 1951, 439 234 471 in 1961, 548 159 652 in 1971, 685 184 692 in 1981, 846 302688 in 1991 and 1028 610 328 in 2001.

During the period 1921-31, the population grew steadily each decade by 11.00 percent or at an average annual rate of 1.05 percent per annum in the decade (1921-31); by 14.22 percent or at an average annual rate of 1.34 percent during 1931-41; by 13.13 percent or at an average annual rate of 1.26 per cent per annum during 1941-51. There was however substantial increase during 1951-81. The increases were by 21.5 per cent or at the rate of 1.98 per cent per annum in the decade 1951-61; by 24.8 per cent or at an average annual rate of 2.24 per cent during 1961-71; and by 25.0 per cent or an yearly average of 2.28 per cent during 1971-81.

The above data can be represented in a tabular form as follows :

Table 3.1 Population of India 1921-2001: percentage decadal increase and Av. Annual growth rates (Simple and Exponential)

Census Year	Population	Decadal growth rate	Average Annual Growth Rate	
			(Simple)	(Exponential)
1921	25,13,21,213			
1931	27,89,77,238	11.00	1.05	1.04
1941	31,86,60,580	14.22	1.34	1.33
1951	36,10,88,090	13.31	1.26	1.25
1961	43,92,34,771	21.51	1.98	1.96
1971	54.81,59.652	24.80	2.24	2.22
1981	68.33.29,097	24.66	2.23	2.20
1991	84,63,02,688	23.85	2.16	2.14
2001	102,86,10,328	21.50	1.96	1.93

Note: Av. Ann. Growth Rate (Simple): $[\{P(1931) \div P(1921)\}^{1/10} - 1] \times 100$ (Exponential): $[(ln\ \{P(1931) \div P(1921)\}/10] \times 100$ and so on.

The data is presented in a table with rows and columns. The use of rows and columns facilitates comparison. The table is much briefer and is more concise.

There should be a title at the head of the table, and the title should indicate the contents of the table. While presenting data in a table, data having some similarity and resemblance should be arranged into groups or classes. This process is known as classification and is the first step in tabulation. The purposes of *classification* are the following:

(1) It facilitates meaningful comparisons.

(2) It helps in condensing the data.

(3) It aids in studying the relationships.

The bases of classification are:

(1) *Geographical:* arranging data according to geographical region.

(2) *Chronological:* arranging data according to the order of time.

(3) *Quantitative:* arranging data according to its numerical magnitude.

For example, in Table 3.1, the basis of classification preceding the tabulation is chronological-data re-arranged in the order of occurrence in points of lime.

Besides helping in condensing information, the table (Table 3.1) facilitated comparisons by bringing out the increasing trend of growth.

One method of presenting quantitative data is to arrange in increasing or decreasing order of magnitude. Such a presentation is known as an *array.* Though this method of presentation is better than presentation as raw data, it does not help in condensing the data. Further it is cumbersome to construct an array. This method, therefore, is not a useful one nor does it have any particular advantage.

3.2. DISCRETE AND CONTINUOUS VARIABLES

Consider the marks secured by students in a class and suppose that marks are given in terms of distinct integral values such as 0, 1, 2, . . . ,100. Here each value differs from the nearest one by a *distinct* and *finite* value 1. Suppose we consider the weekly wages of employees in a particular establishment. Here each value of wage differs from the nearest value by a distinct value Re .01 or 1 paise. We may say the difference of wages between two employees is at least one paise – a finite amount. These are examples of *discrete variables*. Consider now the heights of students of a class. Here the height of one student may vary from that of another of nearest height by an *infinitely* small amount. The values, that is, the heights of students, are capable of infinitely small variations. The heights between two students of nearly equal heights could be a very very small number and it is not possible to specify how small this might be. Here we get example of a *continuous variable*. Weights, ages of persons are also examples of continuous random variables. In the case of continuous variables, we specify a small interval around a value. For example, when we say that a student is 152 cm. in height we shall mean that his height is between say 151.5 cm to 152.5 cm. This is how we shall specify values of a continuous random variable.

In general, when *measurements* are taken, we come across data of a continuous variable while when *counting* is done we get data of a discrete variable.

3.3. FREQUENCY DISTRIBUTION

The most important method of organising and summarising statistical data is by constructing a frequency distribution table. In this method, classification is done according to quantitative magnitude. The items are classified into groups or classes according to their increasing order in terms of magnitude and the number of items falling into each group is determined and indicated.

We shall discuss later questions such as how the classes are to be formed and how many classes are to be taken. We consider now how a frequency distribution table is to be constructed in the case of a discrete variable by taking a particular example.

Example 3 (a). Suppose that the marks secured by 60 students of a class are as follows:

46, 67, 23, 5, 12,	53, 38, 58, 26, 43,
36, 63, 26, 48, 76,	45, 66, 74, 16, 86,
56, 31, 58, 90, 32,	43, 36, 66, 46, 58,
36, 59, 54, 48, 21,	36, 64, 58, 45, 76,
58, 84, 68, 65, 59,	74, 48, 64, 58, 50,
46, 53, 64, 57, 65,	58, 95, 56, 66, 44.

Construct a frequency distribution table.

Marks obtained are divided into 10 groups or intervals as follows: Marks below 10, between 11 and 20, between 21 and 30, and so on, between 91 and 100. Represent each mark by a tally (/), for example, corresponding to the mark 46 we put a tally (/) in the group 41 to 50; similarly we continue putting tallies for each mark. We continue upto four tallies and the fifth tally is put crosswise (\ ) so that it becomes clear at once that the lot contains five tallies, i.e. there are five marks. A gap is left after a lot of five tallies, before starting again to mark the tallies after each lot. The number of tallies in a class or group indicates the number of marks falling under that group. This number is known as the frequency of that group or corresponding to that class interval. Proceeding in this way, we get the following frequency table.

Table 3.2. Frequency distribution of marks secured by 60 students

Class interval	Tally	Frequency (No. of students securing marks which fall in the class interval)
0 to 10	/	1
11 to 20	//	2
21 to 30	////	4
31 to 40	~~////~~ //	7
41 to 50	~~////~~ ~~////~~ //	12
51 to 60	~~////~~ ~~////~~ ~~////~~	15
61 to 70	~~////~~ ~~////~~ /	11
71 to 80	////	4
81 to 90	///	3
91 and above	/	1
Total		60

We shall now consider construction of a frequency distribution table of a continuous variable.

Example 3(b). The heights of 50 students to the nearest centimetre are as given below:

151, 147, 145, 153, 156, 152, 159, 153, 157, 152,
144, 151, 157, 147, 150, 157, 153, 151, 149, 147,
151, 147, 155, 156, 151, 158, 149, 147, 153, 152,
149, 151, 153, 150, 152, 154, 150, 152, 149, 151,
151, 154, 155, 152, 154 152, 156, 155, 154, 150.

Construct a frequency distribution table.

We form the classes as follows: 145-146, 147-148, 149-150, 151- 152, 153-154, 155-156, 157-158, 159-160 and construct the following frequency table:

Table 3.3. Frequency distribution of heights of 50 students

Class interval (Height in cm)	Tally	Frequency (Number of students having height)
145-146	//	2
147-148	~~////~~	5
149-150	~~////~~ ///	8
151-152	~~////~~ ~~////~~ ~~////~~	15
153-154	~~////~~ ////	9
155-156	~~////~~ /	6
157-158	////	4
159-160	/	1
Total		50

We have given heights in cms in whole numbers or heights have been recorded to the nearest centimetre. Thus a height of 144.50 or more but less than 145.5 is recorded as 145; a height of 145.5 or more but less than 146.5 is recorded as 146 and so on. So the class 145-146 could also be indicated by 144.5-146.5 implying the class which includes any height greater than or equal to 144.5 but less than 146.5; the class 147-148 could be indicated by 146.5-148.5, meaning the class which includes any height greater than or equal to 146.5 but less than 148.5. Following this convention, the classes could be represented as: 144.5-146.5, 146.5-148.5, and so on. The above frequency distribution should finally be represented as follows.

Table 3.3a. Frequency distribution of heights of 50 students

Height (in cm)	Frequency (Number of students)
144.5-146.5	2
146.5-148.5	5
148.5-150.5	8
150.5-152.5	15
152.5-154.5	9
154.5-156.5	6
156.5-158.5	4
158.5-160.5	1
Total	50

Class intervals, Class limits and Class boundaries

The interval defining a class is known as a *class interval*. For Table 3.3, 145-146, 147-148, . . . are class intervals. The end numbers 145 and 146 of the class interval 145-146 are known as *class limits*; the smaller number 145 is the *lower class limit* and the larger number 146 is the *upper class limit*.

When we refer to the heights being recorded to the nearest centimetre and consider a height between 144.5 and 146.5 (greater or equal to 144.5 but less than 146.5) as falls in that class and the class is represented as 144.5-146.5, the end numbers are called *class boundaries*, the smaller number 144.5 is known as *lower class boundary* and the larger number 146.5 as *upper class boundary*. The difference between the upper and lower class boundaries is known as the *width* of the class. Here the width is 146.5 − 144.5 = 2 cm and is the same for all the classes. The common width is denoted by c: here c = 2 cm. Note that in certain cases, it may not be possible to have the same width for all the classes (specially the end classes).

Note also that the upper class boundary of a class *coincides* with the lower class boundary of the next class; there is no ambiguity: we have clearly indicated that an observation less than 146.5 will fall in the class 144.5-146.5 and an observation equal to 146.5 will fall in the class 146.5-148.5.

3.4. SELECTION OF THE NUMBER OF CLASSES OR GROUPS AND THE CLASS LIMITS

In Table 3.2 we have taken the class intervals as 0 to 10, 11 to 20 and so on. The class limits for the first class interval are 0 and 10; those for the second class are 11 and 20 and so on. There are 10 classes for 60 observations. In Table 3.3 we have chosen the class intervals as 144.5 or more, but less than 146.5 and so on. The class limits for the first class are 144.5 and 146.4999 (less than 146.5). There are 8 classes for 50 observations. If we choose the class intervals in a somewhat different manner we will find a somewhat different frequency table. For example, if in the case of Example 3(a), we choose the class intervals as 0-20, 21-40, 41-60, 61-80, 81-100 we shall have only five classes and the frequency table will be different. Similarly in case of Example 3(b), if we choose the class intervals as 144.5-147.5, 147.5-150.5, 150.5-153.5, 153.5-156.5, 156.5-159.5, 159.5-162.5 we get six class intervals and we will have a corresponding frequency table.

The question naturally arises as how to select the number of classes to be taken and also how to select the class limits or class intervals.

There is no hard-and-fast rule for the selection of the number of classes or the class intervals in a frequency table. It may be borne in mind that if there are too many classes, many of them will have few frequencies; the distribution may then show certain irregularities which are not really attributable to the behaviour of the variable. On the other hand, if there are too few classes, most of them will contain large frequencies and as a result a lot of information may be lost. The number of classes, so also the class intervals depend on the total number of observations or total frequencies as well as on the nature of the data. If the nature of data is such that the distribution of the frequencies shows marked regularity, then more classes may be taken. Another criterion in the selection is the accuracy of computation desired.

In general, it can be said that the number of classes should not generally be less than six or eight and be greater than 15. As soon as the number of classes is decided upon, the class intervals can be determined from the range of values, that is, difference between the greatest and the smallest observation. For example, if the range is 80 and the number of classes is eight, then the class interval will be 80/8 or 10.

The class *limits* are the lowest and highest values in that class. In selecting these, the point to be borne in mind is that each class is represented by its mid-value and the mid-values of classes are required for computing various descriptive measures of the data. As such the class limits are to be so chosen that the mid-value which is the average between the lower and the upper class limits can be properly determined. In case of continuous variables care should be taken to select the class limits so that there is no ambiguity in assigning a value to a class. In Example 3(b), we have taken class limits as 144.5 and 146.5. The mid value is nearly equal to $\frac{1}{2}(144.5 + 146.5) = 145.5$; it can be properly determined. The class limits are so defined that there is no ambiguity in assigning a value to any particular class.

Again suppose that the data consists of ages of persons to the nearest year. Then the reported age of 25 year would imply an age between 24.5 to 25.5 years, and the average age of such persons will be 25 years. On the other hand if age is reported to nearest completed years, then the reported age of 25 years would imply an age between 25 to 26 years with an average age of 25.5 years.

When dealing with continuous variables, as in the case of Example 3(b), we find that the classes appear to overlap, in that the classes are denoted by 144.5-146.5, 146.5-148.5, . . .; but this should not create any difficulty since we mean that the lower class limits imply any value equal to that value and the upper class limits imply any value just less than that value and any value coming in between falls in that class. We could however make another convention when the class limits overlap: for a value which falls on two overlapping class limits, a frequency of 1/2 may be assigned to each class. This will result in some fractional frequencies in the distribution. This should be avoided, if possible. So it is better to have a clear cut convention as to what the overlapping class limits imply, as we have done in Example 3(b).

3.4.1. General Rules for Construction of Frequency Distribution

First, find the smallest and largest observations in the raw data supplied and find the range i.e., the difference between the largest and the smallest observations.

Secondly, divide the range into a convenient number of class intervals having equal sizes. Sometimes it may be necessary to consider a slightly higher value than the exact range, so as to get a convenient number of class intervals of equal size. The number of class intervals taken should not be less than six or eight and greater than 15. The number may be chosen on the basis of the number of observations and the order of accuracy desired. In choosing class intervals, care should be taken that the *mid point* of the class intervals can be properly calculated. Another point to be borne in mind is that the midpoints coincide with actually observed data. However, whenever class boundaries are considered, it should be seen that no observation falls on the class boundaries. Sometimes the data is such that it is not possible to have all class intervals of equal size; in such case class intervals of unequal size, specially the class intervals at each end, may be conveniently taken.

Thirdly, find the number of observations falling in each class interval (or between corresponding class boundaries). This is best done by using tally marks.

Give tally marks (/) upto four marks (////) and then use a cross mark on them to denote the fifth(~~////~~). Tallies are marked in lots of five or less whenever there is less in the last lot.

Example 4(a). The following observations give the yield of paddy in kg from 50 experimental plots in a research station:

4.4, 3.4, 4.5, 4.8, 5.1,	5.5, 4,6, 4.7, 3.6, 3.5,
4.8, 4.2, 3.4, 5.0, 4.3,	3.6, 3.5, 4.7, 5.3, 5.4,
4.6, 4.0, 5.3, 3.6, 4.3,	5.0, 3.0, 5.8, 4.8, 4.5,
3.6, 5.0, 4.0, 3.7, 4.2,	3.4, 5.3, 5.6, 4.2, 5.8,
4.6, 6.0, 6.2, 6.7, 5.0,	6.2, 6.0, 4.8, 5.6, 6.6.

Form a frequency distribution table.

Here the smallest observation is 3.0 and the largest is 6.7, so that the range is 6.7 – 3.0 = 3.7. Since there are 50 observations, we make classes each of size 5 as follows: 3.0-3.4, 3.5-3.9 and so on. Taking class size or width as .5 we can make 8 classes as 3.0-3.4; 3.5-3.9 and so on. The class limits for the class 3.0-3.4 are 3.0 and 3.4 and the class boundaries are 2.95 and 3.45 and so on. The width of a class c = 3.45 – 2.95 = 0.5.

Any observation between 3.0 and 3.4 will fall in the first class and weights are given to the first decimal point. An observation between 3.45 and 3.55 is recorded as 3.5 i.e. an observation greater than or equal to 3.45 but less than 3.55 is recorded as 3.5. We now get the following frequency table:

Table 3.4. Distribution of yield of paddy in 50 experimental plots

Class interval	*Tally*	*Frequency*
3.0-3.4	////	4
3.5-3.9	~~////~~ //	7
4.0-4.4	~~////~~ ///	8
4.5-4.9	~~////~~ ~~////~~ /	11
5.0-5.4	~~////~~ ////	9
5.5-5.9	~~////~~	5
6.0-6.4	////	4
6.5-6.9	//	2
Total		50

3.5. GRAPHICAL REPRESENTATION

Frequency distributions are easier to visualize and comprehend when they are represented graphically or diagrammatically. There are two ways of graphical representation of frequency distributions: these are (1) histograms, and (2) frequency polygons.

Histogram: A histogram of a frequency distribution is drawn as follows:

(a) The class boundaries are marked on the x-axis starting and finishing at convenient points on the axis; the class intervals are thus marked on the x-axis and are taken as bases.

(b) On each base, a rectangle is drawn whose height is equal to the frequency of that class. If the class intervals are of equal size of width, the areas of the rectangles are proportional to the corresponding class frequencies. Here the vertical axis (or y-axis, as is commonly known) is the frequency axis.

(c) Instead of class boundaries class limits may be used if the frequency distribution is given or constructed in terms of class limit. But it is better to use class boundaries, specially in case of continuous variables. We draw below the histogram corresponding to the frequency distribution given by Table 3.3a.

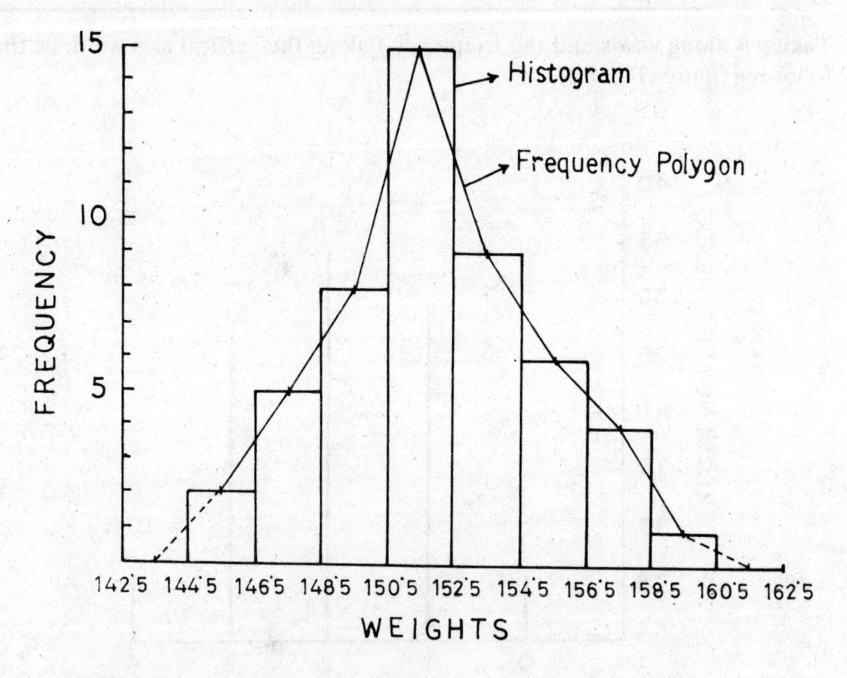

Fig. 3.1 Histogram of Frequency distribution of weight of 50 students.

Remark: Histograms may be used also for discrete variables. However, the histogram in case of such a variable would have a gap between each rectangle. For example, when a histogram is drawn for frequency distribution of Example 3(a), the bases of the rectangles will be 0-10, 11-20, 21-30, . . . and so on. There is a gap of 1 unit between each base of the rectangles. In case of continuous variable, it is desirable to take the class boundaries as bases of the rectangles.

Histogram of discrete variable

Consider a discrete variable which assumes integral values only and each value differs from its nearest value by exactly 1.

Example 4(b). Four similar coins were tossed 100 times. The number of heads x in each of the 100 tosses were noted. It was found that out of 100 tosses (each with four coins), not a single head (and four tails) appeared in six tosses, only one head (and three tails) appeared in 28 tosses, two heads (and two tails) appeared in 36 tosses, three heads (and one tail) appeared in 25 tosses, and all heads (no tail) appeared in five tosses.

To represent the data in a tabular form and to represent the same graphically.

The data can be represented in tabular form as follows:

Table 3.5. Frequency distribution of number of heads in 100 tosses with four coins

No. of heads (n)	0	1	2	3	4	Total
Frequency of heads (f)	6	28	36	25	5	100

Taking n along x-axis and the frequency f along the vertical axis we draw the following figure (Fig. 3.2).

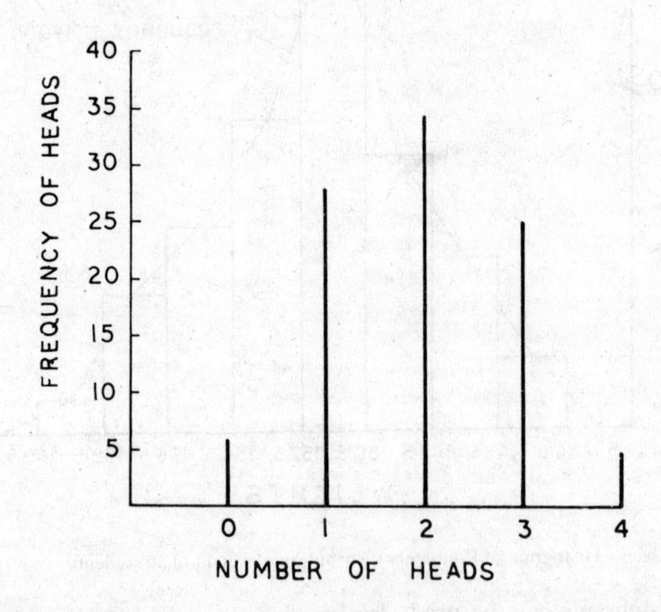

Fig. 3.2. Rod graph of distribution of heads in 100 tosses each with 4 coins.

This is a *natural* way of representing the data by a graph. This is a type of histogram where each rectangle has zero width. It is also called a *rod* graph. However, the data can also be represented by a histogram as indicated in Fig. 3.3.

Draw rectangles of width one having the midpoints of their bases at the values x = 0, 1, 2, 3, 4, and with height of the rectangles equal to the corresponding frequencies. The histogram is as shown in Fig. 3.3.

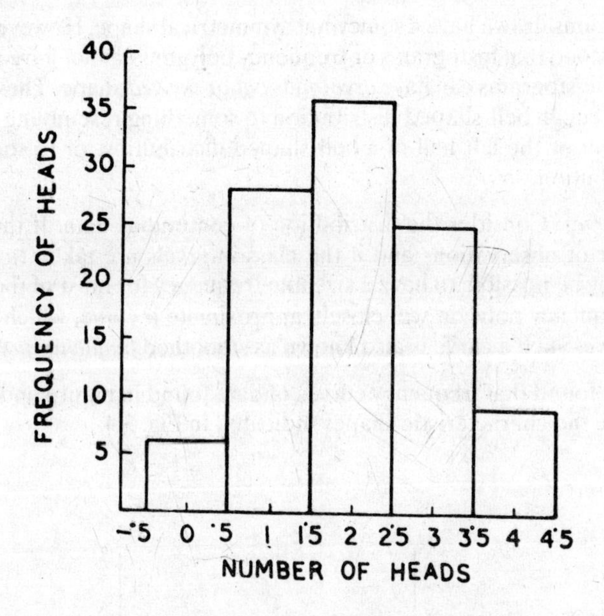

Fig. 3.3. Histogram of distribution of number of heads in 100 tosses each with four coins.

The histogram of Fig. 3.3 is typical of many frequency distributions of data obtained from various sources. For example, frequency distributions of data found in nature and industry have that shape specially when the data are large enough. The corresponding frequency polygons have a rough bell-shaped curve gradually rising of the middle and then tappering off towards the end in a symmetrical manner, with maximum occurring around the mid point. As an example of data found in industry, we may consider the lengths (in cms) of a large sample of wallpins manufactured by a machine.

Frequency Polygon: To draw this, at first, points with ordinates (x-coordinates) as the class mark (or mid point of class limits or boundaries) and abscissae (y-coordinates) as the frequency of the corresponding classes are plotted. These points correspond to the middle points of the upper horizontal side of the corresponding rectangles of the histogram. The neighbouring points are joined by straight lines.

It is customary to add two more classes: the next lower class in case of extreme left class and the next higher class in case of extreme right class, and to take that each of these two classes has zero frequency. The end points of the polygon already drawn are joined to the middle points of these two classes thus added; the middle points lie in the x-axis and are at a distance equal to the class width from the corresponding end mid points or class marks. We thus get the complete frequency polygon. This procedure is indicated in Fig. 3.1 by the two dotted lines.

In this case, it may be noted that the total area (sum of the areas) of the rectangles in the histogram becomes equal to the area of the frequency polygon and the x-axis.

The histograms drawn have a somewhat symmetrical shape. However, it must not be understood that histograms or frequency polygons cannot have any other shape. In fact histograms can have *asymmetrical* or *skewed* shape. These usually range from a rough bell-shaped distribution to something resembling one half, either the right or the left half of a bell-shaped distribution, or to some other type of distribution.

Frequency curves: Consider the distribution of continuous data. If there are a large number of observations and if the class intervals are taken to be small enough, it may be possible to have a sizeable frequency for most of the classes. Then the frequency polygon will closely approximate a curve, which is called *frequency curve*. Such a curve is also known as smoothed *frequency polygon*.

It has been found that frequency curves of data found in nature and industry generally take the characteristic shapes indicated in Fig. 3.4.

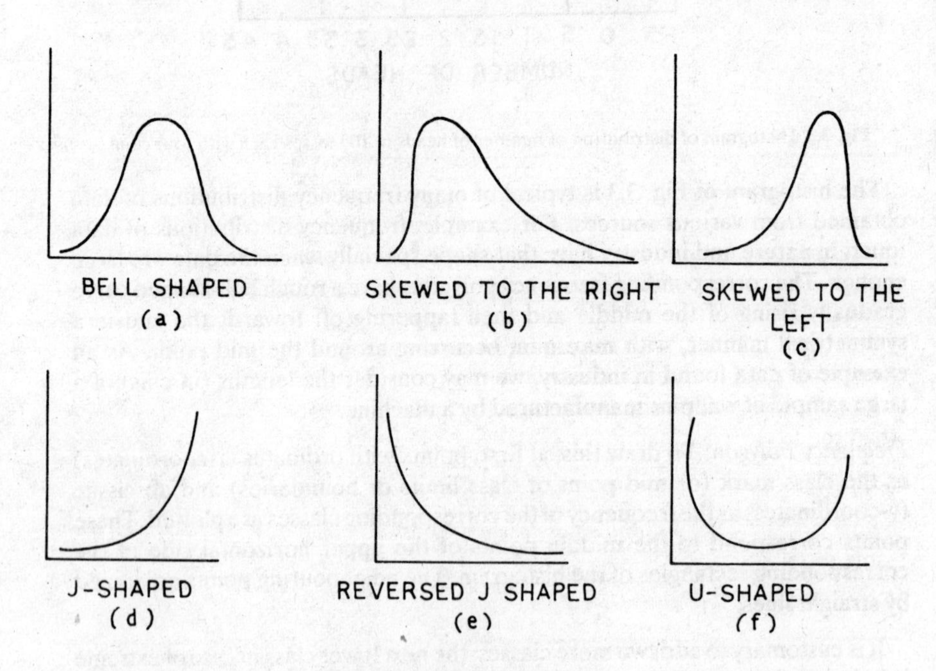

Fig. 3.4. Some frequency curves.

3.6. CUMULATIVE FREQUENCY DISTRIBUTION AND OGIVES

Consider the number of all observations which are *less than the upper class boundary* of a given class interval: this number is the sum of the frequencies upto and including that class to which the upper class boundary corresponds. This sum is known as the *cumulative frequency* upto and including that class interval. For example consider Table 3.3; the cumulative frequency upto and including the class interval 145-146 is 2, that upto and including the next class interval 147-148 is 2 + 5 = 7, that upto and including the next class interval 149-150 is 2 + 5 + 8 = 15 and so on. This implies that two students have heights less than the upper class boundary of the class 145-146, seven students have heights less than the upper class boundary of the class 147-148 and so on. We can thus construct the cumulative frequency table as follows:

Table 3.6. Cumulative frequency (less than) table of heights of 50 students

Class (in cm.) interval	Frequency	Cumulative Frequency (less than)
145-146	2	2
147-148	5	7
149-150	8	15
151-152	15	30
153-154	9	39
155-156	6	45
157-158	4	49
159-160	1	50
Total	50	

The cumulative frequency distribution is represented by joining the points obtained by plotting the cumulative frequencies along the vertical axis and the corresponding upper class boundaries along the *x*-axis. The corresponding polygon is known as cumulative frequency polygon (less than) or ogive. By joining the points by a freehand curve we get the cumulative frequency curve ("less than"). Similarly we can construct another cumulative frequency distribution ("more than" type) by considering the sum of frequencies greater than the lower class boundaries of the classes. For example, the total frequency greater than the lower class boundary 158.5 of the class 159-160 is one (1), while the total frequency greater than the lower class boundary 156.5 of the class 157-158 is 1 + 4 = 5, that of the class 155-156 is 1 + 4 + 6 = 11, and so on. Given below is Table 3.7 of cumulative frequency distribution ("more than") of the same distribution.

Table 3.7. Cumulative frequency (more than) table of heights of 50 students

Class (in cm.) interval	Frequency	Cumulative frequency (more than)
145-146	2	50
147-148	5	48
149-150	8	43
151-152	15	35
153-154	9	20
155-156	6	11
157-158	4	5
159-160	1	1
Total	50	

The graph obtained by joining the points obtained by plotting the cumulative frequencies ("more than") along the vertical axis and the corresponding *lower* class boundaries along the x-axis is known as *cumulative frequency polygon* (greater than) or *ogive*. By joining the points by a free-hand curve, one gets the *cumulative frequency curve* ("more than" type).

These two curves are shown in Fig. 3.5.

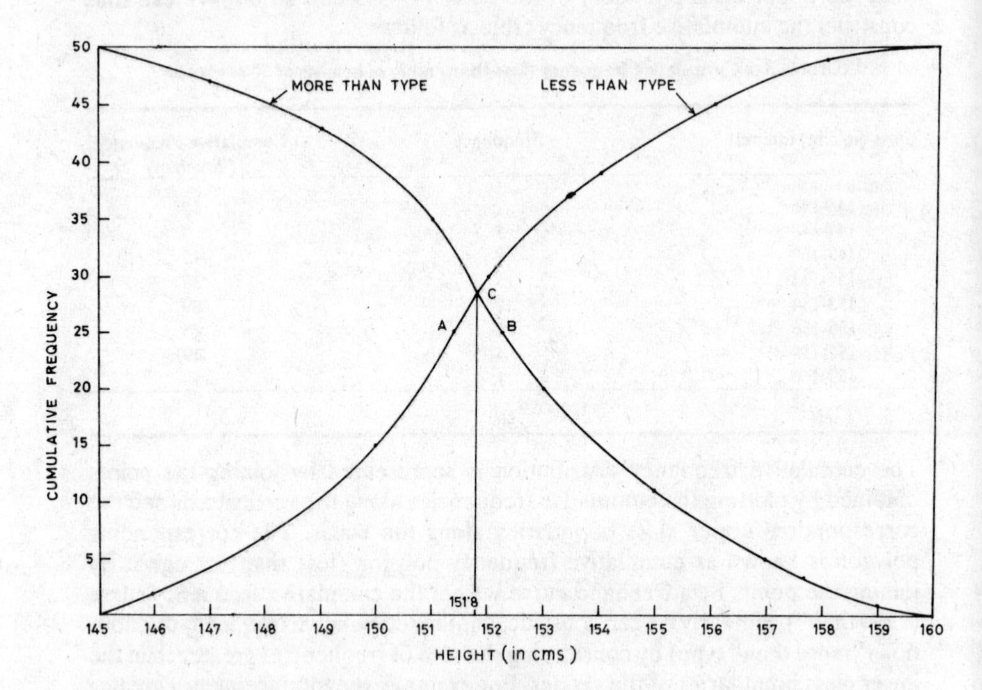

Fig. 3.5. Cumulative frequency curves ('more than' and 'less than' types) of heights of 50 students

3.7. BIVARIATE FREQUENCY DISTRIBUTIONS

Consider the heights and weights of a sample of individuals. Here we have two variables heights (X) and weights (Y). The two variables have some relationship between them. To represent such data involving two variables, a bivariate frequency distribution is constructed by taking appropriate class intervals for each of the variables. The procedure is explained below with the help of an example.

Example 7(a). The marks secured by 50 students in Mathematics and Statistics are given below:

Sl. No.	1	2	3	4	5	6	7	8	9	10
Mathematics	62	45	46	67	23	53	58	38	26	43
Statistics	73	55	42	58	25	62	64	45	31	38

Sl. No.	11	12	13	14	15	16	17	18	19	20
Mathematics	36	63	26	48	76	45	66	72	28	84
Statistics	47	48	35	56	85	54	58	64	36	92

Sl. No.	21	22	23	24	25	26	27	28	29	30
Mathematics	56	38	56	92	34	42	38	68	48	52
Statistics	48	46	65	96	46	54	46	74	56	68

Sl. No.	31	32	33	34	35	36	37	38	39	40
Mathematics	38	57	65	46	38	46	65	56	48	75
Statistics	48	65	76	38	48	52	58	65	64	58

Sl. No.	41	42	43	44	45	46	47	48	49	50
Mathematics	56	84	66	68	58	76	46	76	52	54
Statistics	42	68	74	75	53	84	65	65	58	68

Denote the marks in Mathematics by X and the marks in Statistics by Y. As a first step arrange the values of X (or those of Y) in an array. Arranging the values of X in ascending order we get the following array (the serial numbers used for identification are dropped).

X	23	26	26	28	34	36	38	38	38	38
Y	25	31	35	36	46	47	45	46	46	48

X	38	42	43	45	45	46	46	46	46	48
Y	48	54	38	54	55	38	42	52	65	56

X	48	48	52	52	53	54	56	56	56	57
Y	56	64	58	68	62	68	42	48	65	65

X	58	58	62	63	65	65	66	66	67	68
Y	53	64	73	48	58	76	58	74	58	74

X	68	72	75	76	76	76	76	84	84	92
Y	75	64	58	65	85	85	84	68	92	96

We form the classes 0-10, 11-20, 21-30, . . . both for X and Y, along the horizontal and vertical axes respectively. An observation will fall in the class corresponding to its X and Y values. Give one tally mark for each observation e.g. for the first set (23, 25) we give a tally mark in the class corresponding to 21-30 for X and 21-30 for Y.

Marginal Frequencies

Consider the column with classes 21-30, 31-40, ... and so on indicating columns with the distribution of X. Adding the individual frequencies along a column of X, say 21-30, we get a marginal total $1 + 3 = 4$; along a column 31-40, a marginal total 7 and so on.

The vertical totals constitute a *marginal frequency distribution* of the variable X and the horizontal totals constitute a marginal frequency of the variable Y.

The marginal frequency distribution of X given by the vertical totals under the columns is as follows:

We thus construct the following Table:

Table 3.8. Bivariate frequency distribution of marks secured in Mathematics and Statistics by 50 students

Y \ X	21-30	31-40	41-50	51-60	61-70	71-80	81-90	91-100	Horizontal Total
21-30	/ =1								1
31-40	/// =3		// =2						5
41-50		卌 // =7	/ =1	// =2	/ =1				11
51-60			卌 / =6	// =2	/// =3	/ =1			12
61-70			// =2	卌 / =6		// =2	/ =1		11
71-80					卌 =5				5
81-90						/// =3			3
91-100							/ =1	/ =1	2
Vertical Total	4	7	11	10	9	6	2	1	Grand Total 50

Table 3.9(a) Marginal distribution of X

Class Interval	21-30	31-40	41-50	51-60	61-70	71-80	81-90	91-100	Total
Marginal Frequency	4	7	11	10	9	6	2	1	50

Similarly the marginal distribution of Y is given as follows:

Table 3.9(b) Marginal distribution of Y

Class Interval	21-30	31-40	41-50	51-60	61-70	71-80	81-90	91-100	Total
Marginal Frequency	1	5	11	12	11	5	3	2	50

The marginal frequencies are the horizontal totals along the rows.

It is to be noted that the overall total frequency obtained by adding the marginal frequencies of X is the same as that obtained by adding the marginal frequences of Y. In this example, the overall total frequency is 50.

Example 7(b). The following data (X, Y) indicate the ages (X) and the systolic blood pressures (Y) of 40 individuals between the ages of 45 and 65.

(45, 131), (50, 128), (57, 135), (61, 142), (64, 138), (62, 144), (59, 154), (53, 125)
(61, 138), (55, 135), (45, 140), (61, 144), (59, 150), (50, 152), (46, 135), (52, 146)
(46, 148), (47, 138), (63, 148), (63, 138), (60, 144), (53, 135), (63, 146), (54, 144)
(50, 143), (60, 146), (45, 132), (64, 146), (46, 150), (46, 144), (52, 148), (62, 145)
(63, 144), (63, 144), (64, 150), (65, 152), (49, 144), (57, 150), (60, 135), (45, 132)

Table 3.10(a) Bivariate frequency distribution table of age (X) and blood pressure (Y) of 40 individuals

X (age) / Y (blood pressure)	45-47	48-50	51-53	54-56	57-59	60-62	63-65	Total
125-129		/ =1	/ =1					2
130-134	/// =3							3
135-139	// =2		/ =1	/ =1	/ =1	// =2	// =2	9
140-144	/// =3	/ =1		/ =1		//// =4	// =2	11
145-149	/ =1		// =2			// =2	/// =3	8
150-154	/ =1	/ =1			/// =3		// =2	7
Total	10	3	4	2	4	8	9	40

The smallest value of X is 45 and the largest 65. We can form X classes as follows 45-47, 48-50, 51-53, 54-56, 57-60, 60-62, 63-65. The smallest value of Y is 125 and the largest 154; we can form the Y classes with intervals 125- 129, 130-134, 135-139, and so on. Giving one tally for each observation in the corresponding class, we form the above bivariate frequency distribution Table 3.10(a).

The marginal frequency distribution of X and Y are given as follows:

Table 3.10(b) Marginal frequency distribution of X

Class interval	45-47	48-50	51-53	54-56	57-59	60-62	63-65	Total
Frequency	10	3	4	2	4	8	9	40

Table 3.10(c) Marginal frequency distribution of Y

Class interval	125-129	130-134	135-139	140-144	145-149	150-154	Total
Frequency	2	3	9	11	8	7	40

3.8. TABULATION OF DATA

In a simple frequency distribution table classification of data according to a *single characteristic* is shown. For example, Table 3.1 shows classification of data according to marks, Table 3.2 according to heights, Table 3.3 according to yield of paddy. Such a table is a *one-way table*, in which data is classified according to one characteristic or criterion. In a bivariate frequency table classification of data according to *two characteristic* is shown. Such a table is a *two-way table*. For example, Table 3.8(a) shows classification of data according to two characteristics, marks in Mathematics and marks in Statistics. In Table 3.10(a), we have simultaneous classification of data according to two characteristics, age and blood pressure. If data is classified simultaneously with respect to three characteristics or criteria, we get a *three-way* table. It may be noted that characteristics or criteria need not always be quantitative or expressible in quantitative terms. Some of them may be *qualitative*, for example, one characteristic in classification may be sex, another may be geographical region and so on. We give below examples of classification of data where the characteristics according to which classification is done may be qualitative.

Table 3.11. Birth rates and Death rates and natural growth rates in India, 1911-1981

(Rate per 1,000 population per annum)

Year	Crude Birth Rate	Crude Death Rate	Natural Growth Rate
1901-11	49.2	42.6	6.6
1911-21	48.1	47.2	0.9
1921-31	46.4	36.3	10.1
1931-41	45.2	31.2	14.0
1941-51	39.9	27.4	12.5
1951-61	41.7	22.8	18.9
1961-71	41.1	18.9	22.2
1971-81	37.1	14.8	22.3

Here data are classified according to two characteristics: according to chronology and according to three different types of rates.

Table 3.12. Distribution of districts (in India) according to annual geometric rate of population growth

Decade	Number of Districts					
Annual Rate of growth	1911-21	1921-31	1931-41	1941-51	1951-61	1961-71
less than −2.0	5	1	—	1	2	—
(−2.0)-(−1.0)	28	0	—	1	—	—
(−1.0)-(0.0)	143	6	2	8	—	—
(0.0)-(1.0)	129	149	87	132	22	3
(1.0)-(2.0)	25	154	217	149	154	109
(2.0)-(3.0)	4	19	22	33	118	182
(3.0)-(4.0)	1	5	4	6	26	34
More than 4.0	—	1	3	5	13	7
Total	335	335	335	335	335	335

Note that each vertical marginal total is 335. This gives classification of the same 335 districts according to annual rate of growth for the different decades.

The following is a schematic outline of a three-way table of employees in an establishment according to age, sex and salary drawn.

Salary (Rs.)	Less than 1000		1000-1500		1500-2000		2000-2500		2500-3000		Total
Age (years)	M	F	M	F	M	F	M	F	M	F	
20-30											
30-40											
40-50											
50-60											
Over 60											
Total											

3.9. OTHER FORMS OF REPRESENTATION

3.9.1. Line Graphs

We have discussed about representation of frequency distributions of grouped data through diagrams such as histogram, frequency polygon, frequency curve and

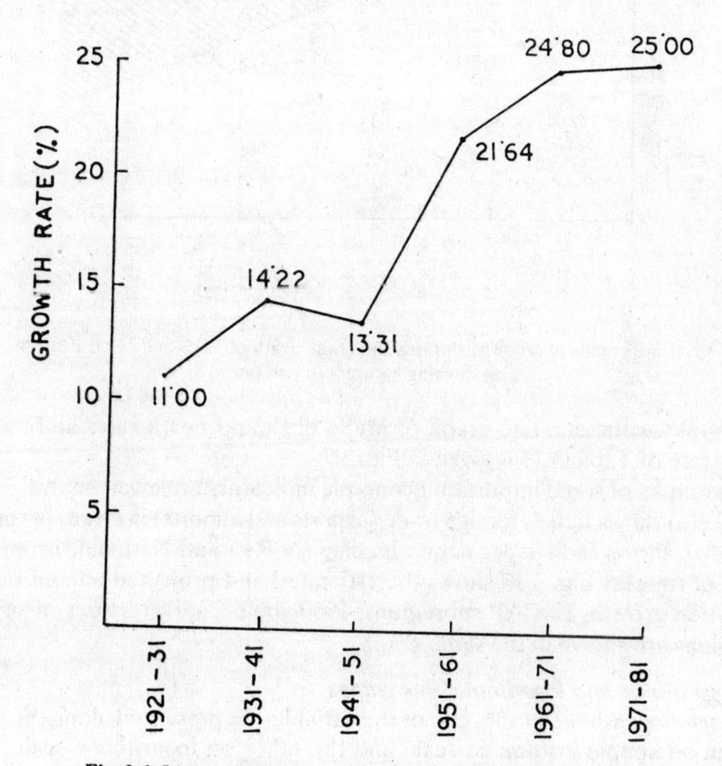

Fig. 3.6 Line graph of decennial growth rate (see Table 3.1)

cumulative frequency distribution curve (or ogive). For drawing these diagrams, first we have to construct the frequency distributions of the raw data.

A line graph of decennial growth of population (Table 3.1) is given in Fig 3.6.

It is found that graphs and diagrams are often used for quick and visual representation of statistical data in the form it is obtained. The main forms of such visual representation are the following:
(1) line graphs or curves, (2) geometric forms, (3) pictures, and (4) maps. Line graphs may be simple arithmetic, semi logarithmic or logarithmic. These are drawn on suitable graph papers. They are suitable for representation of chronological data. The geometric forms include such geometric figures as rectangles (usually represented by bars), circles (or pie), squares, and cubes. Charts consisting of pictures (of the characteristics under observation) are called pictographs. Maps are used for geographical presentation of statistical data.

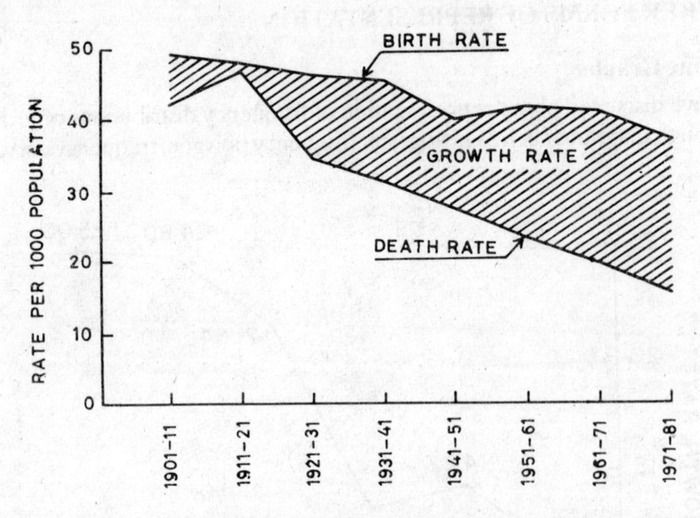

Fig. 3.7. Line graphs of crude birth rates and crude death rates of India (see Table 3.11): also showing natural Growth rate.

A simple arithmetic line graph of crude birth and death rates and natural growth rate of Table 3.11 is given in Fig. 3.7.

Line graphs of some important economic indicators are given below:

Fig 3.8(a) shows India's foreign trade (exports and imports) in crores of rupees.

Fig. 3.9. shows India's per capita income (in Rs.) and National Income (in crores of rupees) Fig. 3.10 shows the estimated and projected annual rate of population growth, ESCAP subregions, 1960-2000. Characteristics of several subregions are shown in the same graph.

Semilogarithmic and logarithmic line graphs

In a *semilogarithmic* graph, one of the variables (represented along the axes) is taken on simple arithmetic scale and the other on logarithmic scale. In a *logarithmic* graph, both the variables are taken on logarithmic scale.

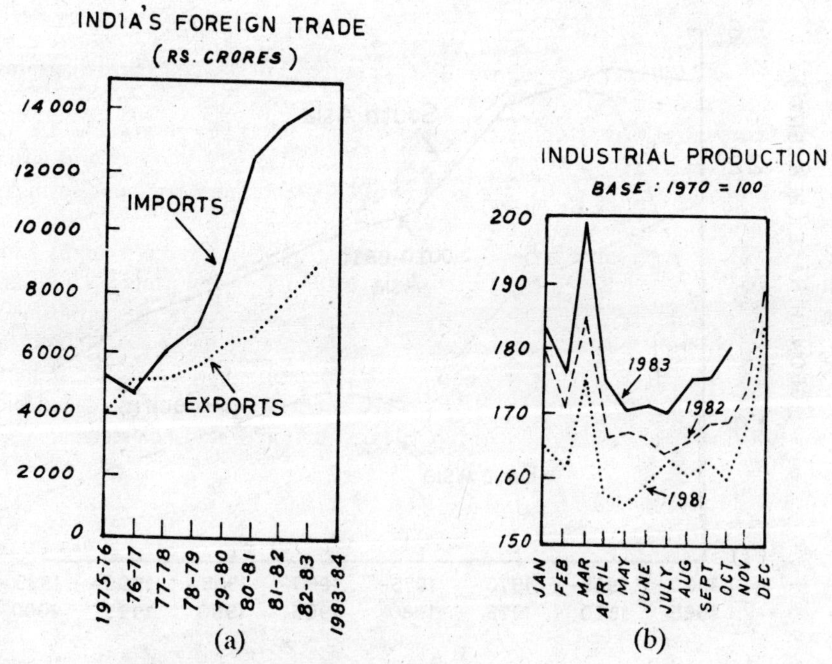

Fig. 3.8(a) India's foreign trade from 1975-76 to 1982-83
(b) Index of Industrial Production 1981, 1982, 1983 (upto October)
(*Courtesy : The Times of India*)

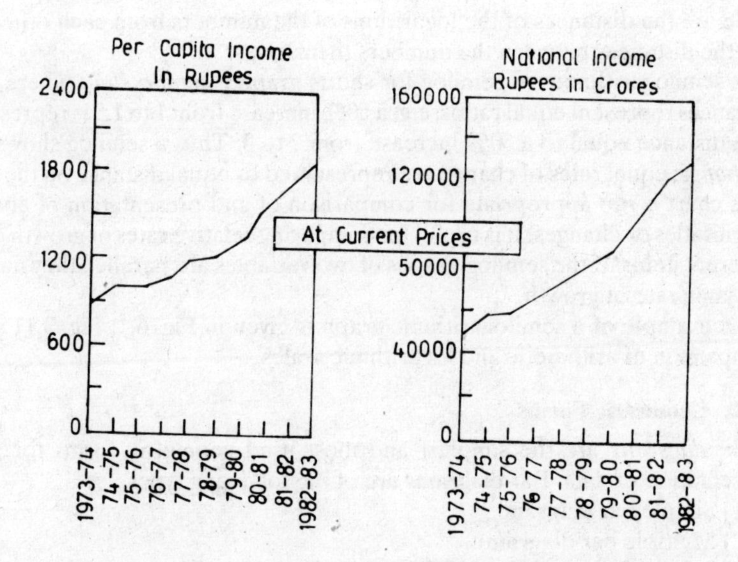

Fig. 3.9. Per capita income (in Rs.) and National Income (in crores of rupees) of India during 1973-74 to 1982-83 (*Courtesy* : The Times of India)

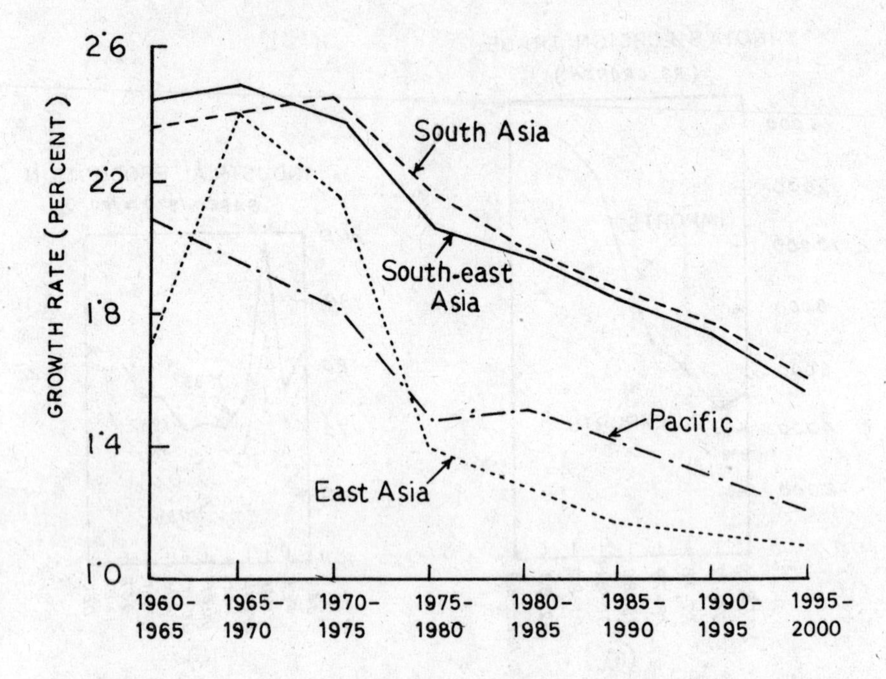

Fig. 3.10. Estimated and projected annual rate of population growth. ESCAP subregions. 1960-2000 (Source : ESCAP Datasheet)

It is to be noted that (1) no zero or negative value can be shown on a logarithmic scales and (2) the distances between the points on the logarithmic scale are the distances of the logarithms of the numbers from each other, and not the distances between the numbers themselves.

A semilogarithmic (or semilog for short) graph is a *ratio chart* where equal distances represent equal ratios, e.g. a 50% increase from 1 to 1.5 is represented by a distance equal to a 50% increase from 2 to 3. Thus a semilog shows *rates of change*; equal rates of change are represented by equal distance on the scale. This chart is *not* appropriate for comparison of and presentation of absolute magnitudes or changes. It is useful for comparing relative rates of growth of two different items. If the semilog graphs of two variables are parallel, they indicate the same rate of growth.

An example of a semilogarithmic graph is given in Fig. 6.1. Fig. 3.11 shows comparison of arithmetic and logarithmic scales.

3.9.2. Geometric Forms

Bar diagrams are the simplest and most used geometric forms for visual presentation of data. Bar diagrams are of the following types :

(1) Simple bar diagram
(2) Multiple bar diagram
(3) Subdivided or component bar diagram
(4) Percentage bar diagram

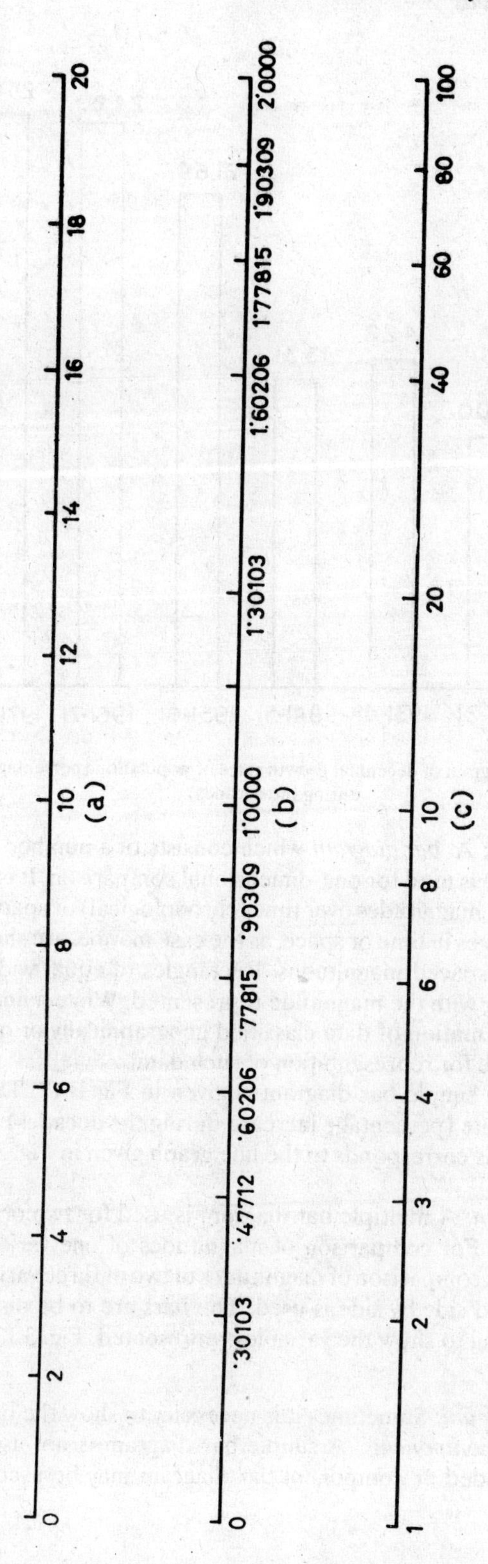

Fig. 3.11. Comparisons of Arithmetic and Logarithmic Scales (a) Arithmetic scale (b) Logarithms on Arithmetic scale (c) Logarithmic scale

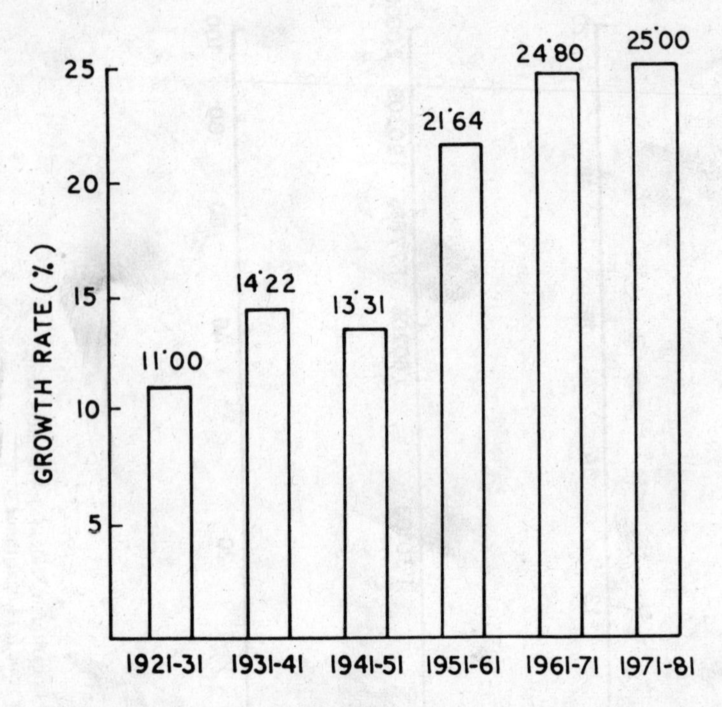

Fig. 3.12. Bar diagram of decennial growth rates of population (percentage increase during the decades)

Simple bar diagram: A *bar diagram* which consists of a number of rectangles, (usually called bars) is used for one-dimensional comparison. It is used to show absolute changes in magnitudes over time (chronological) or space (geographical/regional). Changes in time or space, as the case may be, are shown along the x-axis with equally spaced magnitudes. Rectangles of equal width are drawn with lengths varying with the magnitude represented. While a line graph is not suitable for representation of data classified geographically or qualitatively, a bar graph is suitable for representation of such data.

An example of a simple bar diagram is given in Fig. 3.12. This represents decennial growth rate (percentage increase during the decades) of population (see Table 3.1). This corresponds to the line graph given in Fig. 3.6.

Multiple bar diagram: A multiple bar diagram is used for two or three dimensional comparison. For comparison of magnitudes of one variable in two or three aspects, or for comparison of magnitudes of two or three variables, a group of rectangles placed side by side is used. The bars are to be distinguished by shading or colouring to show the variables represented. Fig. 3.13 is a multiple bar diagram.

Subdivided bar diagram: Sometimes it is necessary to show the breakup of *one variable in several components*. A simple bar diagram is not useful in such a situation; a subdivided or component bar diagram may be used to represent

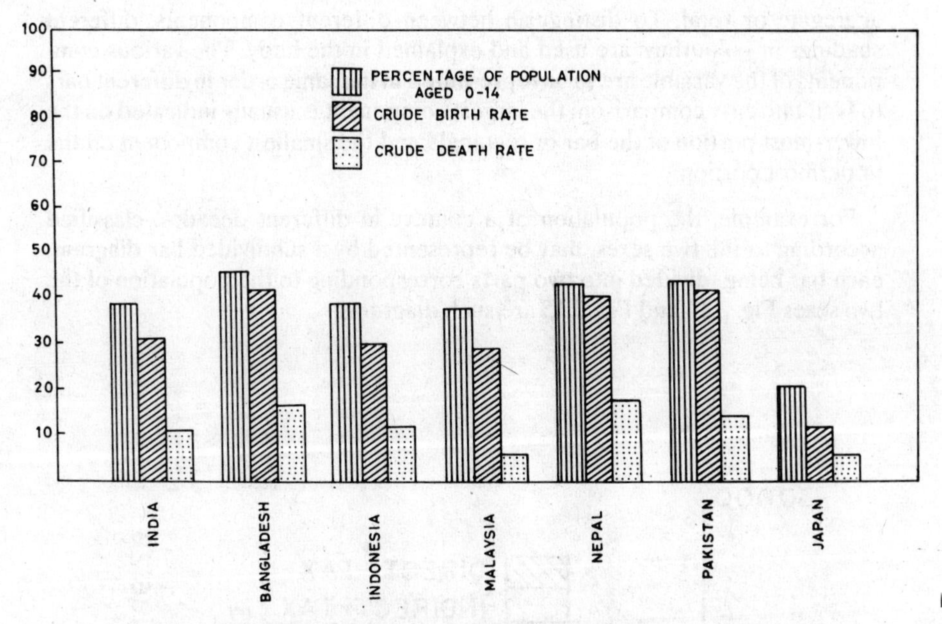

Fig. 3.13. Selected demographic estimates (1985) of some Asian countries.

data relating to such a variable over time or space. Here as in the case of simple bar diagram, bars or rectangles are drawn to represent the total or aggregate magnitudes of the variable; one bar to represent each time period or geographical area or to present each of the different qualitative aspects. Then each bar is divided into several segments, each segment representing a component of the

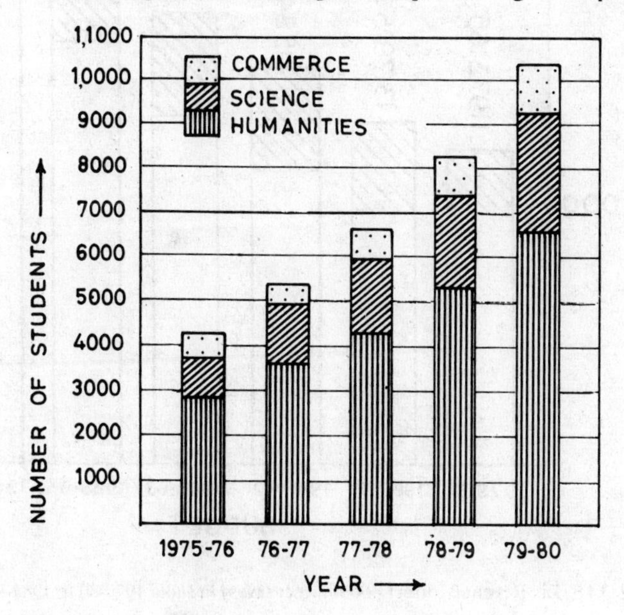

Fig. 3.14. Number of students faculty-wise

aggregate or total. To distinguish between different components, different shadings or colourings are used and explained in the body. The various components of the variable are to be represented in the same order in different bars to facilitate easy comparison; the largest component is usually indicated on the lower-most portion of the bar or rectangle and the smallest component on the uppermost portion.

For example, the population of a country in different decades, classified according to the two sexes, may be represented by a subdivided bar diagram, each bar being divided into two parts corresponding to the population of the two sexes Fig. 3.14 and Fig. 3.15 are such diagrams.

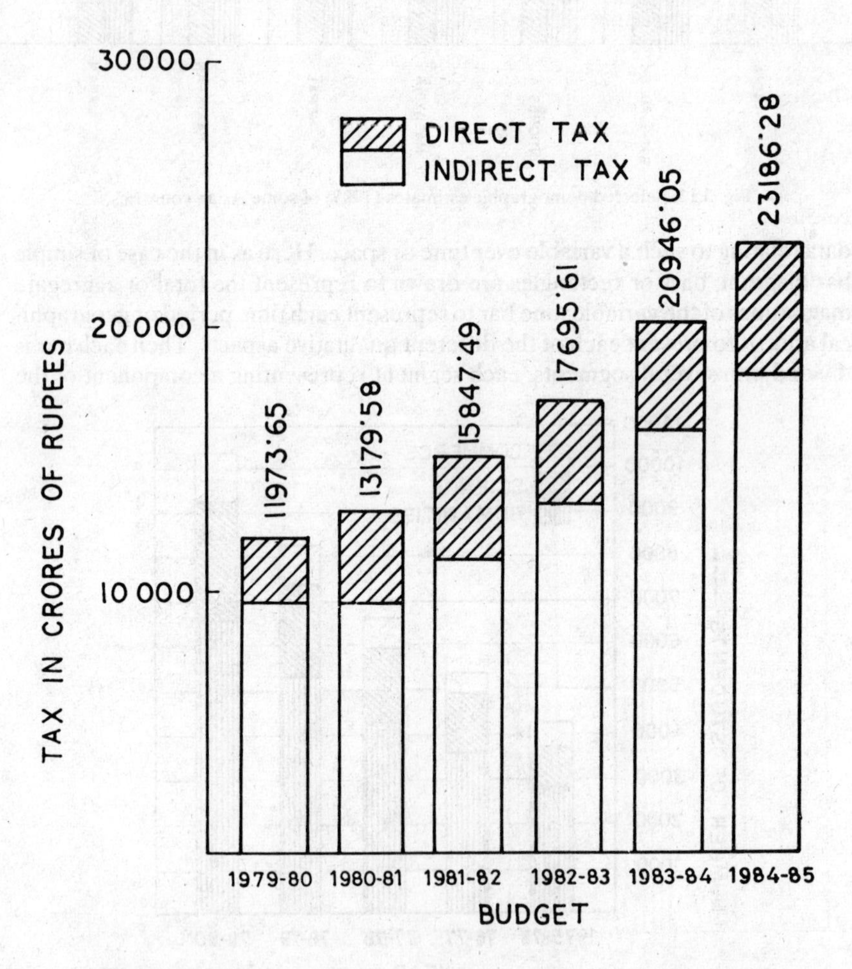

Fig. 3.15. Tax Revenue (direct and indirect taxes) in India 1979-80 to 1984-85

Example 9(a). The following table shows the number of students in different faculties of a University:

Year	Number of students			
	Humanities	Sciences	Commerce	Total
1975-76	2,810	890	540	4,240
1976-77	3,542	1,363	471	5,376
1977-78	4,301	1,662	652	6,615
1978-79	5,362	2,071	895	8,328
1979-80	6,593	2,752	1,113	10,458

Represent the data by a suitable subdivided bar diagram.

Example 9(b). Fig. 3.15 shows gross tax revenue in India (in crores of Rupees) broken up in two components, direct tax and indirect tax.

Percentage bar diagram:

In the subdivided bar diagrams, the heights of the rectangles are proportional to the magnitudes of the variable and in each rectangles drawn in a scale proportional to the magnitudes of the component. Instead of considering absolute magnitudes of the variable, magnitudes of components may be indicated in terms of percentages of the total of the variable for that particular characteristic. Here all the rectangles are drawn with equal heights each representing a total of 100. The rectangles are then subdivided in terms of percentages of the components.

We consider the data as given in Example 9(a). We obtain the percentage of students, faculty-wise, for each year; this is given in Table 3.13 below.

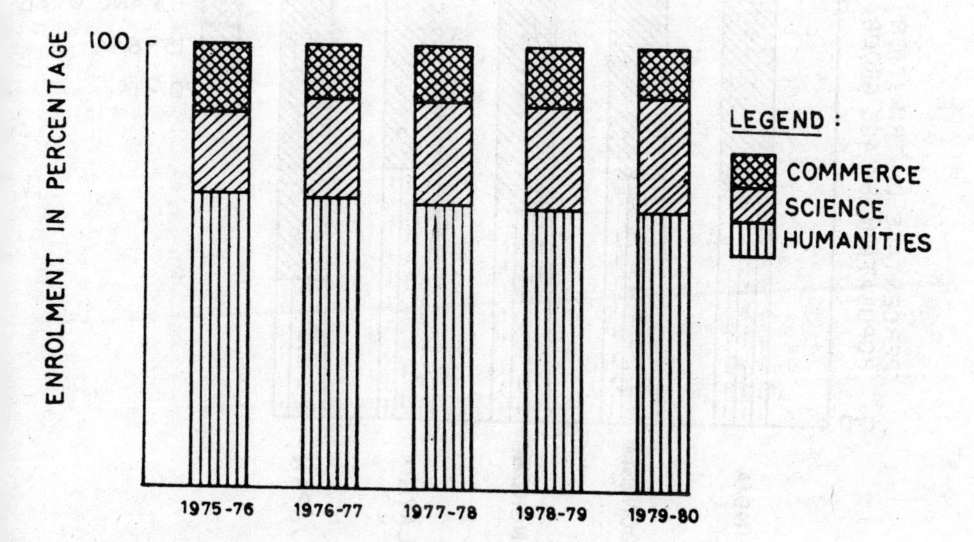

Fig. 3.16 Percentage bar diagram of student enrolment faculty wise

Table 3.13 Percentage of students faculty wise

Year	Percentage of students			
	Humanities	Sciences	Commerce	Total
1975-76	66.27	21.00	12.73	100.00
1976-77	65-89	25.35	8.76	100.00
1977-78	65.02	25.12	9.86	100.00
1978-79	64.38	24.87	10.75	100.00
1979-80	63.04	26.32	10.64	100.00

The above data is represented by Fig 3.16.

The population of some countries by age groups is given below:

Table 3.14 Percentage of population by age group (in percentages)

Countries	Age-group (in years)			Total
	0-14	15-64	65 and Over	
India (1978)	42.1	54.7	3.2	100.00
Indonesia (1975)	44.0	53.5	2.5	100.00
Hongkong (1978)	37.1	58.6	4.3	100.00
Malayasia (1974)	45.0	52.0	3.0	100.00
Japan (1970)	24.0	68.9	7.1	100.00

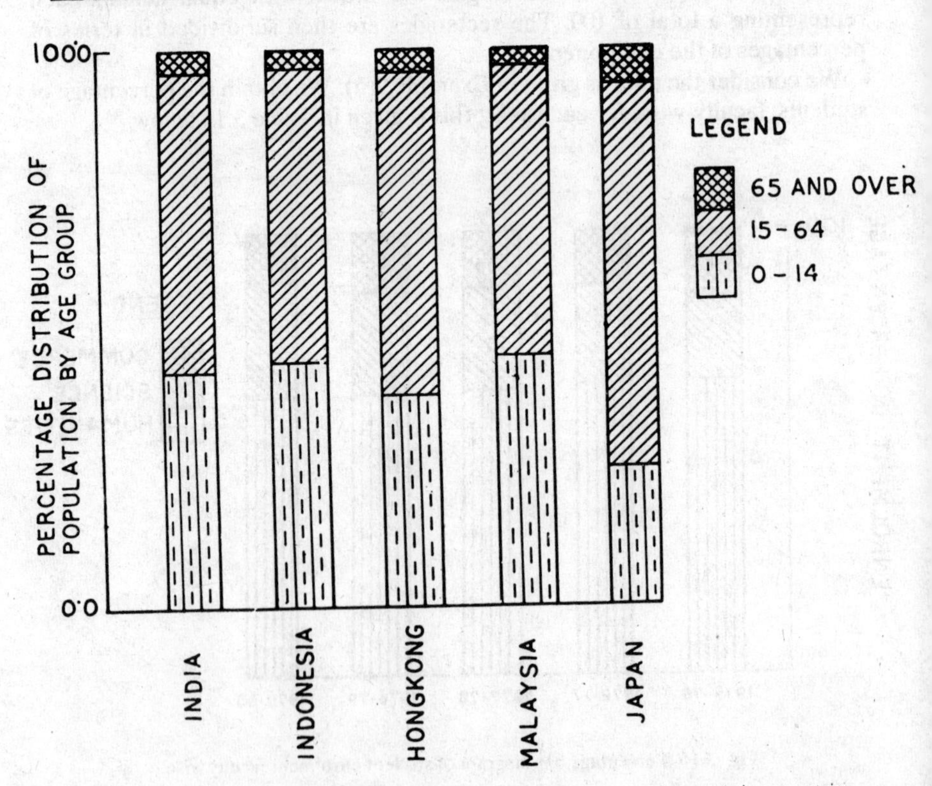

Fig. 3.17 Percentage distribution of population by age-group of some Asian countries

This is represented by the percentage bar diagram in Fig. 3.17
Split bar diagram: For indicating wide variations in magnitudes of data, broken or split bars are used. One of the main types of this diagram is the pyramid, used for representing population by sex and age group. The population pyramids for India for 1961, 1971 corresponding to data in Table 3.15 is given in Fig. 3.18.

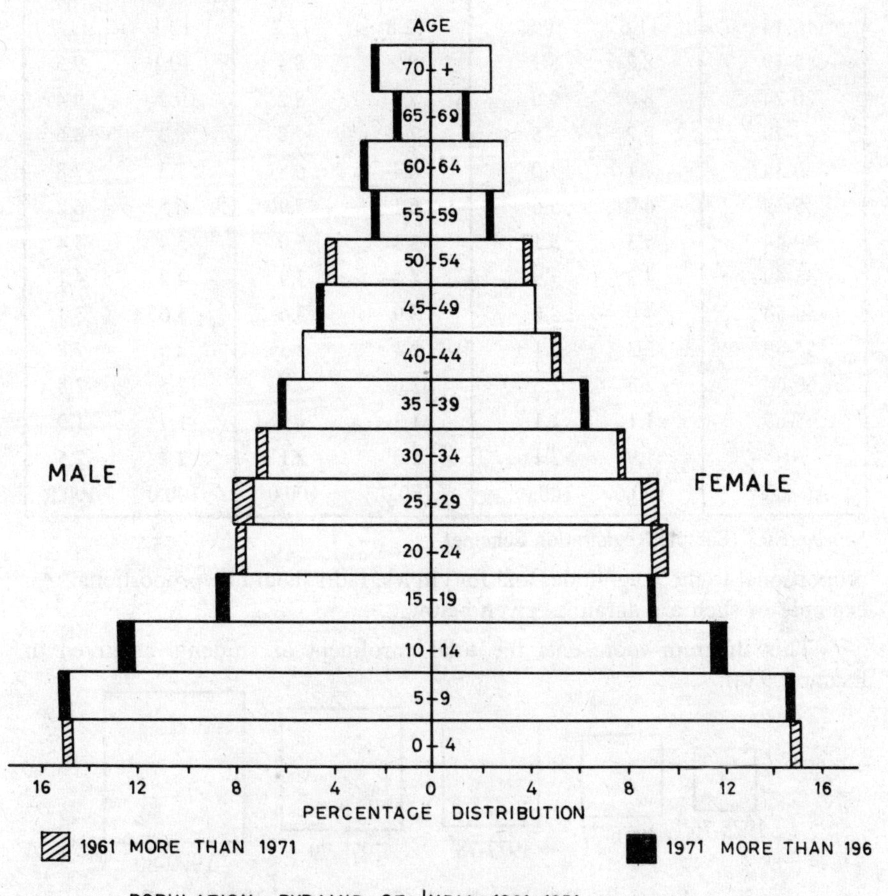

Fig. 3.18 Population pyramid of India 1961, 1971.

Area diagram : For two dimensional comparison, area diagrams in the form of rectangles squares or circles, may be used. In bar diagrams used for one-dimensional comparison, the rectangles are of the same width, the heights of the rectangle indicating the magnitudes of the variable. Rectangles used for area diagram, vary in the heights as well as in the widths, so that the total areas of the rectangles represent the magnitudes of the variable. While using circles or squares for such comparison, the areas are taken into consideration for representing the magnitudes of the variable over time or space, or over some other characteristic of variation. For drawing squares, sides of squares should be

Table 3.15 : Percentage distribution of population of India by age and sex

Age Group	1961 Census		1971 Census		SRS 1997	
	Male	Female	Male	Female	Male	Female
0-4	14.7	15.5	14.2	14.9	11.4	11.1
5-9	14.6	14.9	14.9	15.1	12.8	12.6
10-14	11.6	10.8	12.8	12.2	12.1	11.7
15-19	8.2	8.1	8.9	8.4	10.1	9.5
20-24	8.0	9.0	7.6	8.2	9.2	9.4
25-29	8.2	8.5	7.2	7.8	8.3	8.6
30-34	7.1	7.0	6.4	6.8	7.3	7.8
35-39	6.0	5.6	6.1	5.9	6.5	6.2
40-44	5.3	5.1	5.3	5.0	5.4	5.4
45-49	4.3	3.9	4.4	3.9	4.3	4.1
50-54	4.0	3.8	3.9	3.6	3.6	3.7
55-59	2.3	2.1	2.4	2.3	2.6	2.8
60-64	2.5	2.6	2.6	2.6	2.5	2.8
65-69	1.1	1.1	1.3	1.3	1.7	1.9
70+	1.9	2.1	2.0	2.1	2.2	2.5
All ages	100.0	100.0	100.0	100.0	100.0	100.0

Source: SRS (Sample Registration Scheme).

proportional to the magnitudes and for circles, radii should be proportional. An example of such a diagram is given below.

This diagram represents the total enrolment of students as given in Example 9 (a):

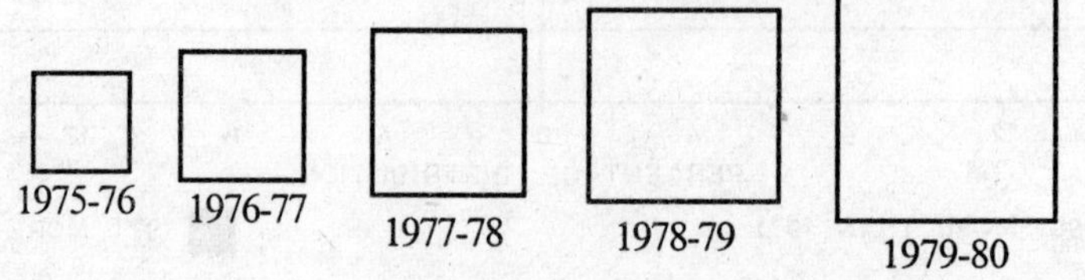

Fig. 3.19. Total enrolment of students, 1975-80.

Year	1975-76	1976-77	1977-78	1978-79	1979-80
Total enrolment.	4,240	5,376	6,615	8,328	10,458
Side of square (square root of total enrolment)	65.1	73.3	81.3	91.3	102.3

Pie diagram: For representing break down of an aggregate into components a pie diagram is used. It is a circle divided into sectors with areas equal to the corresponding components. A pie diagram shows the components in terms of percentages only and not in absolute magnitude. One circle is drawn for one dimensional comparison. It has the same purpose as a percentage bar diagram.

The areas of the sectors are proportional to the angles subtended at the centre of the circle, so that the angles are taken according to the relative proportion (or percentage) of the magnitude of the various components. This is to be found out before drawing a pie diagram.

An example of a pie diagram is given below.

Example 9(c). The tax revenue of India as provided in 1984-85 budget when broken into various sources, is as follows :

Sources	(in crores of Rs.)	Tax Revenue in percentages	Angle (in degrees)
Excise	6526	37.24	134.1
Customs	7108	40.56	146.0
Corporation Tax	2568	14.65	52.7
Income Tax	560	3.20	11.5
Other Taxes	763	4.35	15.7
Total Tax revenue	17525	100.00	360.0

A pie diagram is given in Fig. 3.20

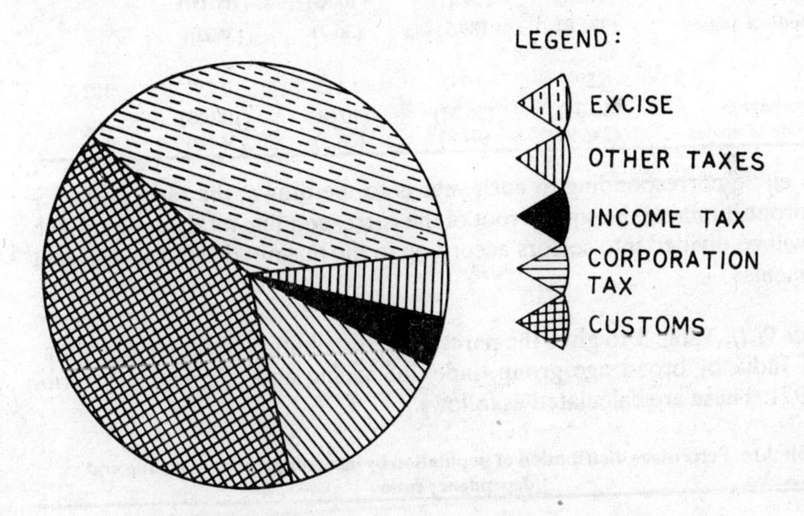

Fig. 3.20. India's tax revenue according to sources (1984-85)

For two-dimensional comparison, where a variable is indicated over time, space or in terms of some other characteristic and the variable is broken up into components, a multiple pie diagram is used. This amounts to drawing as many circles, the areas of circles being proportional to the magnitudes varying with time space or other characteristic, as the case may be. For example, for representing India's tax revenue for a number of years, as many circles may be drawn, the areas of the circles (and therefore their radii) being in proportion to the magnitude for different years. The radii are proportional to the square roots of the magnitudes. A multiple pie diagram has the same purpose as a multiple percentage bar diagram. A multiple pie diagram of data given in Example 9(a) can be constructed as indicated below

Year	Number of students (figures within parentheses are those in percentages) Humanities	Science	Commerce	Total	Radius of circle
1975-76	2,810	890	540	4,240	65.1
(in percentage)	(66.27)	(21.00)	(12.73)	(100.00)	
Magnitude of angles:	(238.6)	(75.6)	(45.8)	(360.0)	
1976-77	3,542	1,363	471	5,376	73.3
(in percentage)	(65.59)	(25.35)	(8.76)	(100.00)	
Magnitude of angles:	(236.1)	(91.3)	(31.5)	(360.0)	
1977-78	4,301	1,662	652	6,615	81.3
(in percentage)	(65.02)	(25.12)	(7.86)	(100.00)	
Magnitude of angles:	(234.1)	(90.4)	(35.5)	(360.0)	
1978-79	5,362	2,071	895	8,328	91.3
(in percentage)	(64.38)	(24.87)	(10.75)	(100.00)	
Magnitude of angles:	(231.8)	(89.5)	(36.7)	(360.0)	
1979-80	6,593	2,752	1,113	10,458	102.3
(in percentage)	(63.04)	(26.32)	(10.64)	(100.00)	
Magnitude of angles	(226.9)	(94.8)	(38.3)	(360.0)	

One circle corresponding to each year is to be drawn, the radii of circles being proportional to the square root of the corresponding total number. Each circle will be divided into sectors according to the magnitudes of the different components.

Example 9(d). Table 3.16 gives the percentage distribution of the total population of India by broad age group, index of ageing and dependency ratio, 1951-1971. These are calculated as follows.

Table 3.16. Percentage distribution of population by age-group, index of ageing and dependency ratio

Census Year	Total population (millions)	Percentage Distribution by age group			Index of Ageing	Dependency ratio		
		0-14	15-59	60⁺		Young	Old	Overall
1951	361-1	37.50	56.85	5.65	15.07	65.96	9.94	75.90
1961	439.2	41.00	53.36	5.64	13.74	76.84	10.55	87.39
1971	548.2	42.02	52.01	5.97	14.21	80.79	11.48	92.27

$$\text{Index of ageing} = \frac{\text{Percentage of population aged } 60^+}{\text{Percentage of population aged } 0\text{–}14} \times 100$$

$$\text{Young dependency ratio} = \left(\frac{\text{Percentage of population aged 0–14}}{\text{Percentage of population aged 15–59}}\right) \times 100$$

$$\text{Old dependency ratio} = \frac{\text{Percentage of population aged } 60^{+}}{\text{Percentage of population aged 15–59}} \times 100$$

Over all dependency ratio = Young dependency ratio + Old dependency ratio.

A multiple pie diagram of the above data is shown in Fig. 3.21

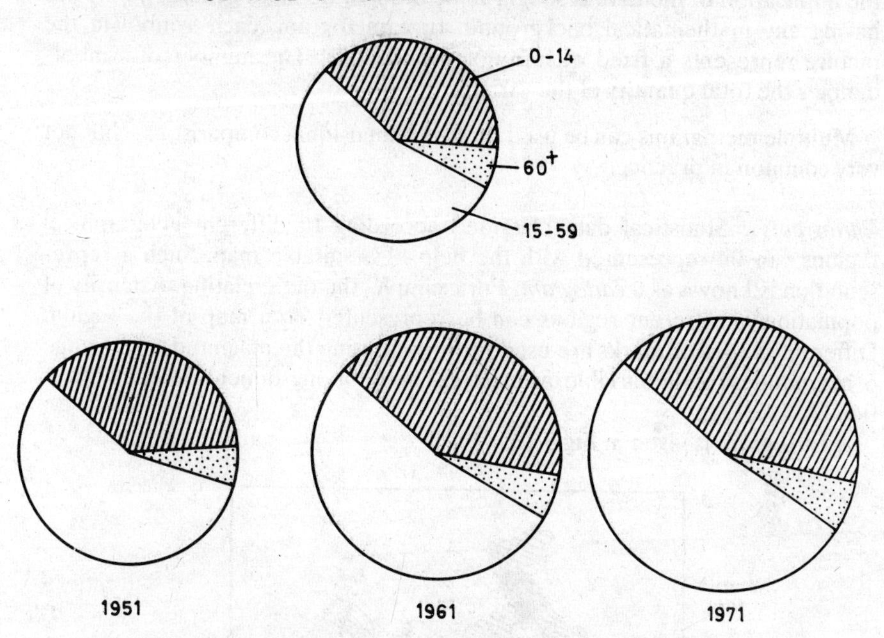

Fig. 3.21 Distribution of Population (of India) by Board Age Group

Table 3.17 shows the comparative position of India, Bangladesh, Indonesia and Japan in 1985.

Table 3.17. Percentage Distribution by age-group and dependency ratio (Source 1985 ESCAP Population Datasheet)

Country	Total population (millions)	Percentage Distribution by age group			Index of Ageing	Dependency Ratio		
		0-14	15-64	65+		Young	Old	Overall
India	762.1	38.1	57.6	4.3	11.29	66.15	7.47	73.62
Bangladesh	101.1	45.7	51.2	3.1	6.78	89.26	6.0.5	95.31
Indonesia	166.4	38.7	57.8	3.5	9.04	66.96	6.06	73.02
Japan	120.8	21.6	68.2	10.2	47.22	31.67	14.96	46.63

Note that 15-64 is taken as the age group of working population.

Note also the difference in the ratios between developing countries and a developed country (Japan).

3.9.3. Pictorial Diagram or Pictogram

Data are presented with the help of pictures also. Such a presentation is known as pictorial diagram or pictogram. Here the magnitudes of quantities of the variable are explained with the help of pictures which depict the variable appropriately. This quick and popular visual presentation of the data is used for one-dimensional comparison over time or space. It is useful specially when the implication of the data is sought to be brought home to general public not having any mathematical background. In a pictogram, each symbol in the picture represents a fixed quantity of the variable. The number of symbols denotes the total quantity of the variable.

Multiple pictograms can be used for two-dimensional comparisons – but not very common in practice.

Cartograms : Statistical data classified according to different geographical regions can be represented with the help of a suitable map. Such a representation is known as a *cartogram*. For example, the data relating to density of population of different regions can be represented on a map of the region. Different shades or marks are used for representing the magnitudes (or range of magnitudes) of the variable over space. This is for one-dimensional classification.

A cartogram is given in Fig 3.22.

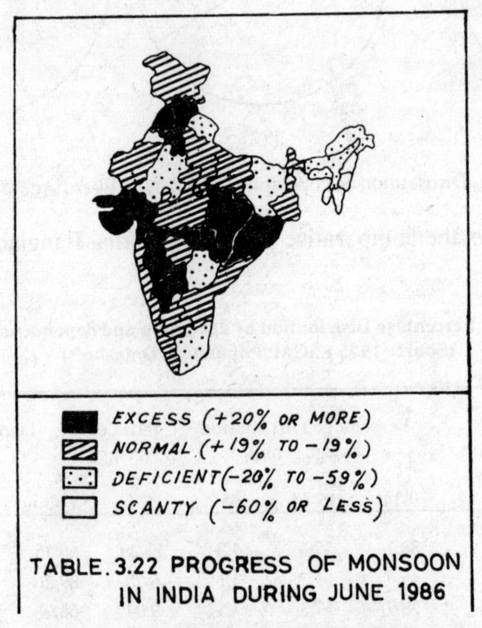

Fig. 3.22 Progress of monsoon in India during June 1986
(Courtesy : The Times of India)

Comparison of tabular and graphic/diagrammatic presentation

Statistical data may be presented in both tabular and graphic/diagrammatic forms. Whether graphs or tables are to be used depends on the purpose as well as the audience for whom such representation is meant. The main advantages of one over the other are as indicated below:

Firstly, a graph gives only a general impression but is more appealing to the general public than a table.

Secondly, a table requires much closer examination and is more difficult to interpret than a graph.

Thirdly, more information can be shown in a table than in a graph.

Fourthly, in a table, exact figures can be indicated while in a graph only approximate figures can be shown.

EXERCISES–3

Section 3.1

1. Distinguish between classification and tabulation of statistical data. What purposes do classification and tabulation serve? Mention the important basis of classification.
2. Prepare a blank table for showing the distribution of population of a country regionwise (four regions) according to sex and to age groups (three groups).
3. Prepare a blank table for presenting the distribution of students of a degree college according to
   (1) class (three classes)
   (2) sex (two sexes)
   (3) type of residence (three types: residing in home, in students' residence and in other type of residence)
   (4) income of guardian (three income groups I, II, III)
4. The following is an extract from the Annual Report of the Indian Oil Corporation Ltd. for the year 1969-70:

   The progress achieved by the Corporation in the three facets of its operations are as follows: The Corporation achieved a crude throughout of 6.25 million tonnes in its own refineries at Jawaharnagar, Barauni and Gauhati (3.40 million tonnes at Jawaharnagar, 2.09 million tonnes at Barauni and 0.76 million tonnes at Guwahati) and a sale of 1046 million kilolitres, thereby recording an increase over the preceding year by 13 percent and 28 per cent respectively. During the year, 2.42 million tonnes of petroleum products were transported through the pipelines of the corporation as against 1.99 million tonnes during the previous year.

   Profit during the year before providing for depreciation and interest amounts to Rs. 36.20 crores as against Rs. 32.41 crores for the previous year. Profit after depreciation and interest amounts to Rs. 20.41 crores during the year as against Rs. 18.46 during the previous year.

   Present the data in a suitable tabular form.
5. The following is an extract from the annual Report of the Life Insurance Corporation of India for 1969-70:

   During 1969-70, new business of Rs. 1,036.08 crores was introduced under 14.01 lakh policies as against Rs. 929.35 crores of business under 14.54 lakh policies in 1968-69. In 1969-70, 4.62 lakh policies were issued to people in rural areas as against 4.77 lakh policies in the previous year. The total income of the Corporation during 1969-70 amounted to Rs. 346.35 crores made up of Rs. 260.41 crores of premium, Rs. 78.43 crores of interest, dividends and rents and Rs. 7.51 crores of miscellaneous income. The overall expense ratio for the year for life business was 27.7 per cent as against 27.5 per cent in the previous year.

   Arrange the above information in a suitable tabular form and represent the same by a suitable diagram or diagrams. Write also an explanatory note on the data, adding your comments thereon.

Section 3.2

6. Distinguish between discrete and continuous variables with suitable examples.

Sections 3.3 and 3.4

7. Explain what do you mean by a frequency distribution. Enumerate the general rules for construction of a frequency distribution.
8. Explain clearly what do you understand by class intervals, class limits, and class boundaries of a frequency distribution.
9. What considerations would you bear in mind in choosing the number of class intervals, the class limits and the class boundaries of a frequency distribution?
10. The following are the weekly wages (in Rs.) of 40 employees of a firm:

160, 85, 148, 144, 95,	139, 94, 68, 135, 148,
125, 135, 85, 114, 63,	105, 100, 75, 125, 105,
85, 145, 132, 60, 135,	75, 115, 98, 105, 74,
76, 84, 110, 114, 115,	125, 126, 84, 148, 86.

Construct a frequency distribution table of the above data.

The following data show the diameters (in cms.) of a sample of 50 items manufactured by a firm.

5.03, 5.08, 5.10, 5.06, 5.02, 5.00, 5.10, 5.04, 5.10, 5.09
4.96, 5.12, 5.08, 5.07, 5.09, 5.16, 5.06, 5.08, 5.07, 5.10
5.09, 5.08, 5.07, 5.10, 5.12, 5.08, 5.07, 5.11, 5.12, 5.08
5.06, 5.10, 5.08, 5.08, 5.10, 5.11, 5.08, 5.10, 5.08, 5.06
5.08, 5.12, 5.06, 5.10, 5.12, 5.13, 5.14, 5.12, 5.08, 5.15.

11. Construct a frequency distribution table and a cumulative frequency distribution table.
12. Draw a histogram and a frequency polygon.
13. The ages (to the nearest years) of 40 workers of a firm are as given below:

24, 28, 26, 35, 46,	52, 26, 48, 59, 46,
32, 36, 38, 42, 53,	40, 42, 46, 29, 26,
36, 40, 35, 57, 48,	56, 58, 34, 36, 42,
30, 35, 42, 46, 50,	52, 54, 50, 56, 38,

Construct a frequency distribution table and draw the histogram.
Prepare a cumulative frequency distribution table ("more than") and draw the ogive.

Sections 3.5 and 3.6

14. Write a note on the representation of a frequency distribution. Explain how you would draw a histogram of data of (1) continuous variable (2) discrete variable.
15. Mention the different types of diagrams used for representation of a frequency distribution.
16. What are frequency polygons and frequency curves? How do you construct them?
17. Indicate the characteristic shapes of frequency curves corresponding to data found in nature and in industry.
18. Represent the data of Table 3.3 by a histogram. Draw a frequency polygon and a frequency curve.
19. Represent the data of Table 3.6 by a histogram, a frequency polygon and a frequency curve.
20. The table below gives the weekly wages (in Rupees) of 200 employees of a factory.

Table 3.18. Weekly wages of 200 employees

Weekly wages (in Rs.)	Number of employees
150-170	10
170-190	24
190-210	36
210-230	58
230-250	45
250-270	15
270-290	12

The class interval 150-170 denotes income of Rs. 150 or more but less than Rs. 170 and so on.

Draw a histogram, frequency polygon and frequency curve of the data.

21. What do you mean by cumulative frequency curve or ogive? What are the different types of ogives? Explain their uses.
22. Prepare "more than" and "less than" type of cumulative frequency distributions of the data relating to distribution of lifetimes of 350 radio tubes:
    Draw the two ogives.

Table 3.19. Distribution of life-times of 350 radio tubes

Life time (in hours)	*No. of tubes with life time.*
300-400	6
400-500	18
500-600	73
600-700	165
700-800	62
800-900	22
900-1000	4

What percentage of tubes have life times:
(i) Greater than 750 hours.
(ii) Between 650 and 850 hours?

23. Toss 3 similar coins together and note the number of heads (x). Repeat the experiment 60 times and count the number of heads $x = 0, 1, 2, 3$. State whether x represents a discrete variable. Construct a frequency table and draw the histogram.
24. Construct the cumulative frequency distributions ("more than" and "less than") of the data of Exercise 20 and draw the ogives thereof. Estimate the percentage of employees who have wages less than or equal to Rs. 200.
25. The distribution of rent of 150 residential houses surveyed in a locality is as follows:

Table 3.20 Distribution of residential houses by rent paid

Rent (in Rs.)	Number of houses
less than or equal to 150	10
151-200	18
201-250	27
251-300	35
301-350	30
351-400	16
401-450	8
451-500	6

Draw a histogram of the data and the frequency polygon.

26. The following table shows the distribution of 785 thousand individuals according to their income.

Table 3.21 Distribution of individuals according to income

Class interval (in Rs.)	Frequency (in thousands)
Below 10,000	110
10,000-20,000	282
20,000-30,000	187
30,000-40,000	82
40,000-50,000	65
50,000-60,000	26
60,000-70,000	14
70,000-80,000	9
80,000-90,000	5
90,000-1,00,000	2
Over 1,00,000	3

Draw a histogram, a frequency polygon and a "more than" type ogive of the above data relating to income of individuals. What kind of graph would the frequency curve show?

If Rs. 18,000 is the minimum *taxable* income, estimate the percentage of individuals liable to pay income tax. Find the percentage if the taxable income level is raised to Rs. 25,000.

Section 3.8

27. What do you mean by bivariate frequency table? The following data show the heights (x) and the weight (y) of 40 students. Represent the data by a bivariate frequency distribution table.

x (in cm):	135	143	175	165	154	146	152	138	148	172
y (in kg):	42	48	60	53	45	48	50	44	42	54
x (in cm):	145	158	163	136	156	148	162	174	138	145
y (in kg):	50	52	54	42	50	46	54	63	41	45
x (in cm):	160	138	153	154	162	164	158	148	146	165
y (in kg):	52	46	50	52	54	52	56	46	48	60
x (in cm):	156	146	154	156	162	143	158	165	167	148
y (in kg):	52	50	53	48	58	45	48	57	60	45

28. What are the various types of diagrams commonly used for representation of data? Explain them and the methods of drawing them.
29. Explain the method of construction of (1) subdivided bar diagram (2) percentage bar diagram (3) pie chart.
    Indicate the situations where they are useful.
30. Mention the advantages and disadvantages of diagrammatic representation over tabular representation of statistical data?
31. The following table gives the gross domestic product at factor cost by industry of origin for India in 1960-61 and 1974-75.

(in crores of Rupees)

Industry of origin	1960-61	1974-75
I Agriculture, etc.	7150	30046
II Mining, Manufacturing, etc.	2720	12790
III Transport, Communication, etc.	2014	10085

IV Finance and Real Estate, etc.	770	2936
V Community and Personal Service, etc.	1414	6040

Represent the above data by a suitable diagram(s).
Write a critical note bringing out the salient features. Add your comments thereto.

32. The following table gives the percentage distribution of Gross Domestic Product (at factor cost) by industry of origin for India and Great Britain in 1965-66:

Industry	Gross Domestic Product (in percentage)	
	India	Great Britain
1. Agriculture etc.	46.1	3.4
2. Mining etc.	1.1	2.3
3. Manufacturing	15.3	34.7
4. Construction, Gas, Electricity etc.	5.7	10.5
5. Transport, Distribution etc.	15.5	20.2
6. Other services	16.3	28.9

Represent the above by some suitable diagram.
Write a note on the data.

33. Selected economic indicators of India are given below:

(A) Per capita income at constant prices	(B) Working Population as % of total	(C) Gross output in Industry per capita	(D) Fertiliser consumption per hectare of cropped area	(E) Electricity consumption per capita (kwh)
(1974-75) (Rs.)	(1971)	(1974-75) (Rs.)	(1974-75) (Rs.)	(1975-76)
989	32.9	445.3	15.8	72.5

Represent the above data by a suitable diagram.

34. The following table gives estimates of reserves of Crude Oil and Carbon in different parts of the world:

	Estimates of Reserves (in percentages)	
	Crude Oil	Carbon
N. America	8.6	18.1
S. America	4.4	0.4
West Europe	1.2	1.8
U.S.S.R. and E. Europe	10.9	64.4
Africa	8.9	1.0
Middle East	60.9	—
China	2.3	11.5
Others	2.8	2.8

Represent the above data by suitable diagrams.

35. The following table gives estimates of percentage of national income going to different income classes in India during 1960-61 to 1970-71 :

Percentage going to	1960-61	1965-66	1970-71
Bottom 20% (of income class)	8.99	9.07	10.94
Next 20% (of income class)	12.14	8.01	13.55
Middle 20% (of income class)	14.25	14.65	15.10
Next 20% (of income class)	26.12	25.98	24.74
Top 20% (of income class)	38.50	42.29	35.67

Represent the above data by suitable diagram(s).
Write a report bringing out the salient features.

36. The following table shows the distribution of marriages by age of groom and age of bride for Hongkong and Singapore for 1977.

Table 3.22. Distribution of marriages by age of groom and age of bride

Age-group

	-19	20-24	25-29	30-34	35-39	40-44	45 and above	Total
Hongkong Groom :	842	10205	16317	5940	2440	1549	4097	40930
Bride :	5057	19005	10694	1956	803	799	2076	40930
Singapore Groom :	345	6849	9157	2716	954	392	408	20821
Bride	3209	10757	5134	1041	411	138	131	20821

Represent the data by suitable diagram(s).

37. Given below is the country-wise (main) breakdown of production of tea. Represent the data by a suitable diagram.

	Production of Tea (in million kg) 1985	1984
India	657.3	645.1
Sri Lanka	214.1	208.1
Bangladesh	43.4	38.2
Indonesia	107.7	102.9
Kenya	147.1	116.2
Uganda	4.6	5.2
Tanzania	16.5	16.5
Malawi	39.9	37.5
Total	1.230.6	1.169.7

Write a note on the above data bringing out the salient features.

Chapter 4

Measures of Location and Dispersion

4.1. INTRODUCTION

The main purpose of classification of data and of giving graphical and diagrammatical representation is to indicate the *nature* of the distribution: to find out the pattern or type of the distribution. Over and over the graphical and diagrammatical represenatation there are certain arithmetical measures which give a more precise description of the distribution. Such measures also enable us to compare two similar distributions. For example, if we draw the histograms of weights of a large number of school students in a developed country and in a developing country, the two histograms may look similar in shape. But if we compare the *average* weights in the two cases we may notice an appreciable difference. There are other arithmetical description measures besides the average which are helpful in comparing two distributions. These measures are helpful for solving some of the important problems of statistical inference.

Measures of Central Tendency: One of the most important aspects of describing a distribution is the central value around which the observations are distributed. Any arithmetical measure which is intended to represent the centre or central value of a set of observations is known as a *measure of central tendency* or *measure of location*.

4.2. THE ARITHMETIC MEAN (OR SIMPLY 'MEAN')

Suppose that n observations are obtained from a sample from a population. Denote the values of the n observations by $x_1, x_2, \ldots, x_n$: x_1 being the value of the first sample observation, x_2 that of the second sample observation and so on. The *arithmetic mean* or *mean* or *average*, denoted by $\bar{x}$, is given by the formula

$$\bar{x} = \frac{x_1 + x_2 + \ldots + x_n}{n} \qquad (2.1)$$

The above mean represents the centre of the set of observations. Suppose that the values $x_1, \ldots, x_n$ are plotted on a thin bar in the form of a straight line and balls representing them are placed on the positions of these points and the bar is cut in the two extreme points, then the value $\bar{x}$ will represent the point on which the bar so cut will balance, viz. $\bar{x}$ is the centre of gravity of the bar. The values of the observations will lie round the mean. This is the geometric interpretation of the mean.

The value of $\bar{x}$ represents an average value. Since there are many other types of averages, it is called the *arithmetic mean* or simply the *mean*. Since this mean is obtained from a *sample* of n observations of a population, it is also known a *sample mean*. The above formula enables us to find the mean when values $x_1, \ldots, x_n$ of n discrete observations are available.

Sometimes the data set are given in the form of a frequency distribution table.

4.2.1. The Arithmetic Mean of Grouped Data

To illustrate the technique of calculation of arithmetic mean from grouped data, consider the frequency table (Table 3.4). The mid point of the class interval 3.0-3.4 is 3.2, that of 3.5-3.9 is 3.7 and so on. We find that the frequency of the class 3.0-3.4, is 4, i.e., four observations lie between 3.0 and 3.4. It will be assumed that these four original observations will be replaced by four observations each equal to the midvalue 3.2. The observations falling in the calss 3.5-3.9 will each be taken to be equal to the mid value of the class 3.7. It is the same with all other class intervals. Table 3.3 is rewritten as follows:

Class interval	Class mid point	frequency
3.0-3.4	3.2	4
3.5-3.9	3.7	7
4.0-4.4	4.2	8
4.5-4.9	4.7	11
5.0-5.4	5.2	9
5.5-5.9	5.7	5
6.0-6.4	6.2	4
6.5-6.9	6.7	2
		50

Here the measurement 3.2 is repeated four times, 3.7 is repeated seven times and so on. The total yield which is the sum of these observations is thus

$$3.2 \times 4 + 3.7 \times 7 + 4.2 \times 8 + 4.7 \times 11 + 5.2 \times 9 + 5.7 \times 5 + 6.2 \times 4 + 6.7 \times 2 = 237.5$$

We divide this by the total number 50 to obtain the arithmetic mean, which equals

$$\bar{x} = \frac{237.5}{50} = 4.75$$

The formula for the arithmetic mean of grouped data can thus be written down as follows:

Suppose that there are k classes or intervals. Let $x_1, x_2, \ldots x_k$ denote the class mid points of these k intervals and let $f_1, f_2, \ldots, f_k$ denote the corresponding frequencies of these classes. Then the arithmetic mean $\bar{x}$ is given by

$$\bar{x} = \frac{\sum_{i=1}^{k} x_i f_i}{\sum f_i} = \frac{\text{Sum of } \{(\text{mid point of class intervals}) \times (\text{frequency of the class})\}}{\text{Total frequency}} \quad (2.2)$$

Here we replace each measurement of a class by its class mid point. The assumed value differs from the actual value of the observation, and thus the formula (2.2) will be an approximation to the correct value obtained by considering the original observations. If we use formula (2.1) to the raw data of Example 4.a(p.20), we find the total yield (sum of the observations) is 235.7 and the arithmetic mean is

$$\frac{235.7}{50} = 4.714$$

This exact value 4.714 differs from the approximate value 4.75 obtained by using the formula (2.2). The difference in this particular case is small. Usually the difference is small and can be neglected unless however the classification is rather crude.

The formula (2.2) is useful as data sets are often given in the form of a frequency distribution table with specified class intervals. The formula is also very convenient to use. However, where exactness is desired, it is advisable to calculate the arithmetic mean from raw data: in these days of high speed computers this should not be a problem. When only a desk calculator is available and approximation can be admitted, formula (2.2) can be used.

4.2.2. Properties of the Arithmetic Mean

The arithmetic mean has some simple but important algebraic properties. Because of such simple properties, the arithmetic mean is more often used as an average, though there are other types of averages. The properties are discussed below:

(a) The sum of the deviations of a set of n observations $x_1, x_2, \ldots, x_n$ from their mean $\bar{x}$ is equal to zero.

Proof: Let d_i denote the deviation of x_i from $\bar{x}$, i.e.,

$$d_1 = x_1 - \bar{x}, d_2 = x_2 - \bar{x}, \ldots, d_i = x_i - \bar{x}, \ldots, d_n = x_n - \bar{x}$$

where $\bar{x} = \dfrac{\Sigma x_i}{n}$.

Then sum of deviations

$$\Sigma d_i = \Sigma(x_i - \bar{x}) = \Sigma x_i - n\bar{x}$$

$$= \Sigma x_i - n \cdot \frac{\Sigma x_i}{n} = 0$$

(b) If $x_1, \ldots, x_n$ are n observations, $\bar{x}$, is their mean and $d_i = x_i - A$, is the deviation of x_i from a given number A, then

$$\bar{x} = A + \frac{\Sigma d_i}{n} \tag{2.3}$$

Proof : We have

$$\bar{x} = \frac{\Sigma x_i}{n} = \frac{1}{n}\Sigma(d_i + A)$$

$$= \frac{1}{n}\Sigma d_i + \frac{1}{n}.nA$$

$$= A + \frac{1}{n}\Sigma d_i$$

(c) If the numbers $x_1, \ldots, x_n$ occur with the frequencies $f_1, \ldots, f_n$ respectively and $d_i = x_i - A$, then

$$\bar{x} = A + \frac{\Sigma f_i d_i}{N} \tag{2.4}$$

where $N = \Sigma f_i$

Proof:

$$\bar{x} = \frac{\Sigma f_i x_i}{N} = \frac{\Sigma f_i(d_i + A)}{N} = \frac{\Sigma f_i d_i}{N} + \frac{NA}{N}$$

$$= A + \frac{\Sigma f_i d_i}{N}$$

(d) If in a frequency distribution all the k class intervals are of the same width c, and $d_i = x_i - A$ denote the deviation of x_i from A, where A is the value of a certain class mid point and $x_i, \ldots, x_k$ are the class mid points of the k classes, then

$$d_i = cu_i \text{, where } u_i = 0, \pm 1, \pm 2, \ldots$$

and

$$\bar{x} = A + \left(\frac{\Sigma f_i u_i}{\Sigma f_i}\right) c \tag{2.5}$$

Proof: Since the class intervals are of equal width c the class midpoints are equally spaced, each with a difference of c from the neighbouring one. Taking A as the value of a *certain midpoint*, say, around the middle of the frequency table, it can be seen that the deviations on the lower side from the middle are positive integral multiples of c and those on the upper side of the middle are negative integral multiples of c. That is, the deviations d_i can be expressed as cu_i where $u_i = 0, u_i = 1, u_i = 2, \ldots$ for the lower side class mid points and $u_i = -1, u_i = -2$ for the upper side class mid points.

Again, writing $\Sigma f_i = N$

$$\bar{x} = \frac{\Sigma f_i x_i}{N} = \frac{\Sigma f_i(d_i + A)}{N}$$

$$= \frac{\Sigma f_i(cu_i + A)}{N} = \frac{c\Sigma f_i u_i}{N} + A$$

$$= A + \left(\frac{\Sigma f_i u_i}{\Sigma f_i}\right) c.$$

Note that A is a certain given class mid point. Whenever grouped data with equal class intervals are given, we would find that this is the most convenient form for computation. We shall illustrate the use of this formula with the help of an example.

Example 2(a). To find the mean of the distribution given by the frequency Table 3.6.

Here the class width of each class is the same, $c = 2$. We have to find the class mid points and choose a class mid point around the middle of the table. Here there are 8 classes, so that 4th and 5th classes are in the middle. We take the 4th class mid point as A (= 151.5). We may write the table as follows:

Class interval (in cm.)	Class mid point (x_1)	$d_i = x_i - A$ (A = 151.5)	$u_i = \frac{d_i}{c}$	f_i	$f_i u_i$
145-146	145.5	−6	−3	2	−6
147-148	147.5	−4	−2	5	−10
149-150	149.5	−2	−1	8	−8
151-152	151.5	0	0	15	0
153-154	153.5	2	1	9	9
155-156	155.5	4	2	6	12
157-158	157.5	6	3	4	12
159-160	159.5	8	4	1	4
	Total			50	13

Thus $$\bar{x} = A + \left(\frac{\Sigma f_i u_i}{\Sigma f_i}\right) c = 151.5 + \frac{13}{50} \times 2$$

$$= 151.5 + 0.52$$

$$= 152.02 \text{ (cm.)}$$

Note:

(1) The value A is taken as the class mid point of a suitably chosen class interval. This interval is around the middle of the table. It is convenient to take this class mid point corresponding to the class which has the largest frequency, though such a class may not be exactly in the middle.

(2) In some cases *all* the class intervals are not of the same width specially the end class intervals. In that case formula (2.4) is to be used taking a suitable value for A preferably as the class mid point of the class with highest frequency.

4.2.3. The Merits and Demerits of the Arithmetic Mean as a Measure of Location

The mean has the following *merits*:

(1) Its definition is precise and exact. It is easy to compute and has a determinate value.
(2) It is based on each and every observation constituting the data.
(3) The calculation of mean does not necessitate arrangement the data in an array as is required in case of calculation of median or grouping.
(4) It has some simple and interesting properties and easy algebraic manipulation is possible with the mean.
(5) The mean is a point of balance.

The mean has the following *demerits*:

(1) It is greatly influenced by the extreme values, the largest and smallest observations. This sometimes reduces its utility as a representative value. Moreover, such extreme values greatly effect the value of the mean.
(2) For its computation, each and every value is necessary and every item counts. Even if one observation is erroneous, then the mean is effected, and if it is missing, it cannot be computed.
(3) Sometimes in grouped data, some of the end intervals are not fully specified, for example, the observations below (or above) a certain value may be grouped in an end interval. In such cases the mean cannot be exactly calculated (the mean can then be approximated from the rest by ignoring the end interval(s).)
(4) It is difficult to locate the mean by inspection or by the graphical method.
(5) The mean computed can be a value which is not assumed by any of the observations.

4.3. THE MEDIAN

The median of a set of n measurements or observations $x_1, \ldots, x_n$ is the middle value when the measurements are arranged in an array according to their order of magnitude. If n is odd, the middle value which is the $(n/2 + 1)$th in the ascending order of magnitude is unique and is the median. If n is even, there are two middle values and the average of these values is the median. The median of the set 2, 3, 5, 6, 7 is 5 and that of the set -3, -1, 0, 1, 2, 3 is 0.5. The median of the data given in Example 3.3 is 4.7, being the mean of the 25th and 26th observation, when the observations are arranged in an array.
The median is the value which divides the set of observations into two equal halves, such that 50% of the observations lie below the median and 50% above the median.

The median is not affected by the actual values of the observations but rather on their positions.

4.3.1. The Median of Grouped Data

Consider the case when data is given in the form of a frequency table. At first the class interval in which the median (i.e. the middle value) lies is to be determined; then the problem is to determine where in this class interval does the median lie. The method is illustrated by means of an example below. Consider the data given in Table 3.6. The median is the 50/2 (= 25)th measurement.

The median lies in the interval 151-152, that is, the interval with class boundaries 150.5 and 152.5. Since the total number of measurements upto the preceding interval is 15, the desired median is the (25 − 15 = 10)th measurement. The 10th one in this interval 150.5-152.5 is to be determined – there are in all 15 measurements in this interval; the 10th will lie in the 10/15 of the way between the lower and upper class boundaries 150.5 and 152.5. Thus the median is

$$150.5 + \frac{10}{15} \times 2 = 150.5 + 1.33$$
$$= 151.83 \text{ (cm)}$$

The formula for obtaining the median of grouped data may be written as follows:

$$\text{Median} = L_m + \left(\frac{\frac{N}{2} - (\Sigma f)_0}{f_m} \right) \times c \qquad (3.1)$$

where N = total frequency

f_m = frequency of the class where the median lies

L_m = lower class boundary of the class where the median lies

$(\Sigma f)_0$ = sum of frequencies of classes below (or lower than) the class where the median lies

and c = width of the median class interval

= difference between the upper and lower class boundaries of the median class.

Let us find the median of the data given in Table 3.4 using formula (3.1).

Here 4.5-4.9 is the class where the median lies. We have $L_m = 4.45$, $(\Sigma f)_0 = 19$, $c = 0.5$, $f_m = 11$ and $N = 50$.

Using (3.1) we get

$$\text{Median} = 4.45 + \frac{\frac{50}{2} - 19}{11} \times 0.5$$
$$= 4.45 + 0.27$$
$$= 4.72 \text{ (kg.)}.$$

Note: The above formula is usable even if all the class intervals are not of the same width. Thus even when the end class intervals are open, this can be used. The median is not effected by the values assumed by the extreme observations.

4.3.2. Graphical Determination of Median

The median can also be obtained graphically, either using the histogram or the ogive. We discuss the method of obtaining the median through the ogive. We indicated the method of drawing the two ogives ("less than" and "more than" types) in section 3.6. Consider the two ogives shown therewith in that section for the data given in Example 3(b). Consider the "less than" ogive: along the x-axis, we have heights and along the y-axis cumulative frequencies "less than" and "more than" in the two graphs given in Fig. 3.5. The median is the abscissa of the point A on the ogive whose ordinate corresponds to 50% of cumulative frequency; in other words A is the point on the graph corresponding to 25 (=50%) of cumulative frequency. From the graph we find that the x-coordinate of the point A is 151.4. Similarly the median can be determined from the "more than" type ogive.

Another method, perhaps a better one is to draw the two ogives in the same figure. The x-co-ordinate of their point of intersection gives the median. Here c is the point of intersection with abscissa 151.8 cm. which gives the median. This tallies with the median found arithmetically.

The graphical method is found to be very useful.

4.3.3. Merits and Demerits of the Median as a Measure of Location

The merits are:

(1) It is easy to compute. It can be calculated by mere inspection and by the graphical method.

(2) It can be found even when the data are not completely known. If the number of items is known then the median can be computed even if the extreme values are not known or are missing.

(3) The median is affected by the middle one or two values but not the extreme values.

(4) In case of grouped data, it can be computed even if the end intervals are open or not clearly specified.

(5) The median is more useful in certain situations, for example, in computing the average income of a group of individuals.

(6) It is useful in the study of certain attributes which cannot be directly measured.

The demerits are:

(1) If the series is long then rearranging the data in an array is tedious and time-consuming.

(2) It does not possess much useful algebraic property. It is *not* amenable to algebraic manipulations.

(3) Its computation does not involve every value or observations. As such it cannot be considered as representative in many situations.

(4) It may not be representative values if the sets of observations lack pattern.

(5) It involves interpolation in case of group data—and the calculation is based on the assumption that all frequencies in a given interval are uniformly spaced or distributed in the class-interval. This may not always hold.

4.4. THE MODE

The *mode* is the observation which occurs most frequently in a set. The word comes from the French word '*la mode*' (which means the fashion).

4.4.1. Calculation of Mode

The computation of mode from a set of observations is simple enough. It is the observation which occurs most frequently. For example the mode of 3, 2, 3, –1, 0, –3, 2, 5 is 3. The mode may not exist for certain sets such as set –1, 0, 1, 5. There may be more than one mode.

For grouped data the mode may be determined as follows: Consider a histogram of grouped data and concentrate on three rectangles – the central rectangle GBCJ, containing the modal class and the two adjacent rectangles as shown in Fig. 4.1.

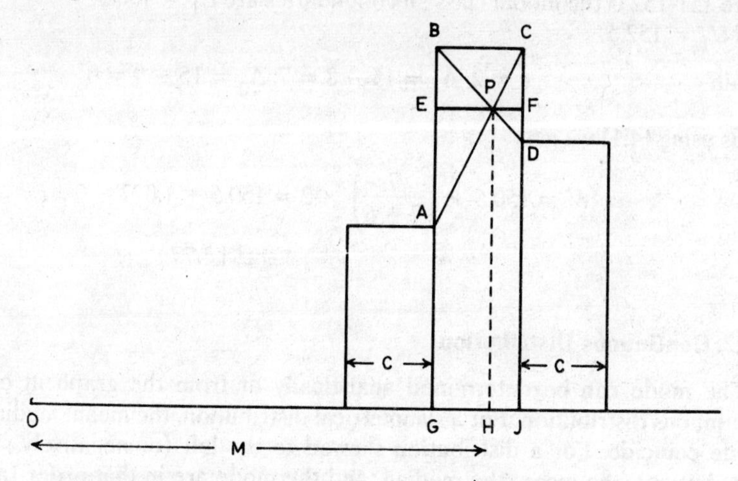

Fig. 4.1 Mode of Grouped Data-determination from Histogram

The mode M is *defined* as the abscissa OH of P, the point of intersection of the two straight lines AC and BD. Let

c = size of the class intervals ($= GJ$)

$OG = L_1$, the lower boundary of the modal class,

$OJ = U_1 = L_1 + c$, the upper boundary of model class,

$\Delta_1 = AB = BG - AG$ = excess of the modal class frequency over the frequency of the class to its left,

and $\Delta_2 = CD = CJ - DJ$ = excess of the modal class frequency over the frequency of the class to its right.

From the similar Δ^s ABP and DPC, we have

$$\frac{EP}{AB} = \frac{PF}{CD}$$

or

$$\frac{M - L_1}{\Delta_1} = \frac{U_1 - M}{\Delta_2} = \frac{L_1 + c - M}{\Delta_2}$$

or

$$M(\Delta_1 + \Delta_2) = L_1\Delta_2 + \Delta_1 L_1 + c\Delta_1$$

$$= L_1(\Delta_1 + \Delta_2) + c\Delta_1$$

Thus
$$M = L_1 + \left(\frac{\Delta_1}{\Delta_1 + \Delta_2}\right)c \quad (4.1)$$

Example 4(a). To calculate the mode for data given in Table 3.6.
Here 151-152 is the modal class : its boundaries are $L_1 = 150.5$ and $U_1 = 152.5$

Again
$$c = 2, \Delta_1 = 15 - 8 = 7, \Delta_2 = 15 - 9 = 6$$

Thus using (4.1) we get

$$M = 150.5 + \left(\frac{7}{7+6}\right) \times 2 = 150.5 + 1.077$$
$$= 151.577$$

4.4.2. Continuous Distribution

The mode can be determined analytically or from the graph in case of continuous distribution. For a symmetrical distribution, the mean, median and mode coincide. For a distribution skewed to the left (or negatively skewed distribution), the mean, the median, and the mode are in that order (as they appear in the dictionary) and for a distribution skewed to the right (or positively skewed distribution) they occur in the reverse order: mode, median and mean. This is shown in Fig. 4.2.

4.4.3. Empirical Formula

There is an *empirical* formula for a moderately asymmetrical skewed distribution. It is given by

$$\text{Mean} - \text{Mode} = 3\,(\text{Mean} - \text{Median}) \quad (4.2)$$

This enables us to find, in case of a moderately asymmetrical distribution, the mode from the mean and the median.

For Example 2(a), Mean = 152.02 cm, median = 151.8 cm

and thus

$$\text{Mean} - \text{Mode} = 3(152.02 - 151.83) = 3 \times 0.19 = 0.57 \text{ (cm.)}$$

and
$$\text{Mode} = \text{Mean} - 0.57 = 152.02 - 0.57 = 151.45 \text{ (cm.)}$$

This also indicates that the distribution is skewed to the right.

4.4.4. Merits and Demerits of Mode as a Measure of Location

The merits of mode are :

(1) It is comparatively easy to understand.
(2) It is the simplest descriptive measure of average.
(3) It is easy to locate in case of ungrouped data.

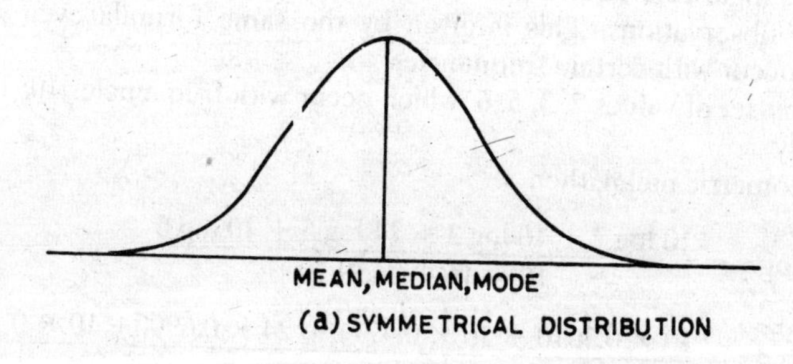

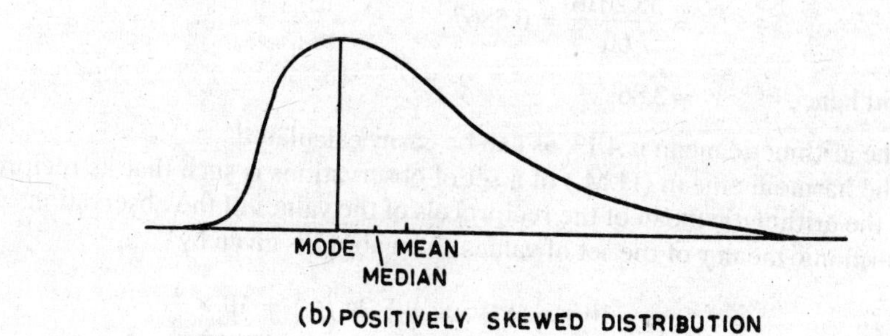

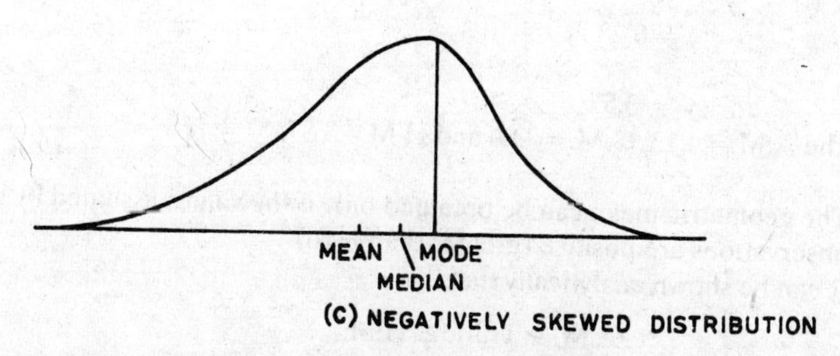

Fig. 4.2. Location of Mean. Median and mode for (a) Symmetrical (b) Positively skewed. (c) Negatively skewed distributions.

(4) The mode is not effected by the extreme values.
(5) It is useful as an average value for many situations, such as the average size of shoe and the average price of a commodity, the average type of dress or cloth and so on.

The demerits are :

(1) It is not precisely defined.
(2) The mode does not exist in many cases and there may be more than one mode – that is, it is not useful as an average in such situations.
(3) It is not based on all the observations of the set.
(4) It is not amenable to algebraic manipulation.

4.5. THE GEOMETRIC AND HARMONIC MEANS

There are two other averages, the geometric mean and the harmonic mean which are sometimes used. The geometric mean (G.M.) of a set of observations is such that its logarithm equals the arithmetic mean of the logarithms of the values of the observations. This is given by the same formula even if the observations occur with certain frequencies.

Consider the set of values 2, 3, 5, 6 which occur with frequencies 10, 16, 24, 10 respectively

If x is their geometric mean, then

$$\log x = \frac{10 \log 2 + 16 \log 3 + 24 \log 5 + 10 \log 6}{10 + 16 + 24 + 10}$$

$$= \frac{10 \times 0.3010 + 16 \times 0.4771 + 24 \times 0.6990 + 10 \times 0.7782}{60}$$

$$= \frac{35.2016}{60} = 0.5867$$

and hence $\quad x = 3.86.$

The arithmetic mean is 4.13, as can be easily calculated.

The harmonic mean (H.M.) of a set of observations is such that its reciprocal is the arithmetic mean of the reciprocals of the values of the observations. The harmonic mean y of the set of values given above is given by

$$\frac{1}{y} = \frac{10 \times \frac{1}{2} + 16 \times \frac{1}{3} + 24 \times \frac{1}{5} + 10 \times \frac{1}{6}}{60}$$

$$= 0.28$$

so that.

$$y = 3.57$$

Thus the A.M. = 4.13, G.M. = 3.86 and H.M = 3.57.

Note :

(1) The geometric mean can be obtained only if the values assumed by the observations are positive (greater than zero).

(2) It can be shown analytically that

$$\text{A.M.} \geq \text{G.M.} \geq \text{H.M.}$$

4.5.1. Merits and Demerits of Geometric Mean and Harmonic Mean

The merits of Geometric Mean are as follows:

(1) It can be precisely determined.

(2) It takes into account every measurement or observation.

(3) It has algebraic properties and can be algebraically manipulated.

(4) It seeks to minimise the effect of extreme values. As logarithm of the values of the measurements are taken, comparatively small weight is given to large values and large weight to small values. If some items are large and some items are small, G.M. will be more representative as an average.

(5) It is useful in dealing with ratios, rates etc.

The demerits are :

(1) It is defined only for positive values of the variables. If there is any negative value or the value 0, G.M. is undefined.

(2) It involves use of logarithms.

The merits and demerits of Harmonic Mean.

The merits are :

(a) It can be precisely found.

(b) It is amenable to algebraic manipulation.

(c) It takes into account the values of each item.

(d) It measures relative changes.

The demerits are :

(a) It is rather complicated to calculate

(b) It gives much more weight to small item of values and this can lead to fallacious results.

The geometric mean is useful in dealing with ratios, rates and percentages. Thus it deals better with relative change rather than absolute change or difference. Index numbers are in fact relative measures and they deal with ratios. Thus the geometric mean is often used for construction of index numbers.

The harmonic mean is more useful in a situation where more weights are to be given to small items. It is useful for certain types of ratios which give measure of relative changes.

4.6. OTHER MEASURES OF LOCATION : QUARTILES, DECILES AND PERCENTILES

The median of a set of measurements is the value which divides the set into two equal halves, each containing 50% of the measurements. In the same way, some other measures of location can be considered. The three quartiles Q_1, Q_2 and Q_3 are such that when the measurements are arranged in increasing order they divide the set of measurements into four equal parts, the first quartile Q_1 contains the 25% of the measurements the second quartile Q_2 contains 50% of the measurements and the third quartile Q_3 contains 75% of the measurements. In fact the second quartile Q_2 is the median. Similarly the 9 deciles $D_1, D_2, \ldots,$ D_9 divide the set of measurements into 10 equal parts, 10% of the measurements being contained between two neighbouring deciles. The first decile D_1 contains 10% of the measurements, the second decile D_2 contains 20% of the measurements, and so on. The fifth decile is the median. In the same manner percentiles can be considered. The 99 percentiles $P_1, \ldots, P_{99}$ divide the set the measurements into 100 equal parts. The first percentile P_1 contains 1% of the measurements, the second percentile P_2 contains 2% of the measurements, and so on, the kth percentile P_k contains k% of the measurements. Thus 1% of the measurements is contained between two neighbouring percentiles. The 50th percentile is therefore the median.

The principle of finding out the quartiles, deciles and percentiles is basically the same as that of finding the median: whereas the median divides the set of observations into two equal halves, each containing 50% of the measurements, the 4th decile, for example, divides the set into two parts the first part being 40% of the set and the other containing 60% of the observations.

Using the same principle as is used in determining the median, we find the first quartile Q_1 of the data given in Table 3.6. The first quartile contains 25% of the measurements, that is, the desired value is $\frac{50 \times 25}{100} = 12.5$. The interval is 149-150. The first two intervals contain 7 measurements, that is, the class with boundaries 148.5 and 150.5 contains this. There are 8 measurements in this interval and the first quartile will lie in the $\frac{(12.5 - 7)}{8}$ of the way between the upper and lower class boundaries, 148.5 and 150.5. Thus the first quartile is

$$148.5 + \frac{12.5 - 7}{8} \times 2 = 148.5 + 1.375$$

$$= 149.875 \text{ (cm.)}$$

Similarly, the first quartile can be obtained from one of the ogives.

Example 6(a). The distribution of fortnightly wages of 280 employees of an undertaking is as follows:

Fortnightly wages (in Rs.)	Frequency
Less than 200	12
200-400	16
400-600	38
600-800	78
800-1000	80
1000-1200	35
1200-1400	14
above 1400	7
	280

Find the first quartile, the median and the third quartile; find D_4, P_{66}, P_{10} and P_{90}.

The cumulative frequency table is as given below:

Wages	Frequency	Cumulative frequency
Less than 200	12	12
200-400	16	28
400-600	38	66
600-800	78	144
800-1000	80	224
1000-1200	35	259
1200-1400	14	273
above 1400	7	280

Here the class 200-400 means wages from Rs. 200 to less than Rs. 400, i.e. Rs. 399.99. The class boundaries coincide approximately with the class limits.

The observation for the first quartile Q_1 corresponds to $\frac{280}{4}$ th = 70th observation, which lies in the interval 600-800, with lower class boundary 600.

This interval contains 78 observations and the intervals preceding this contain 66 observations. Thus

$$Q_1 = 600 + \frac{\frac{280}{4} - 66}{78} \times 200$$

$$= 600 + 10.25$$

$$= 610.25 \text{ (Rs.)}$$

The median which is the second quartile Q_2, is given by

$$Q_2 = 600 + \frac{\frac{280}{2} - 66}{78} \times 200$$

$$= 600 + \frac{74 \times 200}{78}$$

$$= 600 + 189.74 = 789.74 \text{ (Rs.)}$$

The third quartile Q_3 is given by

$$Q_3 = 800 + \frac{\frac{280 \times 3}{4} - 144}{80} \times 200$$

$$= 800 + 165$$

$$= 965 \text{ (Rs.)}$$

The observation for 4th decile corresponds to the $\frac{280 \times 4}{10}$ th = 112 th observation, which lies in the interval 600-800. Thus

$$D_4 = 600 + \frac{112 - 66}{78} \times 200$$

$$= 600 + 117.95$$

$$= 717.95 \text{ (Rs.)}$$

The observation for 66th percentile corresponds to $\frac{280 \times 66}{100}$ th = 184th observation, which lies in the interval 800-1000. Thus the 66th percentile P_{66} is given by

$$P_{66} = 800 + \frac{184.8 - 144}{80} \times 200$$

$$= 800 + \frac{40.8 \times 200}{80}$$

$$= 800 + 102$$

$$= 902 \text{ (Rs.)}$$

In the same manner

$$P_{10} = 400 + 0 = 400 \text{ (Rs.)}$$

$$P_{90} = 1000 + \frac{\frac{280 \times 90}{100} - 224}{35} \times 200$$

$$= 1000 + \frac{28 \times 200}{35}$$

$$= 1160.00 \text{ (Rs.)}$$

Note: The quartiles, deciles and percentiles can also be obtained from the graph of either of the ogives as follows:
The observation for Q_1 corresponds to the 70th observation. In the graph for ogive, we have cumulative frequencies along the y-axis. Mark the line $y^* = 70$. The abscissa of the point where this line meets the ogive is the required value.

This is shown in the graph below, Fig. 4.3.

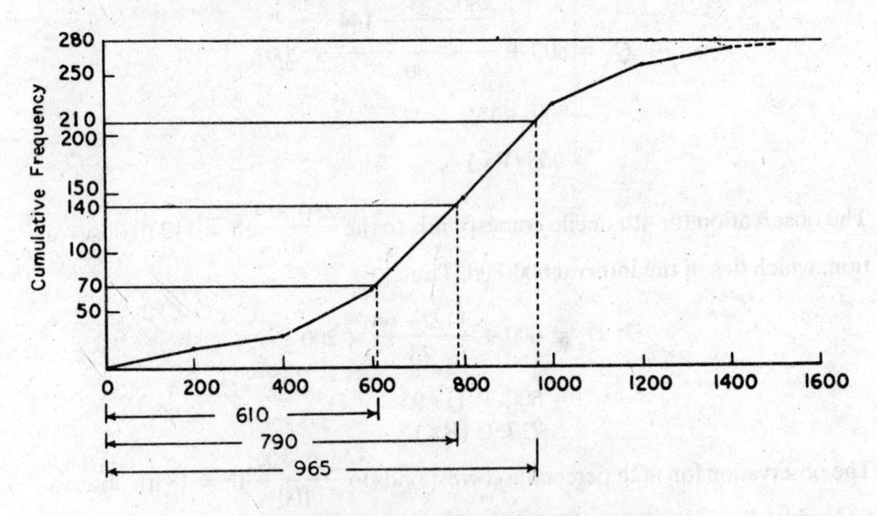

Fig. 4.3 Determination of Quartiles from Ogive

4.7. MEASURES OF VARIATION OR DISPERSION

An average is a measure of central tendency. It tells us where the centre of the set of data lies but does not tell us how the items of the set are distributed around the centre. Two sets may have the same averages but the items in one may scatter wildly around this centre while in the other case, all the items may lie close to the average. For example, the maximum temperature during a week in January in two cities may be as follows:

Max Temperature (in degree celsius)

City A:	10	12	8	9	6	4	8
City B:	0	4	8	12	14	11	8

Both show the same average 8.14. In the case of city B, the scatter around the average 8.14 is much greater than in the case of city A. The average does not enable us to draw a full picture of a set of observations. A further description about the degree of scatter is necessary to get a better description of the data. The degree to which data tend to spread around an average is called the *variation* or *dispersion*. We shall consider how the variation is measured and also the various *measures* of variation or dispersion that are available. Such a measure tells us how the items are spread around the average. Such a measure also permits us to compare two sets with regard to their variability. The most common measures of variation are the range, the semi-interquartile range, the mean deviation and the standard deviation.

These are measures of absolute variation. Some measures of relative dispersion are also available. These are the coefficient of variation, the coefficient of mean deviation and the coefficient of quartile deviation.

4.7.1. The Range

The range of a set of numbers is the difference between the largest and the smallest items of the set. For example, consider the maximum temperature in the example cited above, the range for city A is $12 - 4 = 8$ while for city B it is $14 - 0 = 14$. The range is a very crude measure. It does not tell us about the distribution of the values of the set relative to the average. Further a single extreme value, either the maximum or the minimum can very much effect the value of the range. Nevertheless the range gives us a quick and simple measure. It is used in statistical quality control as a ready measure of variation.

4.7.2. The Semi-Interquartile Range

Leaving aside the values below the first quartile Q_1 (or the lowest 25% of the values of the set) and the values above the third quartile Q_3 (or the highest 25% of the values of the set) we are left with the central 50% of the set.
This does lead to a more refined form of range.
The semi-interquartile range Q given by

$$Q = \frac{Q_3 - Q_1}{2}$$

is a measure of variation.

Consider Example 6(a) on distribution of wages of 280 employees. It is found that Q_1 = Rs. 610.25 and Q_3 = Rs. 965. Hence

$$Q = \frac{965.00 - 610.25}{2} = 177.37 \text{ (Rs.)}$$

4.7.3 The Mean Deviation

Consider a set of observations $x_1, \ldots, x_n$.
The mean or average deviation (M.D.) is defined by

$$M.D. = \frac{\sum_{i=1}^{n} |x_i - \bar{x}|}{n} \tag{7.1}$$

where
$$\bar{x} = \frac{\sum_{i=1}^{n} x_i}{n}$$

is the arithmetic mean and $|x_i - \bar{x}|$ denotes the *absolute* value of the deviation (or difference between x_i and $\bar{x}$.). The absolute value of a quantity is the quantity without the associated sign, that is, $|5| = 5$, $|-5| = 5$, $|-0.35| = 0.35$ and so on.

Note

(1) Since we use absolute deviation, it may be appropriate to call it *absolute* mean deviation, rather than simply mean deviation.

(2) The quantity $\sum_{i=1}^{n} (x_i - \bar{x})$ is always equal to zero.

Example 7(a). Find the mean deviation of the set of measurements 1, 3, 8.

Here the arithmetic mean $= \dfrac{1+3+8}{3} = 4$. Thus

$$Q = \frac{|1-4| + |3-4| + |8-4|}{3}$$

$$= \frac{|-3| + |-1| + |4|}{3} = \frac{3+1+4}{3} = 2.67$$

Mean deviation for grouped data

If $x_1, \ldots, x_k$ occur with frequencies $f_1, \ldots, f_k$, then

$$M.D. = \frac{\sum_{i=1}^{k} f_i |x_i - \bar{x}|}{\sum_{i=1}^{k} f_i} \tag{7.2}$$

where
$$\bar{x} = \frac{\Sigma f_i x_i}{\Sigma f_i}.$$

The above formula is applicable in the case of a frequency distribution whose class intervals have mid points $x_1, \ldots, x_k$ and the classes have frequencies $f_1, \ldots, f_k$.

***Example** 7(b).* Consider the data given in example 2(a), where $\bar{x}$ = 152.02 cm. We have the following Table.

Class mid point (x_i)	Absolute deviation $\lvert x_i - \bar{x} \rvert$	Frequency f_i	$f_i \lvert x_i - \bar{x} \rvert$
145.5	6.52	2	13.04
147.5	4.52	5	22.60
149.5	2.52	8	20.16
151.5	0.52	15	7.80
153.5	1.48	9	13.32
155.5	3.48	6	20.88
157.5	5.48	4	21.92
159.5	7.48	1	7.48
Total		50	127.20

Thus $$\text{M.D} = \frac{127.20}{50} = 2.544 \text{ (cm.)}$$

Note. In the above the arithmetic mean $\bar{x}$ is the quantity from which the deviations are taken. However, deviations from other averages could also be considered. For example, mean deviation about the median is the corresponding measure where deviations are taken from the *median*. Such a measure will be called *mean deviation about the median*. It can be shown that this quantity is less than the M.D. about the mean.

There is an important result which can be analytically proved. It can be stated as follows:

Of all the mean deviations taken about different averages (or any arbitrary value), the mean deviation about the median has the smallest value

Denote the mean deviation (M.D.) about an arbitrary value a by

$$d(a) = \frac{\Sigma f_i \, |x_i - a|}{\Sigma f_i} \tag{7.3}$$

and the M.D. about the median m by

$$d(m) = \frac{\Sigma f_i \, |x_i - m|}{\Sigma f_i} \tag{7.4}$$

It can be shown that ($\Sigma f_i = N$)

$$N[d(a)] = N[d(m)] + A \tag{7.5}$$

where $A \geq 0$ is non-negative; $A = 0$ only when $a = m$,
In other words,

$$d(a) \geq d(m) \tag{7.6}$$

and the equality holds when $a = m$.
[The proof of the relation (7.5) requires algebraic manipulations, see Ex. 4-37.]

4.8. THE VARIANCE AND THE STANDARD DEVIATION

4.8.1. Ungrouped Data

Consider a set of sample observations $x_1, x_2, \ldots, x_n$ with mean $\bar{x} = \frac{\Sigma x_i}{n}$. The quantities $(x_1 - \bar{x}), (x_2 - \bar{x}), \ldots (x_n - \bar{x})$ are deviations from the mean. The average deviation of the set of observations can be obtained by taking a suitable average of the n deviations $(x_i - \bar{x}), i = 1, 2, \ldots, n$.

A simple mean of the n deviations is

$$\frac{(x_1 - \bar{x}) + (x_2 - \bar{x}) + \ldots + (x_n - \bar{x})}{n} = \frac{x_1 + x_2 + \ldots + x_n - n\bar{x}}{n} \tag{8.1}$$

$$= \frac{n\bar{x} - n\bar{x}}{n} = 0.$$

This is always 0 for every set of observations. One measure of dispersion can be obtained by taking the average of the set of absolute deviations $|x_1 - \bar{x}|, |x_2 - \bar{x}|, \ldots, |x_n - \bar{x}|$. The mean of the set of absolute deviations is the mean deviation. This method of considering absolute deviations in place of simple deviations *eliminates the signs* of the deviations before averaging the deviations. Another method of eliminating the signs is to square the deviations before averaging them. Thus by averaging the squared deviations $(x_1 - \bar{x})^2, (x_2 - \bar{x})^2, \ldots, (x_n - \bar{x})^2$ we get a measure of dispersion. This is known as *sample variance* and is the average of the squared deviations, that is, the sum of the squared deviations divided by n, the number of observations. However, it may be indicated that the sum of squared deviations is divided by $(n - 1)$ instead of n, when the observations constitute a sample from a population and not the whole population. This is used as an estimate of the variance of the population of which the set of observations $x_1, \ldots, x_n$ is a sample. In order to obtain an improvement in the estimate of the population variance the sum of squared deviations is divided by $(n - 1)$ instead of n. This is explained below. Thus the sample variance of a set of n observations is defined by

$$s^2 = \frac{\sum_{i=1}^{n} (x_i - \bar{x})^2}{(n - 1)} \tag{8.2}$$

Example 8(a). A *population* consists of 5 numbers 2, 4, 6, 7, 11. Find the population mean μ and the population variance σ^2.

Here $$\mu = \frac{2 + 4 + 6 + 7 + 11}{5} = 6$$

and the mean of squared deviations from the mean μ is the population variance σ^2. Thus

$$\sigma^2 = \frac{(2-6)^2 + (4-6)^2 + (6-6)^2 + (7-6)^2 + (11-6)^2}{5}$$

$$= \frac{16 + 4 + 0 + 1 + 25}{5} = \frac{46}{5} = 9.2$$

Example 8(b). A *sample* of 5 observations drawn from a large population are 2, 4, 6, 7, 11. Find the mean $\bar{x}$ and sample variance s^2.

Here $$\bar{x} = \frac{2 + 4 + 6 + 7 + 11}{5} = 6.$$

and

$$s^2 = \frac{(2-6)^2 + (4-6)^2 + (6-6)^2 + (7-6)^2 + (11-6)^2}{4} = \frac{46}{4} = 11.5$$

Note: Whenever a set of observations is given, then it is to be understood that the set is a sample from a population, unless otherwise stated. Then the variance, would imply sample variance and to calculate the same, formula (8.2) is to be used.

The variance gives a measure in terms of the squared unit – in terms of a unit which is the square of the unit of the original data. For example, if heights of students constitute the data and if heights are given in cms., then the variance will be $(\text{cm})^2$ units. In order to obtain a measure of variability in the same unit as that of data, the + ve square root of the variance is taken. The positive square root of the variance is called the *standard deviation,* which serves as a basic measure of variability or dispersion rather than the variance. The population standard deviation is the square root of the population variance and the sample standard deviation is the square root of the sample variance. These are both in the same unit as the original data as well as the mean. For example, if data of heights are given in cms., then the mean and the standard deviation are also in cms.

The standard deviation is defined by

$$s = \sqrt{variance} = \sqrt{\left\{\frac{\sum_{i=1}^{n} (x_i - \bar{x})^2}{n-1}\right\}} \quad (8.3)$$

In Example 8(b) we found that the variance is 11.5. The standard deviation is

$$\sqrt{11.5} = 3.39$$

Method of computation of variance and standard deviation (s.d.):

We have

$$s^2 = \frac{\sum_{i=1}^{n}(x_i - \bar{x})^2}{n-1} = \frac{\sum_{i=1}^{n}(x_i^2 - 2x_i\,\bar{x} + \bar{x}^2)}{n-1}$$

$$= \frac{\Sigma x_i^2 - 2\bar{x}\,\Sigma x_i + n\bar{x}^2}{n-1}$$

$$= \frac{\Sigma x_i^2 - 2\bar{x}(n\bar{x}) + n\bar{x}^2}{n-1}$$

$$= \frac{\Sigma x_i^2 - n\bar{x}^2}{n-1} \qquad (8.4)$$

and $$s = \sqrt{\left[\frac{\Sigma x_i^2 - n\bar{x}^2}{n-1}\right]} \qquad (8.5)$$

Consider Example 8(b). We have

						Total
x	2	4	6	7	11	$\Sigma x_i = 30$
x^2	4	16	36	49	121	$\Sigma x_i^2 = 226$

Thus

$$s^2 = \frac{226 - 5 \times (6)^2}{4} = \frac{46}{4} = 11.5$$

$$s = \sqrt{11.5} = 3.39$$

4.8.2. Grouped Data

The above method is used for finding the variance and the standard deviation of ungrouped data. For grouped data the variance s^2 and the standard deviation s are defined as follows :

If $x_1, \ldots, x_k$ are the class mid-points of the k classes constituting the classification of the data and if $f_1, \ldots, f_k$ are the frequencies of the respective classes, then the variance s^2 and the standard deviation s are defined as follows:

$$s^2 = \frac{\sum_{i=1}^{k} f_i (x_i - \bar{x})^2}{\sum_{i=1}^{k} f_i - 1}$$

$$s = \sqrt{\left[\frac{\sum_{i=1}^{k} f_i (x_i - \bar{x})^2}{\sum_{i=1}^{k} f_i - 1}\right]}$$

Denote the total frequency = $\Sigma f_i = N$. We have

$$s^2 = \frac{\sum_{i=1}^{k} f_i (x_i - \bar{x})^2}{N-1} = \frac{\Sigma f_i (x_i^2 - 2x_i \bar{x} + \bar{x}^2)}{N-1}$$

$$= \frac{\Sigma f_i x_i^2 - 2\bar{x} \Sigma f_i x_i + \bar{x}^2 \Sigma f_i}{N-1}$$

$$= \frac{\Sigma f_i x_i^2 - 2\bar{x}(N\bar{x}) + N(\bar{x})^2}{N-1} \text{ since } \frac{\Sigma f_i x_i}{N} = \bar{x}$$

$$= \frac{\Sigma f_i x_i^2 - N\bar{x}^2}{N-1} = \frac{\Sigma f_i x_i^2 - N\left(\frac{\Sigma f_i x_i}{N}\right)^2}{N-1}$$

when the total frequency N is large for grouped data, dividing the sum of the squared deviations $\Sigma f_i(x_i - \bar{x})^2$ by $N - 1$ is in practice equivalent to dividing it by N. Thus we may use the formula : for large N

$$s^2 = \frac{\sum_{i=1}^{k} f_i (x_i - \bar{x})^2}{N}$$

$$= \frac{\sum_{i=1}^{k} f_i x_i^2 - N\bar{x}^2}{N} \qquad (8.6)$$

$$= \frac{\sum_{i=1}^{k} f_i x_i^2}{N} - \left(\frac{\Sigma f_i x_i}{N}\right)^2 \qquad (8.6a)$$

Thus

$$s = \sqrt{\left[\frac{\Sigma f_i x_i^2}{N} - \left(\frac{\Sigma f_i x_i}{N}\right)^2\right]} \qquad (8.7)$$

where $N = \Sigma f_i$ is large.

When N is not so large, (8.6) with N in the denominator replaced by $N - 1$ is to be used. Usually N is large and (8.6a) is used.

Example 8(c). Calculate the standard deviation of the data given in Table 3.4.

We construct the following table.

Class mid point x_i	Frequency f_i	$f_i x_i$	$f_i x_i^2$
3.2	4	12.8	40.96
3.7	7	25.9	95.83
4.2	8	33.6	141.12
4.7	11	51.7	242.99
5.2	9	46.8	243.36
5.7	5	28.5	162.45
6.2	4	24.8	153.76
6.7	2	13.4	89.78
Total	50	237.5	1170.25

We have
$$\bar{x} = \frac{\Sigma f_i x_i}{N} = \frac{237.5}{50} = 4.75$$

$$s^2 = \frac{\Sigma f_i x_i^2}{N} - \left(\frac{\Sigma f_i x_i}{N}\right)^2$$

$$= \frac{1170.25}{50} - (4.75)^2$$

$$= 23.405 - 22.5625$$

$$= 0.8425$$

and
$$s = \sqrt{0.8425} = 0.918$$

Division by $(N - 1)$ instead of N would yield

$$s^2 = \frac{1170.25 - 50 \times (4.75)^2}{49} = \frac{1170.25 - 50 \times 22.5625}{49}$$

$$= \frac{42.125}{49} = 0.8597$$

$$s = \sqrt{0.8597} = 0.927$$

The difference would have been still smaller for larger N. The formula (8.6) is applicable wherever the width of the class intervals are equal (as in the above). We can use some simplification as indicated below.

4.8.3. Computation of Variance and Standard Deviation – Short Method

(a) *Ungrouped Data*

Let A be an arbitrary constant, suitably chosen somewhere around the middle of the values of the variable. Let

$d_i = x_i - A$ be the deviation of x_i from this constant A. We have then $x_i = A + d_i$ and the mean of the deviations is given by

$$\bar{d} = \frac{\Sigma d_i}{n} = \frac{\Sigma(x_i - A)}{n} = \frac{\Sigma x_i}{n} - \frac{nA}{n} = \bar{x} - A$$

or $$\bar{x} = A + \bar{d}$$

Thus

$$\sum_{i=1}^{n}(x_i - \bar{x})^2 = \sum_{i=1}^{n}(A + d_i - A - \bar{d})^2 = \sum_{i=1}^{n}(d_i - \bar{d})^2$$

$$= \sum_{i=1}^{n} d_i^2 - 2\bar{d}\sum_{i=1}^{n} d_i + \sum_{i=1}^{n}(\bar{d})^2.$$

$$= \sum_{i=1}^{n} d_i^2 - 2\bar{d}\,.\,n\,\bar{d} + n\bar{d}^2$$

$$= \Sigma d_i^2 - n\,\bar{d}^2$$

Hence $$s^2 = \frac{\Sigma(x_i - \bar{x})^2}{n-1} = \frac{\Sigma d_i^2 - n\,\bar{d}^2}{n-1} \tag{8.8}$$

and $$s = \sqrt{\left(\frac{\Sigma d_i^2 - n\,\bar{d}^2}{n-1}\right)} \tag{8.9}$$

(b) *Grouped data*

If $x_1, \ldots, x_k$ are the class mid points of the k classes with respective frequencies $f_1, f_2, \ldots, f_k$, then

$$s^2 = \frac{\Sigma f_i d_i^2}{N} - \left(\frac{\Sigma f_i d_i}{N}\right)^2 \tag{8.10}$$

where $$N = \Sigma f_i$$

Write $d_i = x_i - A$ or $x_i = A + d_i$, where A is an arbitrary constant;

then $$\bar{x} = \frac{\Sigma f_i x_i}{N} = \frac{\Sigma f_i (A + d_i)}{N} = \frac{A\Sigma f_i + \Sigma f_i d_i}{N}$$

$$= \frac{AN}{N} + \frac{\Sigma f_i d_i}{N} = A + \bar{d}$$

where $$\bar{d} = \frac{\Sigma f_i d_i}{N}$$

Thus

$$\Sigma f_i (x_i - \bar{x})^2 = \Sigma f_i (A + d_i - A - \bar{d})^2$$

$$= \Sigma f_i d_i^2 - 2\bar{d}\, \Sigma f_i d_i + \bar{d}^2 \Sigma f_i$$

$$= \Sigma f_i d_i^2 - 2\bar{d} \quad N\bar{d} + \bar{d}^2 \cdot N$$

$$= \Sigma f_i d_i^2 - N\bar{d}^2.$$

Hence

$$s^2 = \frac{\Sigma f_i (x_i - \bar{x})^2}{N} = \frac{\Sigma f_i d_i^2}{N} - \frac{N\bar{d}^2}{N}$$

$$= \frac{\Sigma f_i d_i^2}{N} - \left(\frac{\Sigma f_i d_i}{N}\right)^2$$

and $$s = \sqrt{\left[\frac{\Sigma f_i d_i^2}{N} - \left(\frac{\Sigma f_i d_i}{N}\right)^2\right]} \tag{8.11}$$

(c) *Grouped Data with Equal Class Widths*

Let all the class intervals have an equal width, c. Then if A is an arbitary constant and if

$d_i = x_i - A$ then d_i can be expressed as

$d_i = cu_i$ where $u_i = 0, \pm 1, \pm 2, \pm 3, \ldots$

Then

$$\bar{d} = \frac{\Sigma f_i d_i}{N} = \frac{\Sigma f_i (cu_i)}{N}$$

$$= c\,\frac{(\Sigma f_i u_i)}{N} = c\bar{u}$$

where $$\bar{u} = \frac{\Sigma f_i u_i}{N}$$

and $d_i^2 = c^2 u_i^2$. Thus from (8.10) we get

$$s^2 = \frac{\Sigma f_i (cu_i)^2}{N} - \left[\frac{\Sigma f_i (cu_i)}{N}\right]^2$$

$$= c^2 \left[\frac{\Sigma f_i u_i^2}{N} - \left(\frac{\Sigma f_i u_i}{N}\right)^2\right] \tag{8.12}$$

and
$$s = c\sqrt{\left[\frac{\Sigma f_i u_i^2}{N} - \left(\frac{\Sigma f_i u_i}{N}\right)^2\right]} \tag{8.13}$$

Note:

(1) For finding variance and standard deviation, (8.10) and (8.11) are convenient in case of grouped data with unequal class intervals while (8.12) and (8.13) are convenient in case of grouped data with equal class width or class intervals. For ungrouped data (8.8) and (8.9) are generally used.

(2) In the case of ungrouped data where the total number of observations n is small we divide the sum of the squares of deviations by $(n-1)$ to find the variance, while in case of grouped data the total frequency $N = \Sigma f_i$ is *usually* large and we divide the sum of the squares of deviations by N.

Example 8 (d). Consider the data given in Table 3.4 for which the variance was obtained in Example 8(c). We shall use (8.12) and (8.13) below.

Let $A = 4.7$, $c = 0.5$ We get the following table

x_i	$d_i = x_i - A$	$u_i = (d_i/c)$	f_i	$f_i u_i$	$f_i u_i^2$
3.2	−1.5	−3	4	−12	36
3.7	−1.0	−2	7	−14	28
4.2	−0.5	−1	8	−8	8
4.7	0	0	11	0	0
5.2	0.5	1	9	9	9
5.7	1.0	2	5	10	20
6.2	1.5	3	4	12	36
6.7	2.0	4	2	8	32
		Total	50	5	169

We have
$$\bar{u} = \frac{\Sigma f_i u_i}{N} = \frac{5}{50} = 0.1$$

$$s^2 = c^2\left[\frac{\Sigma f_i u_i^2}{N} - \left(\frac{\Sigma f_i u_i}{N}\right)^2\right]$$

$$= c^2\left[\frac{169}{50} - (0.1)^2\right] = (0.5)^2[3.38-0.1]$$

$$= 0.8425$$

and
$$s = \sqrt{0.8425} = 0.918$$

Calculations have been very much simplified here.

Example 8 (e). The distribution of heights of 160 students is as follows:

Height (in cm)	Frequency
150.0-154.0	15
155.0-159.0	38
160.0-164.0	64
165.0-169.0	25
170.0-174.0	15
175.0-179.0	3
	160

Find the variance and standard deviation of the distribution. Here take $A = 162.0$, the class mid point of the class interval 160.0-164.0, the class with highest frequency. The class interval 150.0-154.0 has boundaries 149.5 and 154.5 and the width of the class is 5.0. So also is the width of all other classes. Thus $c = 5.0$. Thus we get the following table:

(x_i) Class mid point	$d_i = x_i - A$	$u_i = \frac{d_i}{c}$	f_i	$f_i u_i$	$f_i u_i^2$
152.0	−10	−2	15	−30	60
157.0	−5	−1	38	−38	38
162.0	0	0	64	0	0
167.0	5	1	25	25	25
172.0	10	2	15	30	60
177.0	15	3	3	9	27
		Total	160	−4	210

We have, the variance s^2, given by

$$s^2 = c^2 \left[\frac{\Sigma f_i u_i^2}{N} - \left(\frac{\Sigma f_i u_i}{N} \right)^2 \right]$$

$$= 25 \left[\frac{210}{160} - \left(-\frac{4}{160} \right)^2 \right]$$

$$= 25\,[1.3125 - 0.000625]$$

$$= 32.796875;$$

the standard deviation s is given by

$$s = 5.727$$

Incidentally the mean

$$\bar{x} = A + \bar{d} = A + c\bar{u} = 162 + 5 \times \frac{\Sigma f_i u_i}{N}$$

$$= 162 + 5 \times \left(\frac{-4}{160} \right) = 162 - 0.125 = 161.875.$$

4.8.4. Interpretation of the Mean, Variance and S.D.

There is a simple physical interpretation of the arithmetic mean $\overline{x}$: it is the point about which a light rod with frequencies as weights along different points of the rod will balance. It is analogous to the concept of ***centre of gravity*** in mechanics. The concept of variance is analogous to that of ***moment of intertia.***

If two sets of data have both the same mean, but the first set has a standard deviation twice as large as the second set, we may say that the second set of data reflects smaller variation of data but nothing beyond. The situation is something like this. To the question 'Who is Amal?', if the answer is, 'Brother of Bimal' and then to the question 'Who is Bimal', if the answer is, 'Brother of Amal', we know that Amal and Bimal are brothers but we do not have their true identity.

A Russian mathematician P. Chebyshev (1821-94) gives a clear meaning of the value of s in relation to the proportion of data points located in the intervals surrounding the mean. We state the result below.

4.8.5. Chebyshev's Lemma or Rule (for sample)

If $\overline{x}$ and s are the arithmetic mean and standard deviation respectively of a set of observations, then

(1) the interval $(\overline{x} - 2s, \overline{x} + 2s)$ contains *at least*

$$\left(1-\frac{1}{2^2}\right) = \frac{3}{4} \text{ of the total number of observations;}$$

(2) the interval $(\overline{x} - 3s, \overline{x} + 3s)$ contains *at least*

$$\left(1-\frac{1}{3^2}\right) = \frac{8}{9} \text{ of the total number of observations;}$$

(3) in general, for any k (> 1), the interval $(\overline{x} - ks, \overline{x} + ks)$ contains at least $\left(1-\frac{1}{k^2}\right)$ fraction of the total number of observations.

Note: While the above rule is for sample data, a similar rule exists for probability distributions.

Example 8(f). Consider the data of Example 8(*d*). Its mean $\overline{x} = 4.75$ and standard deviation $s = 0.92$.

The interval $(\overline{x} - 2s, \overline{x} + 2s) = (4.75 - 1.84, 4.75 + 1.84) = (2.91, 6.59)$ contains at least $\frac{3}{4}$ of the total number of points, i.e. at least about $37\left(\frac{3}{4}\times 50 = 37.5\right)$ observations.

By drawing a cumulative frequency curve we can actually find out the percentile values corresponding to the values 2.91 and 6.59 and find out what percentage of data lies in this interval.

Similarly, the interval $(\overline{x} - 1.5s, \overline{x} + 1.5s) = (4.75 - 1.38, 4.75 + 1.38) = (3.37, 6.13)$ contains at least $\left(1-\frac{1}{(1.5)^2}\right) = 0.56$ fraction of the data.

By drawing an ogive we can check the percentage or proportion of points lying within 3.37 and 6.13.

In case of a specific distribution (normal distribution) it can be precisely indicated as to what proportion of observations will lie between $(\bar{x} - s, \bar{x} + s)$, $(\bar{x} - 2s, \bar{x} + 2s)$, $(\bar{x} - 3s, \bar{x} + 3s)$ and so on. This has been considered in Chapter 7.

Note : While the mean can be thought of as the origin (reference point), the standard deviation can be considered as the scale parameter.

4.8.6. Properties of Standard Deviation

(1) The standard deviation is the square root of the mean of the squares of deviations $(x_i - \bar{x})^2$ from the arithmetic mean $\bar{x}$. Instead of taking deviations from the arithmetic mean $\bar{x}$, deviations from some other central measure or any other point can also be considered. For example, to measure dispersion, we could define a quantity

$$\frac{\Sigma f_i (x_i - a)^2}{N}$$

where a is an arbitrary point, say, the median or the mode or any other average or any arbitrary point. This is root mean square deviation. When a is the arithmetic mean, the root mean square deviation is standard deviation.

Of all the deviations $\frac{\Sigma f_i (x_i - a)^2}{\Sigma f_i}$, where a is any point, it can be *shown* that the *minimum* is that for which $a = \bar{x}$, i.e. the quantity is the standard deviation.

The standard deviation is the least possible root mean square deviation:

We shall now prove this: Now

$$D^2 = \frac{\Sigma f_i (x_i - a)^2}{N}, \; N = \Sigma f_i$$

is the mean square deviation about an arbitrary point a. We have

$$D^2 = \frac{\Sigma f_i \left\{(x_i - \bar{x}) + (\bar{x} - a)\right\}^2}{N}$$

$$= \frac{\Sigma f_i (x_i - \bar{x})^2}{N} + \frac{2\Sigma f_i (x_i - \bar{x})\,(\bar{x} - a)}{N} + \frac{\Sigma f_i (\bar{x} - a)^2}{N}$$

$$= s^2 + \frac{(\bar{x} - a)^2 \Sigma f_i}{N} \quad \text{since } \Sigma f_i (x_i - \bar{x}) = 0$$

$$= s^2 + (\bar{x} - a)^2.$$

The quantity $(\bar{x} - a)^2 \geq 0$, so that $D^2 \geq s^2$ and the equality holds when $\bar{x} = a$. Thus D^2 is least when the deviations are taken from the mean $\bar{x}$. That is, variance is the least possible mean square deviation.

It follows that the standard deviation is the least possible root mean square deviation.

(2) Chebyshev's lemma helps one determine the minimum proportion of observations lying between $\bar{x} \pm ks$, where $k > 1$ i.e. in the interval $(\bar{x} - ks, \bar{x} + ks)$.

(3) Suppose that two sets of data having N_1 and N_2 observations have the same mean $\bar{x}$ and standard deviations s_1 and s_2. Then the variance s^2 of the combined set of $(N_1 + N_2)$ observations is given by

$$s^2 = \frac{N_1 s_1^2 + N_2 s_2^2}{N_1 + N_2} \qquad (8.14)$$

If the two sets have different means $\bar{x}_1$ and $\bar{x}_2$, then

$$s^2 = \frac{N_1 s_1^2 + N_2 s_2^2}{N_1 + N_2} + \frac{N_1 N_2}{(N_1 + N_2)^2} (\bar{x}_1 - \bar{x}_2)^2 \qquad (8.15)$$

4.8.7. Uses of Standard Deviation

These are as given below:

(1) We can compare the degree of variation of two or more sets by comparing their standard deviations. The one with smaller s.d. (if in the same unit) will have lesser variation or dispersion. More observations will lie closer to the mean in this case.

(2) The mean will be more representative in the case of a distribution having smaller standard deviation.

(3) The standard deviation will give a good idea of the proportion of observations lying in an interval around the mean. This follows from Chebyshev's lemma.

(4) The standard deviation can also be used to locate appproximately the place of an item in the distribution.

(5) The standard deviation together with the arithmetic mean gives a fairly good idea of the distribution. These help in the summarization of the data.

4.8.8. Sheppard's Correction for Variance

When the observations are grouped into classes, all the observations in a class are taken to be equal to the mid point of the class. This introduces some error known as grouping error. Sheppard suggests a correction, known as Sheppard's correction. It is given by $c^2/12$ where c is the common class width.
The corrected variance is as follows:

$$\text{corrected variance} = \text{variance from grouped data} - c^2/12 \qquad (8.16)$$

It is generally used for distributions of continuous variables where the tails go towards zero gradually in both the directions.

4.8.9. Absolute and Relative Dispersion

Coefficient of Variation : The mean deviation the standard deviation and other deviations are the measures of absolute dispersion. This however does not give a correct picture of the situation. For example, a standard deviation of 5 in the marks of one paper of 100 marks is not of the same degree as the standard deviation of 5 in the total of 1000 marks. In fact, it is of much smaller degree in the latter case. To obviate such a situation a measure of relative dispersion is defined. It is given as a proportion of average and is defined by

$$\text{Relative dispersion} = \frac{\text{absolute dispersion}}{\text{Average}}$$

In particular, the standard deviation is taken as a measure of absolute dispersion and the arithmetic mean as a measure of average. It is generally expressed in percentages, and is known as *coefficient of variation.* Thus

$$\text{C.V.} = \text{Coefficient of variation} = \frac{\text{s.d}}{\text{mean}} \times 100$$

$$= \frac{s}{\bar{x}} \times 100$$

The coefficient of variation is independent of the unit of the observations. It can therefore be used for comparison of distributions having observations with different units.

Since $\bar{x}$ appears in the denominator, the measure cannot be used when $\bar{x}$ is zero or close to zero.

Example 8(g). Two factories producing similar products are such that the mean daily wage of workers of one factory is Rs. 100 with a standard deviation of Rs. 10; whereas in the other factory the mean wage is Rs. 150 and standard deviation of Rs. 12. Find which of these have the large variation.

The coefficient of variation of factory *I* equals

$$\frac{10}{100} \times 100 = 10\%$$

and that of factory II equals

$$\frac{12}{150} \times 100 = 8\%$$

Thus the factory I has a slightly greater variation.

Example 8(h). The screws produced in machine I has mean length 5.2 cm with s.d. 0.4 cm while that produced by machine II has mean length 6.4 cm with s.d. 0.6 cm. Compare their variations.

The coefficient of variation in case of machine I

$$= \frac{0.4}{5.2} \times 100 = 7.7\%$$

and that in case of machine II

$$= \frac{0.6}{6.4} \times 100 = 9.4\%$$

Thus products in machine II have more variation.

4.9 MOMENTS OF HIGHER ORDER

4.9.1. Ungrouped Data

We have discussed the importance of the mean and the variance and the standard deviation in the process of summarization of data. We have

$$\text{mean} = \frac{\Sigma(x_i)}{N}$$

and

$$\text{variance} = \frac{\Sigma(x_i - \bar{x})^2}{N}$$

The mean, denoted by m_1, is the mean of the quantitites; the variance, denoted by m_2 is the mean of the squares of the deviations from the mean. The mean m_1 is the first moment and m_2 is the second moment about the mean. By considering the mean of the cubes of the deviations from the mean we get the third moment m_3 about the mean, and in general, by considering the mean of the rth powers ($r = 2, 3, 4, \ldots$) of the deviations from the mean, we get rth moment about the mean, denoted by m_r.

Thus

$$m_3 = \frac{\Sigma(x_i - \bar{x})^3}{N} \tag{9.1}$$

$$m_4 = \frac{\Sigma(x_i - \bar{x})^4}{N} \tag{9.2}$$

and

$$m_r = \frac{\Sigma(x_i - \bar{x})^r}{N}, \quad r = 1, 2, 3, \ldots \tag{9.3}$$

Similarly we can find the moment of any order about an arbitary origin, for example, A. We have

$$m'_r = \frac{\Sigma(x_i - A)^r}{N}, \quad r = 1, 2, 3, \ldots \tag{9.4}$$

Grouped data

If $x_1, x_2, \ldots, x_k$ occur with frequencies $f_1, f_2, \ldots, f$ or *if* $x_1, x_2, \ldots, x_k$ are the class midpoints of k classes with respective frequencies $f_1, f_2, \ldots, f_k$, then

$$m_2 = \frac{\sum_{i=1}^{k} f_i(x_i - \bar{x})^2}{N} \tag{9.5}$$

$$m_3 = \frac{\sum_{i=1}^{k} f_i(x_i - \bar{x})^3}{N} \tag{9.6}$$

and

$$m_r = \frac{\sum_{i=1}^{k} f_i(x_i - \bar{x})^r}{N}, \quad r = 1, 2, 3, \ldots \tag{9.7}$$

m_r is the rth moment about the mean; similarly we can consider

$$m'_r = \frac{\Sigma f_i(x_i - A)^r}{N}, \quad r = 1, 2, \ldots \tag{9.8}$$

as rth moment about an arbitrary origin A.

4.9.2. Relation Between Moments m_r and m'_r

We have

$$m'_1 = \frac{\Sigma f_i(x_i - A)}{N} = \frac{\Sigma f_i x_i}{N} - \frac{A \Sigma f_i}{N} \quad (\Sigma f_i = N) \tag{9.9}$$

$$= \bar{x} - A$$

$$m_2 = \frac{\Sigma f_i(x_i - \bar{x})^2}{N} = \frac{\Sigma f_i\{(x_i - A) + (A - \bar{x})\}^2}{N} = \frac{\Sigma f_i\{(x_i - A) - m_1'\}^2}{N}$$

$$= \frac{\Sigma f_i\{(x_i - A)^2 - 2m'_1(x_i - A) + (m'_1)^2\}}{N}$$

$$= \frac{\Sigma f_i(x_i - A)^2}{N} - 2m'_1 \cdot \frac{\Sigma f_i(x_i - A)}{N} + (m'_1)^2 \frac{\Sigma f_i}{N}$$

$$= m'_2 - 2m'_1 m'_1 + (m'_1)^2$$

Thus $\quad m_2 = m'_2 - (m'_1)^2, \tag{9.10}$

$$m_3 = \frac{\Sigma f_i(x_i - \bar{x})^3}{N} = \frac{\Sigma f_i\{(x_i - A) - (m'_1)\}^3}{N}$$

$$= \frac{\Sigma f_i(x_i - A)^3}{N} - (3m'_1)\frac{\Sigma f_i(x_i - A)^2}{N} + (3m'_1)^2 \frac{\Sigma f_i(x_i - A)}{N} - (m'_1)^3 \frac{\Sigma f_i}{N}$$

$$= m'_3 - 3m'_1 m'_2 + 3(m'_1)^3$$

Thus
$$m_3 = m'_3 - 3m'_1 m'_2 + 2(m'_1)^3 \qquad (9.11)$$

and so on.

These formulae enable us to find the moments about the mean from moments from any suitable origin A. Calculations are further simplified for frequency distributions with equal class width c.

Then
$$x_i - A = cu_i\,,\ u_i = 0\,, \pm 1, \pm 2\,, \ldots$$

and

$$m'_1 = \frac{\Sigma f_i(cu_i)}{N} = (c)\frac{\Sigma f_i u_i}{N}$$

$$m'_2 = \frac{\Sigma f_i(cu_i)^2}{N} = (c^2)\frac{\Sigma f_i u_i^2}{N}$$

$$m'_3 = \frac{\Sigma f_i(cu_i)^3}{N} = (c^3)\frac{\Sigma f_i u_i^3}{N}$$

and these can be easily calculated.

Example 9(a). Find m_3 for data given in Table 3.4.
Here let $A = 4.7$ and $c = 0.5$; we can rewrite the table of Example 8 (a) as follows.

x_i	$u_i = (x_i - A)/c$	f_i	$f_i u_i$	$f_i u_i^2$	$f_i u_i^3$
3.2	–3	4	–12	36	–108
3.7	–2	7	–14	28	–56
4.2	–1	8	–8	8	–8
4.7	0	11	0	0	0
5.2	1	9	9	9	9
5.7	2	5	10	20	40
6.2	3	4	12	36	108
6.7	4	2	8	32	128
	Total	50	5	169	113

We have

$$m'_1 = \frac{(0.5)\cdot 5}{50} = 0.05$$

$$m'_2 = (0.5)^2\,\frac{169}{50} = 0.845$$

$$m'_3 = (0.5)^3 \frac{113}{50} = 0.02825$$

Thus

$$\bar{x} = A + m'_1 = 4.7 + 0.05 = 4.75$$

$$m_2 = m'_2 - (m'_1)^2 = 0.845 - (0.05)^2 = 0.8425$$

$$m_3 = m'_3 - 3m'_1m'_2 + 2(m'_1)^3 = 0.02825 - 3 \times 0.05 \times 0.845 + 2(0.05)^3$$

$$= 0.02825 - 0.12675 + 0.00025$$

$$= -0.09825.$$

4.10. OTHER DESCRIPTIVE MEASURES: SKEWNESS AND KURTOSIS

4.10.1. Skewness and its Measurement

Measures of location and dispersion are good descriptive measures of masses of data. These help us to have a better picture of data which we can not get from raw data alone. The central tendency and variability are two important characteristics of the distribution of the mass of data. Besides these two there are two other characteristics of a distribution. These are *Asymmetry,* also known as *Skewness,* and *Peakedness*, or *Kurtosis.*

Mean, median, mode are measures of central tendency and mean deviation and standard deviations are measures of variability. We shall discuss measures of the characteristics of skewness and kurtosis, which will indicate the direction and extent of these characteristics.

In a symmetrical distribution, the mean, median and mode coincide. The more the mean moves away from the mode the greater is the departure from symmetry. That is, the larger the distance between the mean and the mode, the greater is the skewness or asymmetry. Karl Pearson defines a measure of skweness based on this distance.

The *coefficient* of *skewness* (Pearsonian) is defined by

$$S_p = \frac{\text{Mean} - \text{Mode}}{\text{Standard Deviation}} \qquad (10.1)$$

A positively skewed distribution (Fig. 4.2b) has a long tail to the right: here the mean is greater than the mode and thus S_p is positive. On the other hand a negatively skewed distribution (Fig. 4.2c), has a long tail to the left and the mean is smaller than the mode and thus S_p is negative. For a symmetrical distribution, S_p is equal to 0.

Thus departure from 0 will imply the extent of asymmetry, a positive value of S_p positive skewness and negative value negative skewness.

For a moderately asymmetrical distribution, we get from (4.2) that mode can be expressed in terms of the median and mode by the formula

$$\text{Mean} - \text{Mode} = 3\,(\text{Mean} - \text{Median})$$

Using this result we get

$$S_p = \frac{3(\text{Mean} - \text{Median})}{\text{Standard Deviation}}$$

$$= \frac{3(\bar{x} - \text{Median})}{s} \qquad (10.2)$$

For example, for data given in Table 3.4

mean = 4.75, median = 4.72 and s.d = 0.92. Thus

$$S_p = \frac{3(4.75-4.72)}{0.92} = 0.098$$

The distribution has a slight positive skewness – a slight long tall to the right. There is another measure of skewness based on moments which we define below. The *Moment coefficient of skewness* S_m is defined by

$$S_m = \frac{m_3}{s^3} = \frac{m_3}{(\sqrt{m_2})^3} = \frac{\text{third moment about the mean}}{(\text{s.d.})^3} \qquad (10.3)$$

For symmetrical distribution, such as normal distribution, the third order moment about the mean is 0. Hence

$$S_m = \frac{0}{s^3} = 0$$

For Example, for data given in Table 3.4 (Example 9 (a))

$$m_2 = 0.8425$$

and $$m_3 = -0.09825$$

Hence the moment coefficient of skewness is given by

$$S_m = \frac{m_3}{(\sqrt{m_2})^3} = \frac{-0.09825}{0.77331} = -0.0127$$

Note : The measure S_m can be zero without the distribution being symmetrical.

4.10.2. Kurtosis and its Measurement

It may be observed that some distributions have a high peak, some are flat-topped and so on. Kurtosis is the degree or extent of *peakedness* of a distribution. Distributions can be broadly divided into three types according to the nature of their kurtosis. A distribution which has a relatively high peak is called as *leptokurtic*, and one which is flat-topped is called as *platykurtic* while one which is not very peaked or flat-topped is called *mesokurtic* (see Fig. 4.4).

Kurtosis is measured in terms of quartiles and percentiles as follows. The *coefficient of Kurtosis* is defined by

$$K_p = \frac{\frac{1}{2}(Q_3 - Q_1)}{P_{90} - P_{10}} \tag{10.4}$$

Another measure is in terms of the moments of the distribution. The *moment coefficient of Kurtosis* is defined by

$$K_m = \frac{m_4}{m_2^2} = \frac{m_4}{s^4} \tag{10.5}$$

which is also denoted by β_2.

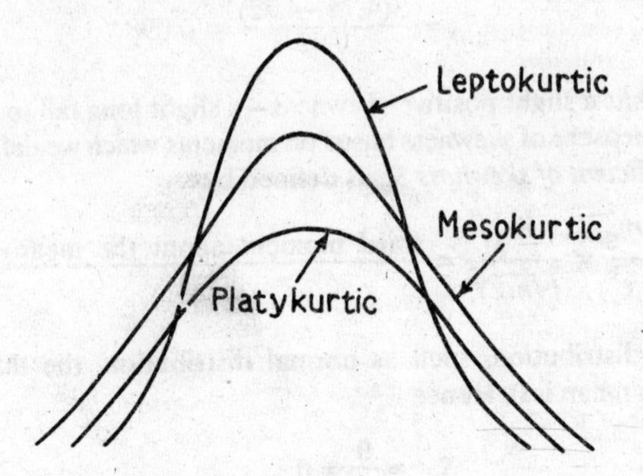

Fig. 4.4 Curve types according to kurtosis

For a very important distribution, known as normal distriution $k_m = 3$ and hence taking normal distribution as the standard, kurtosis is sometimes measured by the quantity $(\beta_2 - 3)$. This quantity is zero for normal distribution. A distribution is

mesokurtic if $K_m = 3$ (or $\beta_2 - 3 = 0$)
platykurtic if $K_m < 3$ (or $\beta_2 - 3 < 0$)
leptokurtic if $K_m > 3$ (or $\beta_2 - 3 > 0$)

For normal distribution, $K_p = 0.263$. A distribution is

mesokurtic if $K_p = 0.263$
platykurtic if $K_p < 0.263$
leptokurtic if $K_p > 0.263$

Note that the coefficients defined above are dimensionless and are pure numbers.

For data given in Example 6(a), we have

Q_1 = Rs. 610.25, Q_3 = Rs. 965, P_{10} = Rs. 400, P_{90} = Rs. 1182.86

Thus the coefficient of kurtosis is given by

$$K_p = \frac{\frac{1}{2}(965.00 - 610.25)}{1182.86 - 400.00} = \frac{177.375}{782.86} = 0.227$$

The coefficient of kurtosis $K_p = 0.227$. Since the kurtosis here 0.227 is slightly lower than that of normal distribution, the distribution is *mesokurtic*.

EXERCISES-4

Sections 4.1-4.6

1. Explain what you mean by the central tendency of a distribution. Mention the measures of central tendency. Which ones are computational and which ones are positional? Discuss the merits and demerits of the three most important measures.
2. State the important properties of the arithmetic mean. Give the physical interpretation of the arithmetic mean.
3. When would you consider the median to be a better representative of the data? Discuss.
4. What are the prerequisites of a good average? Discuss them with particular reference to the three important averages, the mean, the median and the mode.
5. Which one of the averages would you consider to be the most suitable representative of the following:
   (1) the size of ready made shirts,
   (2) the weekly change in price level of a commodity,
   (3) the annual birth rate of a country over 5 years, and
   (4) the average annual rate of interest when it is known that a sum doubles itself in 6 years?
6. Show that
   A.M. ≥ G.M. ≥ H.M.
   where A.M. refers to arithmetic mean, G.M. to geometric mean and H.M. to harmonic mean of a set of observations $a_1, \ldots, a_n (a_i > 0)$.
7. Explain how you can determine the median graphically with the help of (1) ogive. (2) histogram.
8. How can you say about the location of the mean, median and the mode if the nature of the distribution as regards its symmetry is given?
9. (a) Prove the formula
   $$\bar{x} = A + \left(\frac{\Sigma f_i u_i}{\Sigma f_i}\right) c$$
   where the symbols have their usual significance.
   (b) Explain how you would find the mode of a distribution with (1) ungrouped data (2) grouped data.
10. What is a geometric mean? When would you use it? In a newly industrialised country, the consumer expenditure has grown as follows during 1963 to 1968.

Year	1963	1964	1965	1966	1967	1968
Expenditure (million units)	1,995	2,291	2,692	2,951	3,467	4,074

   Find the average annual rate of growth.
11. Find the mean, the median and the mode of the frequency distribution considered in Exercise 3.12.
12. Find the mean and median weekly wages of the employees as given in Exercise 3.20. Draw the cumulative frequency curve and estimate the percentage of employees drawing income between Rs. 200 and Rs. 265
13. Find the mean (m), the median and the two quartiles – the first and the third, of the lifetime of tubes given in Exercise 3.22. Find also the mode of the distribution.
14. Find the median rent of 150 residential houses as given in Exercise 3.25.
    Find Q_1, Q_3 and P_{75}.
15. Find the mean, median and the mode of the distribution of income as given in Exercise 3.26. Which one of the three averages do you consider to be more representative and why? Comment on the accuracy of the mean.
    (a) Draw the "more than" type ogive. Find P_{10}, P_{90}, Q_1 and Q_3.
    (b) Estimate the number of persons with incomes between Rs. 24,000 to Rs. 66,000 who are required to pay a surcharge on tax.
16. What other positional measures of location are you familiar with besides the median? Describe them and explain how to determine them.

Sections 4.7-4.8

17. What do you mean by dispersion? What are the important measures of dispersion? Discuss the requirements of a good measure of dispersion.
18. Find the mean deviation and the semi-inter-quartile range of data given in Table 3.4.
19. Define standard deviation. Explain its uses.
20. State the properties of standard deviation. Discuss the relative merits of standard deviation over other measures in the light of the properties of standard deviation.
21. Explain what is meant by absolute and relative dispersion. How are they measured?
22. Find the variance and the standard deviation of the data of Exercise 3.11.
23. Find the standard deviation of data of Exercise 3.13. Estimate the percentage of tubes having a life time of between $m - 2s$ and $m + 2s$ hours.
24. Find the standard deviation of data of Exercise 3.12. If there is an increase in wage at the flat rate of Rs. 10 what will be the change in the mean, median and mode, the standard deviation and the variance.
25. Find the variance and the standard deviation of data given in Exercise 3.14.
26. The daily maximum temperature of a town for January and February 1984 are as given in table below

Temperature (in centigrade)	Number of days
−14 to −10	5
−9 to −5	7
−4 to 0	24
1 to 5	12
6 to 10	8
11 to 15	4

Find the mean and the median daily maximum temperature. Find also the variance and the standard deviation of the distribution.

27. Find the coefficient of variation of the data of Exercise 3.11 (see also 3.22)
28. The following are the number of orders for a certain product obtained by two salesmen of a firm for 12 months.

x	60	40	20	35	40	20	10	15	70	40	15	20
y	30	20	25	35	30	20	15	20	40	45	25	15

Who is more consistent as regards receipt of orders?

29. The mean and standard deviation of weekly wages of the 65 day shift workers is Rs. 125 and Rs. 12 respectively and those of 35 evening shift workers are Rs. 156 and Rs. 18 respectively. Compare the two sets of workers as regards relative variations of their wage structure. Find the mean and the standard deviation of wages of the workers of the two shifts combined.
30. Find the mean and standard deviation of the data given in (a) Exercise 3.20 (b) Exercise 3.22

Section 4.9

31. The first three moments of a distribution about the origin are 5, 35 and 25. Find the first three moments about the mean.
32. Find the first three moments about the mean of the distribution given in Example 2(a).
33. Find the first three moments about the mean of the distribution given in Exercise 3.12.

Section 4.10

34. Find the coefficient of skewness of the data considered in Exercise 3.33.
35. Describe the distribution considered in Exercise 3.33 as regards skewness and peakedness. Draw the frequency curves and verify your finding.
36. Find the moment coefficients of skewness and kurtosis of the distribution of Example 2(a).
37. Consider n observations $x_1, \ldots, x_n$ with respective frequencies $f_1, \ldots, f_n$. $\Sigma f_i = N$. Let $d(a)$, $d(m)$ denote the mean deviation (M.D.) about an arbitrary value a, and about the median m respectively. Then show that :

(i) for $a > m$

$$N[d(a)] = N[d(m)] + 2 \sum_{x_i > m}^{x_i \leq a} f_i(a - x_i) + (a - m)\left[\sum_{x_i \leq m} f_i - \sum_{x_i > m} f_i\right]$$

$$= N[d(m)] + A$$

where A, the sum of the last 2 terms is greater than 0; thus

$$N[d(a)] > N[d(m)].$$

(ii) for $a < m$

$$N[d(a)] = N[d(m)] + 2 \sum_{x_i > a}^{x_i \leq m} f_i(x_i - a) - (a - m)\left[\sum_{x_i \leq m} f_i - \sum_{x_i > m} f_i\right]$$

$$\geq N[d(m)] + \sum_{x_i > a}^{x_i \leq m} f_i(x_i - a)$$

$$= N[d(m)] + B \text{, where } B > 0 \text{.}$$

so that $N[d(a)] > N[d(m)]$.

The relations hold in the general case.

Chapter 5

Index Numbers

5.1. INTRODUCTION, MEANING AND DEFINITION

An *index* is something that indicates some state of affairs with reference to a phenomenon. An *index number* is an expression of what is to be indicated. An index number is a statistical device designed to measure changes or differences in magnitudes in a variable or a group of related variables. These changes may pertain to the price of a commodity or a group of commodities, the physical quantity of a product or a group of products, (produced or consumed) or to the quantitative measure of such characteristics as income and intelligence. The changes or differences may concern periods of time (such as monthly prices or production of a commodity over a number of months), or different geographical locations or again different characteristics, such as persons, objects or institutions, etc. Index numbers are descriptive measures or changes. We shall also say that index numbers are special types of averages.

5.2. USES OF INDEX NUMBERS

Index numbers are constructed with various aims and purposes. Index numbers constructed are put into several uses. Some of them are described below:

(1) The most commonly used types of index numbers are those that relate to changes in level in prices of a commodity or a group of commodities. Such index numbers are price index numbers. Price index numbers are useful in studying price movements with a view to analysing their causes, as well as to determine their effect on the economy. To determine economic relationships, it is often useful to compare changes in the market price level with changes in other series, such as gold, bank deposits, bank loans, and the physical volume of production. Changes of the later types are sometimes helpful in controlling the market price level, changes in interest and mortgage rates are often taken into consideration. As one of the ways to raise the general price in the U.S.A., the official price of gold was increased in 1933-34 (depression period). Changes in the price level of a commodity is intimately connected with changes in the volume of production or the volume of the commodity made available in the market.

(2) Indices of physical changes over a period of time in production, in marketing, in sales, in imports and exports, and so on have important uses. Such index numbers over a period of time, known as *index time series* are extremely useful in the study of the movement of the characteristics over time, in the study of seasonal and cyclic behaviour, and in the study of business cycles, etc.

(3) Index numbers can be used for forecasting future trend in demand of commodities or in volume of production etc. Index numbers of industrial and agricultural production not only reflect the trend but also are useful in forecasting production. Index numbers of unemployment which reflect the trend of unemployment are useful in determining factors leading to unemployment and in forecasting future trends.

(4) Index numbers which reflect the general price level, and in particular, the consumer price index numbers are useful in determining the quantum of additional wages or dearness allowances to compensate for the change in the cost of living. The Government of India and most of the State Governments have index linked formulae for grant of dearness allowances to their employees. For example, the Govt. of India grants an additional dearness allowance for a point rise in the consumer price index. Several business houses in India and abroad also use cost of living index numbers to provide for automatic increases in wages or allowances. Such index linked salary structures are fairly common now.

(5) The changes in consumer price index can be used to determine the real income over a period. Suppose that a person's income in 1980 was 150 times than that in 1970, that is there is an increase of 50% in money income. Suppose that the cost of living index number in 1980 was 175 times than that in 1970, that is, there is an increase of 75%. Then the person's real income in 1980 is $\frac{150}{175} = 85.7\%$ what was his income in 1970. Similarly real changes in per capita or G.N.P are determined. Thus index numbers are useful in determining the real income or real G.N.P. The process of dividing the index of apparent or physical change by the index of the relevant cost or price is known as *deflation* of time series.

5.3. PRICE RELATIVES, QUANTITY RELATIVES AND VALUE RELATIVES

5.3.1. Price Relatives

One of the simplest types of index numbers is a *price relative*. It is the ratio of the price of a *single* commodity in a given period or point of time to its price in another period or point of time, called the *reference period* or *base period*. If prices for a period, instead of a point of time, are considered, then suitable price average for the period is taken and these prices are expressed in the same units. If p_0 and p_n denote the price of a commodity during the base period or reference period (0) and the given period (n) then the *price* relative of the period n with respect to (w.r.t) the base period 0 is defined by

$$\text{price relative in percentage (of period } n \text{ w.r.t. period 0)} = \frac{p_n}{p_0} \times 100 \qquad (3.1)$$

and is denoted by $p_{0/n}$.
For example, if the retail price of fine quality of rice in the year 1980 was Rs. 3.75 and that for the year 1983 was Rs. 4.50 then

$$p_{1980\ |\ 1983} = \frac{\text{Rs. } 4.50}{\text{Rs. } 3.75} \times 100 = 120\%$$

For example, the exchange rate of a U.S. dollar was Rs. 10.00 in July 1984 (period J) and was Rs. 12.50 in December 1984 (period *D*) then the price relative of a dollar in December w.r.t. that in July is given by

$$p_{J\,|\,D} = \frac{\text{Rs. } 12.50}{\text{Rs. } 10.00} \times 100 = 125\%$$

5.3.2 Quantity Relatives

Another simple type of index numbers is a quantity relative, when we are interested in changes in quantum or volumes of a commodity such as quantities of production or sale or consumption. Here the commodity is used in a more general sense. It may mean the volume of goods (in tonnes) carried by roadways, the number of passenger miles travelled, or the volume of export to or import from a country. In such cases we consider of quantity or volume relatives. If quantities or volumes are for a period instead of a point of time, a suitable average is to be taken and the quantities or volumes are to be expressed in the same units. If q_0 and q_n denote the quantity or volume produced, consumed or transacted during the base period (0) and the given period (n) then *quantity relative* of the period n w.r.t. the base period 0 is defined by

$$\text{quantity relative in percentage (of period } n \text{ w.r.t. period 0)} = \frac{q_n}{q_0} \times 100 \qquad (3.2)$$

and is denoted by $q_{0|n}$.

5.3.3. Value Relatives

A value relative is another type of simple index number, usable when we wish to compare changes in the money value of the transaction, consumption or sale in two different periods or points of time. Multiplication of the quantity q by the price p of the commodity produced, transacted or sold gives the total money value pq of the production, transaction or sale. If instead of point of time, period of time is considered, a suitable average is to be taken and is to be expressed in the same units.

If p_0 and q_0 denote the price and the quantity of the commodity during the base period (0) and if p_n and q_n denote the corresponding price and quantity during a given period (n), then the total value $v_0 = p_0q_0$ and $v_n = p_nq_n$, and the value relative of the period n w.r.t. the base period 0 is defined by

$$\text{value relative in percentage (of period } n \text{ w.r.t. period 0)} = \frac{v_n}{v_0} \times 100$$

$$= \frac{p_nq_n}{p_0q_0} \times 100 \qquad (3.3)$$

and is denoted by $v_{0|n}$.

5.3.4. Properties of Relatives

Let $p_a, p_b, p_c, \ldots$ denote the prices, in the periods a, b, c, . . . respectively,

$q_a, q_b, q_c, \ldots$ denote the quantities and $v_a, v_b, v_c, \ldots$ the volumes for the corresponding periods.

The relatives satisfy some properties which are directly obtained from the definitions. We shall state the results for price relatives and write similar results for the quantity and value relatives:

1. Identity property:
   The price relative of a given period w.r.t. the same period is 1, that is $P_{a/a} = 1$.
2. Time reversal property:
   If the base period and the reference period are interchanged, then the product of the corresponding relatives is unity (one is the reciprocal of the other). That is: $P_{a|b} \times P_{b|a} = 1$

   or
   $$P_{b|a} = \frac{1}{P_{a|b}}$$
   Here
   $$P_{a|b} = \frac{p_b}{p_a} \text{ and } P_{b|a} = \frac{p_a}{p_b} \text{ and so the result follows.}$$
3. Circular or Cyclic property:
   We have
   $$P_{a|b} = \frac{p_b}{p_a}, P_{b|c} = \frac{p_c}{p_b}, P_{c|a} = \frac{p_a}{p_c}$$
   and so
   $$P_{a|b} \times P_{b|c} \times P_{c|a} = 1$$
   That is, if the periods *a, b, c,* are in cyclic order then the product of the three relatives w.r.t. the preceding period as base period is unity.

   This holds for *any* number of periods in cyclic order.
4. Modified Circular or Cyclic property:
   We have
   $$P_{a|b} \times P_{b|c} = \frac{p_b}{p_u} \times \frac{p_c}{p_b} = \frac{p_c}{p_a} = P_{a|c}$$
   More generally
   $$P_{a|b} \times P_{b|c} \times P_{c|d} = P_{a|d}.$$

5.4. LINK AND CHAIN RELATIVES

Consider a series of successive periods of time, denoted by $t_1, t_2, t_3, \ldots$ and let $p_1, p_2, p_3, \ldots$ denote the prices of a commodity at times t_1, t_2, t_3 respectively. Then $p_{t_1|t_2} = (p_2|p_1)$, is the price relative of t_2 relating to the preceding period t_1, $p_{t_2|t_3} = (p_3|p_2)$ is the price relative of t_3 relating to the preceding period t_2 and so on. The series of price relatives of one period relating to the preceding period are given by

$$p_{t_1|t_2}, p_{t_2|t_3}, p_{t_3|t_4}, \ldots$$

are called *link relatives*. The price relatives are also expressed as percentages, e.g. $p_{t_1|t_2} = \frac{p_2}{p_1} \times 100$. For example, if the prices of a commodity during January (J), February (F), March (M), April (A), . . . were Rs. 2.50, Rs. 2.75, Rs. 3.00, Rs. 2.80, . . . then the link relatives are

$$p_{J|F} = \frac{2.75}{2.50} = 1.100 = 110.0\%$$

$$p_{F|M} = \frac{3.00}{2.75} = 1.091 = 109.1\%$$

$$p_{M|A} = \frac{2.80}{3.00} = 0.933 = 93.3\%$$

and so on.

The price relative of any period w.r.t. any other period as base:

It follows from the circular property that any price relative can be expressed as the product of link relatives. For example,

$$p_{t_1|t_3} = p_{t_1|t_2} \times p_{t_2|t_3}.$$

That is, the price relative with respect to a fixed period as base period can be obtained from link relatives. Thus the relatives of one period to the preceding one are chained together by successive multiplication and form a chain index to give a price relative with respect to a fixed base period. It is therefore called *chain index* or *chain relative.*

The above concepts of link and chain relatives are also true in case of quantity and value relatives. So far we have considered one single commodity separately and we have found the price relative of one commodity. This would lead to a rather unweildy number of indexes, one corresponding to each item. It would be difficult to use these series of indexes and they may not serve any useful purpose. Nor can we concentrate on one or two commodities leaving aside all others. For obviously, only one or two items or commodities could not ordinarily claim to represent the whole series of items or commodities. In general we are interested in *general price level* rather than on the price level of any particular item. So we have to consider a number of items at the same time. If each commodity is of different qualities then all such qualities have also to be taken into consideration. Sometimes we are interested not in the general price level but in the change in cost of living of a given class of people, such as daily wage earners, office workers, etc. Then we have to take into consideration the type of goods and services that the particular group uses and also their weightage. Similarly, we may consider the change of production in either agriculture or industry. Different commodities are to be taken into consideration in different cases. In all cases, however, we have to consider simultaneously a number of items or commodities together for measuring change in price or production. Further, the quantities to be reckoned of each item have also to be considered.

5.5. PROBLEMS INVOLVED IN THE CONSTRUCTION OF INDEX NUMBERS

There are several problems that a statistician encounters in the process of going for construction of an index number. These relate to the following:

(1) A clear definition of the purpose for which the index is constructed.
(2) Selection of items for inclusion in the process.
(3) Data for the index number–their sources, their representativeness and their collection.
(4) Selection of the base period.
(5) Adoption of a suitable formula for construction of the index–the system of weights to be given, items and the method of combining data.

We discuss the problems below. The problems are not entirely independent nor are they of equal importance.

(1) Before collection of data for construction of an index number, it is important to know what kind of changes we are trying to measure and also how we intend to use the index constructed. If it is designed and geared to a specific purpose, then the index can be very useful. Otherwise it can serve no useful purpose and can even lead to faulty conclusions. For example, if we wish to measure the trend in price changes in items of domestic consumption for a family, we have to take retail prices and not wholesale prices of the items into consideration.

(2) The next important question is about the selection of items. It is to be carefully examined what items are relevant for the particular type of changes to be measured. For example, in computing the cost of living index number of a middle class family, which reflects the trends in the change in the cost of living, gold will not be a relevant item, whereas family clothing will be, in such a case. Items must be so chosen that they reflect that change in the trend that we wish to measure.

(3) Data constitute one of the important ingredients in the construction of an index number. First of all, data must be reliable, accurate and homogeneous. These must be drawn from the right type of sources. The samples selected must be representative of the class to which they belong. A sufficiently large sample of items relevant to the study must be taken to obtain a reliable index number.

The sources of data must be reliable and care should be taken in selecting the sources of data. For example, official publications as well as other reliable unofficial reports from producers and merchants should be considered for computing index for production. If retail food prices are being considered, then prices in stores, not in wholesale markets, are to be taken into consideration.

Care should be taken to ascertain how the data are collected. There should be a proper basis of obtaining representative samples. To get such a sample, it is often necessary to stratify the original data into suitable homogeneous groups and subgroups before drawing a representative sample. For example, in considering price changes, items which have elastic demand are to be grouped separately from items which have inelastic demand.

(4) It is customary to select a base period for comparison. The base period is the period at which the index is taken to be 100. Index numbers of other selected periods can then be compared with the 100 of the base period. How is this period to be chosen? Since a month is considered too short a period, during which there may be seasonal variation, a whole year is chosen as the base period or base year. However, it is to be seen that the year chosen is a sufficiently "normal" year to be a good basis for comparison. It should not be a period during which production is unusually low if we are considering index of production or a period during which the prices of commodities are unusually high because of drought or flood, if we are considering price index. The base period should be more or less stable and should be free from unusual movement of the characteristic we are studying. A year ravaged by war or famine is to be avoided. A particular base period may be suitable for a number of years that follow the base period, but the base period should not be sufficiently away from the period we are considering. It is then desirable to shift the base period. For example, it may be meaningless to consider 1949 as base period if prices of present times are considered. It may be necessary to shift the base period to 1970 in that case and then again shift the base period to a later year.

The main reasons for shifting the base year are the following: (1) because of growth of population or notable technological development the pattern of production, income prices may be very much altered; (2) because of new ideas the patterns of consumption may change to such an extent that aggregate of commodities may be quite different at different periods; (3) because of change in quality of good or change in tastes of individuals, there may be appreciable change in the pattern of consumption, prices and production of commodities.

(5) Adoption of a suitable formula for construction of an index number poses some problems. Which will be the most suitable way of combining the data under the particular situation? Suppose we are considering price indexes. The combining of prices for each year can be done by two methods, either by aggregating (or totaling) or by averaging. Averaging prices over a year leads to simple average of price relatives. For a simple aggregate of prices the price of each particular item is given in usual units. This may lead to dominance of a particular commodity because the price of that commodity being expressed in the unit chosen.

If the price of the same item is expressed in a different unit, its effect may not be so dominant in the aggregate of actual prices. Rather than the economic importance of the commodity, its price in a particular unit may effect the index. An attempt may be made to express the prices of the items in the same unit but that may also lead to unsatisfactory comparisons. For example, it may not be worthwhile to combine the price of potatoes per kg. with the price of butter per kg.

This difficulty is obviated if price relatives are considered, because then the price of each item is expressed in the same unit and the relative refers the price of an item in one unit for a base year to the price of the same item in the same unit for another year. But when we average the price relatives for different items of consumption, there is a concealed assumption that in the base year each item is purchased for an equal amount of money. That is, all the items of the same value is purchased in the base year. This is also an unsatisfactory assumption. Thus both the methods, of taking the simple aggregate of actual prices and the method of taking the simple average of price relatives, are defective in the construction of an index.

A better method of construction is to bring out the relative importance of individual items by giving them due emphasis or weightage. A suitable method will give due appropriate weights to different items according to their importance. The questions then arise: (1) By what do we allocate weight? (2) What type of weight to be used?, and (3) From what period are the weights to be taken. These questions are to be carefully examined before we can determine the weights to be used in indexing. For the first question we have to find the weights that bring out the economic importance of the item. The weights may be production figures or consumption figures. For the second question we have to see what type of weights, (quantity weight or value weight) is to be used. A value weight corresponds to the value (price X quantity) of the commodity. For meaningful results, quantity weights are used in the method of totaling actual prices and value weights are used in the method of averaging price relatives. Though weights may refer to any period usually they relate to the base period.

After all these points have been determined, the question then remains; which formula is to be used? There are several formulae for construction of an index. Consideration of base year quantities as weights leads us to *Laspeyre's formula*

$$\frac{\Sigma p_n q_0}{\Sigma p_0 q_0}$$

Consideration of given period quantities as weights leads us to *Paasche's formula*

$$\frac{\Sigma p_n q_n}{\Sigma p_0 q_n}$$

Whereas the average (or total) quantities of base and given years leads us to *Marshal-Edgeworth's formula*

$$\frac{\Sigma p_n (q_0 + q_n)}{\Sigma p_0 (q_0 + q_n)}$$

If we take the quantities q_t of another typical year as weights, we get the formula

$$\frac{\Sigma p_n q_t}{\Sigma p_0 q_t}$$

The geometric mean of Laspeyre's and Paasche's formulas leads us to *Fisher's Ideal Index Number formula*

$$\sqrt{\left[\frac{\Sigma p_n q_0}{\Sigma p_0 q_0} \cdot \frac{\Sigma p_n q_n}{\Sigma p_0 q_n}\right]}$$

Note : We have seen that the price relatives possess some interesting properties. Index number formulae also have some desirable properties. We shall examine later the properties that a good index number formula should possess. The choice of a particular formula must involve such considerations.

We have considered above the general problems encountered in the construction of various types of index numbers. In constructing the particular type special problems arise. These have also to be taken into consideration. We shall examine later the special problems that arise in the construction of the cost of living index numbers.

5.6 COST OF LIVING INDEX NUMBERS (CLI)

One of the main types of index numbers in use is the cost of living index number (CLI). This is also known as consumer price index number (CPI). Gradually the expression CLI is being replaced by CPI; it is a special index number of retail prices in which only prices of selected commodities are considered which enter into the consumption pattern of a particular group of people. The commodities of selected items for the group constitute what is known as "market basket of goods" for that group. Thus different items enter into the "market basket of goods", of different groups. Different groups of people have different CLI numbers. The market basket of goods includes goods and services needed for maintaining a certain standard of living for that group over a period of time. The CLI measures changes in the cost of maintaining the standard of living for that group.

In India CLI numbers are being constructed for three groups of people. These index numbers are

(1) The working class cost of living index numbers
(2) The middle-class cost of living index numbers
(3) The cost of living index numbers of the Central Government employees.

We shall describe them later. The basket of goods is divided into five major groups–food, housing, fuel and light, clothing and other goods and services.

In the U.S.A. a "Consumer Price Index for Urban Wage Earners and Clerical Workers" is constructed regularly. The commodities are divided into 8 major groups – food, housing, dress, transportation, medical care, personal care, reading and recreation and other goods and services. The special problems that arise in the construction of cost of living index numbers for a group lie in determining the market basket of goods and services needed for a person of the group for maintaining a certain standard of living. While transport may be an item for city dwellers, this may not be so in the case of villagers in a developing country.

The following points need be considered in the selection of items for a group of people: (1) items taken should be such as to represent the habits, tastes and traditions of the average person in the group; (2) the economic and social importance of the goods and services are also to be examined; (3) items should be such that they are not likely to vary in quality in appreciable degree over two different places or different periods of time; and (4) items should be fairly large in number so as to represent adequately the standard of living for the groups). A fairly reasonable number should be selected. Again after determining the items to be included in the basket, the question that arises is the determination of suitable weights for different items in the basket.

To determine the weights, a proper study of consumption habits of persons of that group is to be made. The usual procedure is to conduct "family budget surveys". Such surveys help in determining the items to be entered into the consumption pattern of the group and also help in determining the weights to be assigned to different items. One problem may occur regarding different *qualities* of the same type of commodity. Another problem may concern items of common use which do not occur in both the base period and the given period.

For a detailed account of this topic refer to Banerjee (1975).

5.7. METHODS OF CONSTRUCTION OF INDEX NUMBERS: FORMULAS

As already indicated there are two methods of constructing index numbers: (1) by computing aggregate values, and (2) by taking averages of relatives. However, under certain conditions the two methods are only alternative methods for obtaining the same result. The aggregate method is more direct. But it cannot be applied in certain situations, which is when the other method of averaging relatives has to be used.

Consider a group of k commodities with prices $p_0^{(i)}$, and $p_n^{(i)}$, $i = 1, 2, \ldots k$ for the prices in the base year and the year under reference n respectively. For simplicity we shall drop the upper suffix (i) and write simply p_0 and p_n. A simple aggregate formula for index number could be obtained by

$$P = \frac{\Sigma p_n}{\Sigma p_0} \equiv \frac{\Sigma p_n^{(i)}}{\Sigma p_0^{(i)}} \tag{7.1}$$

where the summation extends from $i = 1$ to $i = k$. Whether the prices are taken in the same units for all the commodities or in some suitable units of the commodities not much useful purpose is served. Commodities are consumed in varying quantities, a fact which has to be taken into account to make any meaningful comparison. Thus it is necessary to consider *weighted aggregate* rather than a simple aggregate of prices of commodities. Weights should be chosen to adequately emphasize the varying quantities consumed of different commodities.

5.7.1. Weighted Aggregates

A reasonable assumption is to consider quantities consumed or produced as weights. The price multiplied by the corresponding quantity gives the value of the particular commodity considered. The general formula for the aggregate price index is then given by

$$P = \frac{\Sigma p_n q}{\Sigma p_0 q} \equiv \frac{\Sigma p_n^{(i)} q^{(i)}}{\Sigma p_0^{(i)} q^{(i)}} \tag{7.2}$$

Base year quantities as weights

We can take the quantity of the base year 0 or the year under reference (n) or the average of these two years or any other year.

Taking base year quantities as weights we get the formula

$$P_L = \frac{\Sigma p_n q_0}{\Sigma p_0 q_0} \tag{7.3}$$

This is *Laspeyre's* method, formulated in 1864. P_L is known as *Laspeyre's price index number.*

Given year quantities as weights

Taking the quantities of the given period as weights, we get

$$P_p = \frac{\Sigma p_n q_n}{\Sigma p_0 q_n} \tag{7.4}$$

This is *Paasche's method, formulated in 1874. It gives Paasche's price index number.*

Average of base year and given year quantities as weights

Taking the arithmetic mean (or total) of the quantities of the base and given year as weights, we get

$$P_{ME} = \frac{\Sigma p_n (q_0 + q_n)/2}{\Sigma p_0 (q_0 + q_n)/2} = \frac{\Sigma p_n (q_0 + q_n)}{\Sigma p_0 (q_0 + q_n)}$$

$$= \frac{\Sigma p_n q_0 + \Sigma p_n q_n}{\Sigma p_0 q_0 + \Sigma p_0 q_n} \tag{7.5}$$

This is *Marshall-Edgeworth's price index number.*

Taking as weights the geometric mean $\sqrt{(q_0 q_n)}$ of base and given year quantities, we get

$$P_W = \frac{\Sigma p_n \sqrt{q_0 q_n}}{\Sigma p_0 \sqrt{q_0 q_n}} \tag{7.6}$$

This is *Walsh's index number.*

Taking the average quantities of a number of years as weights, we get a formula for price index number.

Average of two index numbers

Take Laspeyre's and Paasche's index numbers and use the average of these two index numbers.

Taking the arithmetic mean, we get

$$P_B = \frac{1}{2}\left(\frac{\Sigma p_n q_0}{\Sigma p_0 q_0} + \frac{\Sigma p_n q_n}{\Sigma p_0 q_n}\right) \tag{7.7}$$

This was recommended by *Bowley* and is known as *Bowley's index numbers.* Taking the geometric mean, we get

$$P_F = \sqrt{\left(\frac{\Sigma p_n q_0}{\Sigma p_0 q_0} \times \frac{\Sigma p_n q_n}{\Sigma p_0 q_n}\right)} \tag{7.8}$$

This is known as *Fisher's price index number.*

We thus find that there are a number of formulae for construction of index numbers. The index numbers under different weighting systems would be markedly different when both the relative magnitudes of prices and quantities vary greatly. If the change in prices is in the same ratio and in the same direction, then all the weighting systems would agree and all the formulae would give the same result.

Denote $r_i = \dfrac{p_n^{(i)}}{p_0^{(i)}}$; r_i is the price relative of the commodity i (of the price in the given year (n) relative to the base year (0)).

Taking the base year values $p_0^{(i)} q_0^{(i)} = v_0^{(i)}$ as weights the weighted average of price relatives is obtained as follows:

$$\frac{\Sigma r_i v_0^{(i)}}{\Sigma v_0^{(i)}} = \frac{\Sigma \dfrac{p_n^{(i)}}{p_0^{(i)}} \cdot p_0^{(i)} q_0^{(i)}}{\Sigma p_0^{(i)} q_0^{(i)}}$$

$$= \frac{\Sigma p_n^{(i)} q_0^{(i)}}{\Sigma p_n^{(i)} q_0^{(i)}} = \frac{\Sigma p_n q_0}{\Sigma p_0 q_0} = P_L \tag{7.9}$$

Thus the Laspeyres price index, is the weighted average of price relatives with values of base year as weights. For an alternative form of weighted price average

denote
$$w_i = \frac{p_0^{(i)} q_0^{(i)}}{\sum\limits_i p_0^{(i)} q_0^{(i)}}$$

We have $\Sigma w_i = 1$ and w_i is the proportion of cost that commodity i would bear on the total cost for the base year (0).

Thus P_L can be written as

$$P_L = \frac{\Sigma p_n q_0}{\Sigma p_0 q_0} = \frac{\Sigma p_n^{(i)} q_0^{(i)}}{\Sigma p_0^{(i)} q_0^{(i)}}$$

$$= \frac{\Sigma \dfrac{p_n^{(i)}}{p_0^{(i)}} \cdot p_0^{(i)} q_0^{(i)}}{\Sigma p_0^{(i)} q_0^{(i)}}$$

$$= \Sigma \frac{p_n^{(i)}}{p_0^{(i)}} \cdot \left[\frac{p_0^{(i)} q_0^{(i)}}{\Sigma p_0^{(i)} q_0^{(i)}} \right] \qquad (7.10)$$

$$= \Sigma r_i w_i \quad \textit{where} \quad \Sigma w_i = 1.$$

This is *Laspeyre's index* considered as weighted average of price relatives.

5.7.2 Aggregate of Price Relatives

Let us consider an alternative method of taking aggregates of price relatives. The price relative of the commodity i is given by the quantity $p_n^{(i)} / p_0^{(i)}$ and the price relative in percentage by $\dfrac{p_n^{(i)}}{p_0^{(i)}} \times 100$ of price relatives. The quantity

$$\Sigma \frac{p_n^{(i)}}{p_0^{(i)}}$$

does not convey much meaning, as there is no emphasis on the importance of different commodities. To give due emphasis, we weight the relatives and find instead the weighted average of price relatives. How are weights to be taken? Consider weighting the price relatives by the values of the commodities $p^{(i)} q^{(i)} = v^{(i)}, i = 1, 2, \ldots$

Similarly taking the year values $p_0^{(i)} q_n^{(i)}$ (product of quantity of given year with price of base year) as weights we get

$$\frac{\Sigma \frac{p_n^{(i)}}{p_0^{(i)}} \cdot p_0^{(i)} q_n^{(i)}}{\Sigma p_0^{(i)} q_n^{(i)}} = \frac{\Sigma p_n^{(i)} q_n^{(i)}}{\Sigma p_0^{(i)} q_n^{(i)}} = \frac{\Sigma p_n q_n}{\Sigma p_0 q_n} = P_P \qquad (7.11)$$

which is Paasche's formula for price index number.

Thus the two methods give the same formula.

Note: Though both the methods of computing index numbers lead essentially to the same result, the weighted aggregate method may be preferred to the weighted price relative method, because of its ease of computation and explanation. In certain situations, however, the aggregative method cannot be used and then the aggregate of relative methods is to be used. When an index of business cycles is to be computed, the aggregative method can be used since the data to be averaged is expressed as percentages of trend and seasonal fluctuation.

5.7.3. Comparison of Laspeyre's and Paasche's Index Numbers

The advantage of Laspeyre's index number formula is that once the weights $q_0^{(i)}$ for the base year are determined these remain the same and only the changes in prices are to be considered. For Paasche formula weights $q_n^{(i)}$ have to be computed afresh for each given year (n). As base year is selected as a typical year when conditions are more or less normal, the use of base year quantities as weights gives more stability to the index number series. On the other hand, if the conditions change rapidly then a given year's quantities afford a more realistic picture. Sometimes it may be advantageous to take the average of quantities of several years as weights.

Laspeyre's price index tends to overestimate price changes

Suppose that there is no change in taste or consumption habit during the period beginning with the base year to the given year. That is, the group of people considered would go for the same commodities. According to the economic law of demand, people tend to purchase in lesser quantity the commodities whose prices increase and to purchase in greater quantity the commodities whose prices decrease. Thus the same amount of satisfaction may be derived by an individual buying less of the commodities whose prices increase relative to the base year. In case of Laspeyre's index we consider $\Sigma p_n q_0$, q_0 being the quantity of the base year. Since if prices rise, less will be bought and the actual total prices paid will be somewhat less than $p_n q_0$ the ratio

$$L_P = \frac{\Sigma p_n q_0}{\Sigma p_0 q_0}$$

tends to be higher than it should be in the actual pattern of consumption. The price change indicated by Laspeyre's formula will be the upper limit of price change and therefore it tends to overestimate price change.

Paasche price index tends to underestimate price changes

In the case of Paasche's price index we consider $p_n\, q_n$ (q_n being the quantity of the given year) as being the value of goods supposed to be purchased in the base year. By taking the quantity of the given year, one may take lesser quantity of the commodities which may have increased in price and higher quantity of those which may have decreased in price. Thus $p_n\, q_n$ will be higher than what it should have been in the actual pattern of consumptions. This is the denominator in the Paasche price index formula. Thus the price change indicated by Paasche's formula will tend to be low and it will be the lower limit of price change. This formula thus tends to underestimate price change.

Note : (1) It does not however imply that the Laspeyre index is always higher than the Paasche index but that it *tends* to be higher. Laspeyre's index can be greater than, equal to, or less than Paasche's index in actual cases. In practice, the two index numbers are usually fairly close when the periods (base and given years) are not very far. (2) A question now arises as to which of the index number formula should be accepted as the best. This will be judged on the basis of certain tests of index numbers (which are being discussed in section 5.9). A detailed discussion on this subject has been given by Fisher (1922). (3) The Laspeyre and the Paasche formula of the price index numbers should be multiplied by 100 to express the index in percentage. So also the other index numbers. Marshall-Edgeworth's, Bowley's and Fisher's are to be multiplied by 100 to express them in percentages.

5.8 QUANTITY INDEX NUMBERS

Instead of comparing prices we may be interested in comparing quantity (of consumption or of production and so on) from year to year. We then obtain the corresponding quantity index numbers.

A simple aggregative quantity index is given by

$$Q = \frac{\Sigma q_n^{(i)}}{\Sigma q_0^{(i)}} = \frac{\Sigma q_n}{\Sigma q_0}.$$

Using base-year *prices* as weights we get Laspeyre's quantity index as follows

$$Q_L = \frac{\Sigma q_n p_0}{\Sigma q_0 p_0}. \tag{8.1}$$

Using given-year *prices* as weights, Paasche's quantity index is given by

$$Q_P = \frac{\Sigma q_n p_n}{\Sigma q_0 p_n} \tag{8.2}$$

Similarly we can get the corresponding quantity index numbers. Fisher's and Marshall-Edgeworth's quantity index numbers are given by

$$Q_F = \sqrt{Q_P \cdot Q_L} = \sqrt{\left[\left(\frac{\Sigma q_n p_n}{\Sigma q_0 p_n}\right)\left(\frac{\Sigma q_n p_0}{\Sigma q_0 p_0}\right)\right]} \tag{8.3}$$

and $$Q_{ME} = \frac{\Sigma q_n (p_0 + p_n)}{\Sigma q_0 (p_0 + p_n)} = \frac{\Sigma q_n p_0 + \Sigma q_n p_n}{\Sigma q_0 p_0 + \Sigma q_0 p_n}. \tag{8.4}$$

These aggregative quantity index numbers can also be obtained as weighted averages of quantity relatives. For example

$$Q_P = \frac{\Sigma q_n p_0}{\Sigma q_0 p_0} = \frac{\Sigma \frac{q_n}{q_0}(p_0 q_0)}{\Sigma q_0 p_0} = \frac{\Sigma \frac{q_n}{q_0} \cdot v_0}{\Sigma v_0}$$

Q_p is the weighted average of the quantity relatives $\dfrac{q_n}{q_0} = \dfrac{q_n^{(i)}}{q_0^{(i)}}$ with weights $v_0 = v_0^{(i)}$ equal to the value for the base year.

The quantity index number measures quantity change. While the price index number measures the change in value of a *fixed aggregate of goods at varying prices*, the quantity index number measures the change in value of a *varying aggregate of goods* at *fixed price.* The price index number tells us how much we shall spend in the given year (as percentage to that spent in the base year) if the *same quantity* or assortment of commodities at *varying* prices are bought. The quantity index number, on the other hand tells us how much we shall spend in the given year (as percentage to that spent in the base year) if *varying quantities* of commodities are bought at the same price.

5.9. TESTS FOR INDEX NUMBERS

We have seen in section 5.3.4 that the relatives have certain important properties. What is true for an individual commodity should also be true for a group of commodities. The index number as an aggregative relative should also satisfy the same set of properties. We shall examine the properties that a good index number should have and also examine the index number formulae in the light of these properties.

Time Reversal Test or Property

If the two *periods*, the base period and the reference period are interchanged, the product of the two index numbers should be unity. In other words, one should be the *reciprocal of the other.*

Denote the index number of the given year (n) relative to the base year (0) by $I_{0/n}$. Then $I_{n/0}$ is the index number of the year (0) relative to the year (n), i.e. with the years interchanged time reversal test requires that

$$I_{0|n} \times I_{n|0} = 1 \tag{9.1}$$

Or,

$$I_{n|0} = \frac{1}{I_{0|n}} \tag{9.2}$$

An index number which satisfied the above is said to satisfy time reversal test. Let us examine which of the index number formulae satisfy this property. Consider Laspeyre's index number

$$\frac{\Sigma p_0 q_n}{\Sigma p_0 q_0}.$$

Interchanging the time subscripts, we get

$$\frac{\Sigma p_n q_0}{\Sigma p_n q_n}.$$

Their product

$$\frac{\Sigma p_0 q_n}{\Sigma p_0 q_0} \times \frac{\Sigma p_n q_0}{\Sigma p_n q_n} \neq 1 \tag{9.3}$$

Consequently, Laspeyre's index number does not satisfy the time reversal test. So also is the case with Paasche's index number.

Consider Fisher's price index number (7.8) Interchanging the time subscripts we find another index number. The product of the two is given by

$$\sqrt{\left(\frac{\Sigma p_n q_0}{\Sigma p_0 q_0} \cdot \frac{\Sigma p_n q_n}{\Sigma p_0 q_n}\right)} \cdot \sqrt{\left(\frac{\Sigma p_0 q_n}{\Sigma p_n q_n} \cdot \frac{\Sigma p_0 q_0}{\Sigma p_n q_0}\right)} \tag{9.4}$$

which equals 1. Thus Fisher's index number satisfies the time reversal test.

Again consider Marshall-Edgeworth's price index number given by (7.5). Interchanging the time subscripts we get

$$P'_{ME} = \frac{\Sigma p_0(q_n + q_0)}{\Sigma p_n(q_n + q_0)}$$

So that

$$P_{ME}P'_{ME} = \frac{\Sigma p_n(q_0 + q_n)}{\Sigma p_0(q_0 + q_n)} \cdot \frac{\Sigma p_0(q_n + q_0)}{\Sigma p_n(q_n + q_0)} = 1 \tag{9.5}$$

Thus Marshall-Edgeworth's index number satisfies time reversal test. It can be seen that Walsh's index number given by (7.6) also satisfies this test.

Factor Reversal Test or Property

If the two factors p and q in a price index formula are interchanged, so that a quantity index number is obtained, then the product of the two index numbers should give the true value ratio

$$\frac{\Sigma p_n q_n}{\Sigma p_0 q_0} \tag{9.6}$$

In other words, the price index number multiplied by the corresponding quantity index number should give the true ratio of value in the given year (n) to the value in the base year (0). This property holds for a single commodity, since

$$\frac{p_n}{p_0} \times \frac{q_n}{q_0} = \frac{p_n q_n}{p_0 q_0} \tag{9.7}$$

but it does not hold for most of the index numbers. For example,

$$P_L \times Q_L = \frac{\Sigma p_n q_0}{\Sigma p_0 q_0} \times \frac{\Sigma q_n p_0}{\Sigma q_0 p_0}$$

and this product is not equal to the ratio of the values $\frac{\Sigma p_n q_n}{\Sigma p_0 q_0}$.

Consider Fisher's index number. We have from (7.8) and (8.3)

$$P_F \times Q_F = \sqrt{\left(\frac{\Sigma p_n q_0}{\Sigma p_0 q_0} \times \frac{\Sigma p_n q_n}{\Sigma p_0 q_n}\right)} \times \sqrt{\left(\frac{\Sigma q_n p_0}{\Sigma q_0 p_0} \times \frac{\Sigma q_n p_n}{\Sigma q_0 p_n}\right)}$$

$$= \frac{\Sigma p_n q_n}{\Sigma p_0 q_0} \qquad (9.8)$$

so that the factor reversal test holds for Fisher's index number.

Fisher's index number is one of the rare index numbers which satisfy both the time reversal and factor reversal tests. So it is called "Fisher's *Ideal* Index Number."

Circular Test

This test is based on the shifting of the base period.

If $p_{i|j}$ denotes the index number the given period j with respect to the base period i then this test requires that

$$P_{a|b} \times P_{b|c} \times P_{c|a} = 1$$

for a, b, c all different.

For Laspeyre's index

$$L_{a|b} = \frac{\Sigma p_b q_a}{\Sigma p_a q_a}, L_{b|c} = \frac{\Sigma p_c q_b}{\Sigma p_b q_b}, L_{c|a} = \frac{\Sigma p_a q_c}{\Sigma p_c q_c}.$$

It can be seen that

$$L_{a|b} \times L_{b|c} \times L_{c|a} \neq 1.$$

Hence, this index number does not satisfy circular test. In fact, none of the weighted index numbers satisfies this circular test.

Note : The arithmetic or geometric mean of two numbers will lie between the numbers. Thus Fisher's ideal number being the geometric mean between Laspeyre's and Paasche's index numbers will lie between those two index numbers. Now Laspeyre's index number tends to overtimate price changes while Paasche's index numbers tends to underestimate price changes and Fisher's ideal index number lying between them should be a better estimate than either Laspeyre's or Paasche's index numbers.

5.10. EXAMPLES OF INDEX NUMBER COMPUTATION

Example 10 (a). Find Laspeyre's, Paasche's and Fisher's price and quantity index numbers for the following data.

Commodity/Item	Base Year		Given or Current Year	
	Price (Rs.)	Quantity (kg)	Price (Rs.)	Quantity (kg)
	p_0	q_0	p_n	q_n
A	5	25	6	30
B	10	5	15	4
C	3	40	2	50
D	6	30	8	35

We construct the following table:

Commodity	p_0	q_0	p_n	q_n	p_0q_0	p_0q_n	p_nq_0	p_nq_n
A	5	25	6	30	125	150	150	180
B	10	5	15	4	50	40	75	60
C	3	40	2	50	120	150	80	100
D	6	30	8	35	180	210	240	280
					Σp_0q_0 = 475	Σp_0q_n = 550	Σp_nq_0 = 545	Σp_nq_n = 620

Price Index numbers are as given below; using (7.3) we get

$$P_L = \frac{\Sigma p_n q_0}{\Sigma p_0 q_0} = \frac{545}{475} = 1.1474$$

and in percentage $$P_L \times 100 = 114.74$$

Using (7.4)

$$P_P = \frac{\Sigma p_n q_n}{\Sigma p_0 q_n} = \frac{620}{550} = 1.1273$$

and in percentage $$P_P \times 100 \doteq 114.74$$

Using (7.8) we get

$$P_F = \sqrt{(1.1474 \times 1.1273)} = \sqrt{1.293464}$$

$$= 1.1373$$

and in percentage $$P_F \times 100 = 113.73$$

Quantity index numbers in percentages are as follows:

$$Q_P \times 100 = \frac{\Sigma q_n p_n}{\Sigma q_0 p_n} \times 100 = \frac{620}{545} \times 100 = 1.1376 \times 100 = 113.76$$

$$Q_L \times 100 = \frac{\Sigma q_n p_0}{\Sigma q_0 p_0} \times 100 = \frac{550}{475} \times 100 = 1.1579 \times 100 = 115.79$$

$$Q_F \times 100 = \sqrt{(1.1376 \times 1.1579)} \times 100 = \sqrt{1.3172} \times 100$$

$$= 1.1477 \times 100 = 114.77$$

Example 10(b). Using the following data, find Laspeyre's Paasche's and Fisher's price and quantity index numbers for 1980, with 1970 as the base year.

Commodity	1970 Quantity (q_0)	1970 Value ($v_0 = p_0q_0$)	1980 Quantity (q_n)	1980 Value ($v_n = p_nq_n$)
I	50	350	60	420
II	120	600	140	700
III	30	330	20	200
IV	20	360	15	300
V	5	40	5	50

We construct the following table

Commodity	q_0	p_0q_0	p_0	q_n	p_nq_n	p_n	p_0q_n	p_nq_0
I	50	350	7	60	420	7	420	350
II	120	600	5	140	700	5	700	600
III	30	330	11	20	200	10	220	300
IV	20	360	18	15	300	20	270	400
V	5	40	8	5	50	10	40	50
		Σp_0q_0 = 1680			Σp_nq_n = 1670		Σp_0q_n = 1650	Σp_nq_0 = 1700

Price index numbers (in percentages) are as given below:

$$\text{Laspeyer I.N.} = P_L \times 100 = \frac{\Sigma p_nq_0}{\Sigma p_0q_0} \times 100 = \frac{1700}{1680} \times 100 = 101.1904$$

$$\text{Paasche I.N.} = P_P \times 100 = \frac{\Sigma p_nq_n}{\Sigma p_0q_n} \times 100 = \frac{1670}{1650} \times 100 = 101.2121$$

$$\text{Fisher I.N.} = \sqrt{(101.1904 \times 101.2121)} \times 100 = 1.0241694 \times 100 = 101.20$$

Quantity index numbers (in percentages) are as given below :

$$\text{Paasche I.N.} = Q_P \times 100 = \frac{\Sigma q_np_n}{\Sigma q_0p_n} \times 100 = \frac{1670}{1700} \times 100 = 98.2353$$

$$\text{Laspeyer I.N.} = Q_L \times 100 = \frac{\Sigma q_np_0}{\Sigma q_0p_0} \times 100 = \frac{1650}{1680} \times 100 = 98.2143$$

$$\text{Fisher I.N.} = \sqrt{(Q_L \times Q_P)} \times 100 = 98.22.$$

Example 10(c). Calculate Marshall-Edgeworth's price and quantity index numbers from the data given in Example 10(b).

Using (7.5) we get Marshall-Edgeworth's index in percentages,

$$P_{ME} \times 100 = \frac{1700 + 1670}{1680 + 1650} \times 100 = \frac{3370}{3330} \times 100 = 101.20$$

Using (8.4) we get the quantity index

$$Q_{ME} \times 100 = \frac{1650 + 1670}{1680 + 1700} \times 100 = \frac{3320}{3380} \times 100 = 98.22.$$

Example 10 (d). If the prices of all the items considered in an index number construction change (increase or decrease) in the same ratio, then Laspeyre's and Paasche's index price and quantity numbers will be equal. Similarly if the quantities change in the same ratio the two numbers will be equal.

Suppose that $$\frac{p_n^{(i)}}{p_0^{(i)}} = \frac{p_n}{p_0} = \text{constant for all commodities } i$$

$$= c \text{ (say)} \qquad (10.1)$$

then $p_n = cp_0$. We have

$$P_L = \frac{\Sigma p_n q_0}{\Sigma p_0 q_0} = \frac{\Sigma c p_0 q_0}{\Sigma p_0 q_0} = c$$

$$P_P = \frac{\Sigma p_n q_n}{\Sigma p_0 q_n} = \frac{\Sigma c p_0 q_n}{\Sigma p_0 q_n} = c$$

so that $$P_L = P_P = c = \frac{p_n^{(i)}}{p_0^{(i)}}$$

Similarly it can be shown that $Q_L = Q_P$ under (10.1).

Suppose now that $$\frac{q_n^{(i)}}{q_0^{(i)}} = \frac{q_n}{q_0} = \text{constant for all commodities } i$$

$$= k \text{ (say)} \qquad (10.2)$$

then $q_n = kq_0$

Then we have

$$Q_P = \frac{\Sigma q_n p_n}{\Sigma q_0 p_n} = \frac{\Sigma k q_0 p_n}{\Sigma q_0 p_n} = k$$

and $$Q_L = \frac{\Sigma q_n p_o}{\Sigma q_o p_o} = \frac{\Sigma k q_o p_n}{\Sigma q_o p_n} = k$$

Thus $Q_L = Q_P$ under (10.2)

It can also be shown that $P_L = P_P$ under (10.2).

Example 10 (e). The following table gives the price relatives (with 1952 as base) of a certain year and the weights for computation of wholesale price index. Calculate the price index.

Groups	*Group Weights* (w_i)	*Price Relatives (%)* $(100\, r_i)$
I. Food Articles	50.4	260
II. Liquor & Tobacco	2.1	233
III. Fuel, Light etc.	3.0	125
IV. Industrial Raw Material	15.5	160
V. Manufactured Article	29.0	250
	100.0	

Using (7.10), we get $P_L = 236.698$.

Example 10 (f). The following table gives the price relatives and the weights for different groups of commodities of consumption. Find Laspeyres price index number of 1975 with 1970 as base

	Prices		
Commodity groups	*Base year*	*Given Year*	*Weights (in %)*
I. Food	450	600	58.55
II. Clothing	250	425	5.37
III. Fuel and Light	50	80	6.15
IV. Housing	300	400	9.61
V. Miscellaneous	300	450	20.32
			100.00

We construct the following table

Commodity	*Price relative* $r_i\ (p_n/p_o)$	*Weights* (w_i)	$r_i w_i$
I.	1.33	58.55	77.8715
II.	1.70	5.37	9.1290
III.	1.60	6.15	9.8400
IV.	1.33	9.61	12.7813
V.	1.50	20.32	30.4800
		$\Sigma w_i = 100$	$\Sigma r_i w_i = 140.1018$

Using (7.10), we get $P_L = 140.10$.

5.11. SOURCES OR COMPONENTS OF ERROR IN AN INDEX NUMBER

An index number has the following components of error associated with it.

1. *Formula error.* It is seen that none of the formulas satisfies all the tests. Any formula used introduces an error, which is called formula error.

2. *Homogeneity error.* Because of change of taste, consumption habit, the commodities consumed in the two periods, (the base year and the given year) may not be the same. There are some commodities, called unique commodities, which are included in one particular period. Thus the total expenditure of two periods, are based not exactly on the same set of commodities, and as such are not strictly comparable. Thus one must consider only these commodities called binary commodities which are in both the periods. This introduces an error, called homogeneity error.

The homogeneity error is measured as follows. Let N_0 and N_1 be the total number of commodities in the two periods, the base period and the given period, and let M be the number of binary commodities. Then the total number of *unique* commodities in the two periods is

$$(N_0 - M) + (N_1 - M) = N_0 + N_1 - 2M$$

A measure of heterogeneity or homogeneity is given by the ratio of the total number of unique commodities to the total number of commodities in the two periods. Homogeneity is measured by

$$R = \frac{N_0 + N_1 - 2M}{N_0 + N_1}$$

Now $$R = 0 \text{ when } N_0 + N_1 = 2M,$$

that is, there is no unique commodity and all the commodities are binary and there is complete homogeneity; $R = 1$ when $M = 0$ when there is no binary commodity and all the commodities are unique and there is no homogeneity. R lies between 0 and 1 and the greater the value of R, the greater is the heterogeneity. R is thus a measure of homogeneity.

3. *Sampling error.* Sampling has to be taken recourse to in the selection of items and this introduces sampling errors.

4. *Miscelleneous types of error.* Errors may crop up due to faulty selection of items, inadequate information about price quotation and due to lack of representative character of price quotations etc.

5. In the construction of cost of living index number of a class of people, error may occur due to *incorrect classification.*

5.12. LIMITATIONS OF INDEX NUMBERS

Some of the limitations of index numbers are mentioned below:

1. Index numbers are approximate measures of relative changes in two periods. Such numbers can measure changes in characteristics which are quantifiable and vary with time.
2. Index numbers are subject to various sources of error as noted in section 5.10. Index numbers are often not representative as they are not based on correct and complete data.
3. Selection of the base year is crucial in index number construction. Unless the base year selected is 'normal', no useful purpose can be served by an index number.

4. Index number construction depends on the selection of representative items and the collection of price quotations. To be useful, index numbers have to be free from error arising out of them.
5. Index number construction procedures have to take into account the quality of items or products as well. Study of change in price without taking into account quality may be quite meaningless.
6. An index number is useful for the purpose for which it is constructed and thus its use is limited only to the phenomenon under study. Index numbers constructed for one purpose may not be appropriate for other purposes.

Because of its limitations, the use of index numbers has also been criticised though not very seriously. Note the following comment!

"It is really questionable whether we should be any worse off if we scrapped all index numbers–so many are so ancient and so out of date, so out of touch with reality, so devoid of practical value, that regular calculation must be regarded as nonsense, only serving to mislead the housewife, the businessman and many others."

This is however a very extreme view. The index numbers do serve a very useful purpose in many situations. That is why their construction is taken regularly by Government and other agencies in most countries of the world.

5.13. SOME IMPORTANT INDEX NUMBERS

A number of index numbers are constructed and released regularly, through Ministries/Departments, by the Government of India and state governments. Some of the important ones constructed by the central government are discussed below:

1. *Index Number of Agricultural Production*

This series has the triennium ending 1969-70 as base. The index covers a number of crops divided into groups and subgroups, as mentioned below, with their weights given in parentheses:

A. Foodgrains (68.1)
   (a) Cereals (60.1), (b) Pulses (8.0)

B. Non-Foodgrains (31.9)
   (a) Oilseeds (11.0),
   (b) Fibres (Cotton, Jute, Mesta) (4.0)
   (c) Plantation Crops (Tea, Coffee, Rubber) (2.3)
   (d) Sugarcane, Tobacco (8.1)
   (e) Others (6.5)

In addition to this series, index numbers of areas under principal crops and of yields of principal crops are also constructed annually.

2. *Index Number of Industrial Production (IIP)*

The first series called the Interim Index with 1946 as base covered only 35 items. The revised series with base 1970 = 100 covers a very large number of items grouped as under:

A. Manufacturing (Food manufacturing, Chemicals, Textiles, Machinery, Transport Equipment etc., etc) with group weight 81.08.

B. Mining and Quarrying with weight 9.69.
C. Electricity with weight 9.23.

A revised series with base 1980-81 = 100 which has now been constructed reflects the present state of industrial growth better than the existing series with base 1970 = 100. It was observed that the "industrial structure in India has undergone major changes since 1970. Many of the present critical growth areas like chemicals, petrochemicals, garments, gem-cutting and electronics do not have commensurate weight in the current index with base 1970 = 100 (some do not even find any place at all), whereas many of the traditional (and currently stagnant) industries, such as the mill sector and cotton textiles, command disproportionately high weight in relation to their share in the current industrial production." This is sought to be improved with the revised index.

The weights assigned in the revised series (1980-81 = 100) are as follows:

A. Manufacturing: 77.11
B. Mining and Quarrying: 11.46
C. Electricity: 11.43

Further Revision of 1980-81 IIP series:

The revised IIP (base 1980-81 = 100) covers 352 items. It is generally agreed that since 1980-81 Indian Industry has undergone significant structural changes and the 1980-81 weights do not fully represent some of the sectors such as automobiles, leather goods and electronics, which showed substantial dynamism in the 1980s and 1990s. It is also felt that small scale industries (SSI) sector, which is a dynamic and fast moving sector of the Indian Economy, remains under-represented in the weighting diagram due to non-availability of relevant data). It was therefore decided, to shift the base to 1993-94 for obtaining a more representative weighting diagrams. The new series covering 543 items and having 1993-94 as base has been released in May 1998.*

*Economic Survey : 1998-99, Government of India.

3. *Wholesale Price Index Number (WPI)*

The earlier series has for base 1970-71 = 100. The number of commodities included in the basket is 360; these are classified into groups and subgroups as indicated below, with weights in parenthesis.

A. Primary Articles (41.67)
   (a) Food Articles (29.80)
   (b) Non-Food Articles (10.62)
   (c) Minerals (1.25)
B. Fuel, Power, Light and Lubricants (8.46)
C. Manufacturing Products (49.87)
   (a) Food Products (13.32)
   (b) Textiles (11.03)
   (c) Chemicals etc. (5.55)
   (d) Basic Metals etc. (5.97)
   (e) Machinery & Transport Equipment etc. (6.72)
   (f) Miscellaneous (7.28)

New Series of WPI (1993-94 = 100)

The WPI is the most widely used price index in India. It is a general index that captures movements of prices in a comprehensive way and is available on a weekly basis. It is also taken as an indicator of the rate of inflation in the country.

The Government of India released a new series of WPI with 1981-82 as the base year in July 1989 to replace the old series with lease 1970-71.

With a view to reflecting properly the changes in the economy that have taken place since 1981-82, the Government of India introduced yet another new series with 1993-94 as the base year; a large number of commodities have been added and some items with diminished importance have been dropped. In all, 136 new items (13 in Group l, 1 in Group 2 and 122 in Group 3) have been added. An idea of the changes made can be gauged from the following table :

Weighting diagram, No. of items & No. of quotations of WPI (1981-82) & (1993-94)

Groups of Items	Weights		No. of items (Commodity Basket)		No. of Quotations	
	1981-82	1993-94	1981-82	1993-94	1981-82	1993-94
1. Primary Articles	32.30	22.02	93	98	519	
2. Fuel, Light, Power & Lubricants	10.66	14.23	20	19	73	
3. Manufactured Articles	57.04	63.75	334	318	1779	
All commodities	100.00	100.00	447	435	2371	1918

Source: *Economic Survey:* 1989-90 & 1999-2000, Government of India.

As can be seen more and more emphasis has been given in Manufactured Articles to reflect the pace of industrialization. The WPI at the end of first week of May, 2004 was 181.8 (after correction). Inflation is measured as an exactly one year point to point (percentage) change in WPI; as on week ended August 28, 2004, it stood at 8.33%.

There are four series of Consumer Price Index (CPI). The CPI for industrial workers; the CPI for Urban non-Manual employees; the CPI for agricultural labourers (base 1986-87) and the CPI for rural workers (base 1986-87). We describe below the first two of them.

4. *All India Consumer Price Index (CPI) for Industrial Workers*

Consumer Price Index Numbers for industrial workers on base 1960 = 100 are compiled by the Labour Bureau in respect of 50 important centres (32 factory, 10 plantation and 8 mining) selected from all over the country on the basis of their relative importance. The items of consumption are grouped as under:

I. Food
II. Fuel and Lighting
III. Clothing
IV. House Rent
V. Miscellaneous

The weighting diagrams for these centre indices were derived from the expenditure data shown by the Family Budget Enquiries, conducted among, sampled working class households at each of these centres. CPI's for each of these 50 centres are being published regularly by the Labour Bureau, Govt. of India. The All-India index (CPI) is a weighted average of the 50 constituent centre indices.

The Government of India introduced a new series with 1982 as base year. It would cover 70 centres and would be based on a 1981-82 consumption expenditure survey. The sample size would be increased to 32,616 families against 23,460 families in the 1960 series. The new series with 1982 as base has been introduced w.e.f. October, 1988. The conversion factor from the new to the old series is 4.93. The weights assigned to the 5 item-groups are: I (60.15), II (6.28), III (8.54), IV (8.67) & V (16.36).

5. *All India Consumer Price Index (CPI) for Urban Non-manual Employees*

This series with base 1960 = 100 present CPI for urban non-manual employees based on Family Budget Survey of 45 selected urban centres of the country. A non-manual employee family was defined as one which derived 50% or more of its income from occupation in non-manual work in 'non-agricultural sector.' The weighting diagram at each centre was derived from the expenditure pattern of the same families covered. CPI's for the centres are being published regularly by the Central Statistical Organization, Govt. of India.

The All India index (CPI) is a weighted average of the indices of the constituent centres.

A new series of CPI for Urban Non-manual Employees with 1984-85 as base was introduced w.e.f. November, 1987. The conversion factor from the new to the old series is 5.32.

The CPI's are being used by the Government, Central and State, and industrial organizations to regulate the fixation of D.A. etc. of employees. *Escalation clauses* are also included in Pay Committee Reports, Collective Bargaining Agreements etc., for automatic adjustment of wages/D.A. in terms of rise of CPI.

The State Governments also construct similar indices. The RBI regularly compiles the Index of Ordinary Shares with base 1980-81 = 100. Even some non-official organisations prepare such special types of Indices, e.g. The share Price Index with 1984-85 = 100 of *The Economic Times.*

These indices serve as important indicators of the economy of the country.

Note: As for other countries, mention may be made of the Dow Jones Industrial Average (Index) of the USA, which reflects the general economic condition of the USA and other countries having wide trade links with USA.

A sharp fall in the Dow Jones Average in October, 1987 was a serious cause of concern in the Business Circles and sent shock waves throughout the world.

EXERCISES-5

Sections 5.1-5.2

1. What are index numbers? What purposes are these expected to serve? Discuss.
2. "Index numbers are economic barometers". Why are index numbers so called? Explain.
3. Enumerate and explain the uses of index numbers.

Sections 5.3-5.4

4. What are price relatives and quantity relatives? Discuss.
5. Define price, quantity and value relatives. What do they measure?
6. State the important properties of relatives.
7. Explain what is meant by link and chain relatives.
8. Explain the fixed base and chain base methods of construction of an index number. Describe their relative merits and demerits.

Sections 5.5-5.6

9. Enumerate and explain the problems involved in construction of index number of prices.
10. Discuss how you would choose a base year for the construction of an index number.
11. What is a cost of living index number? What purpose does such a number serve? Discuss.
12. Discuss the problems that arise in construction of a cost of living index number.
13. Explain the method of constructing a cost of living index number for the unskilled labourers in an industrial area.
14. You are required to construct a cost of living index number for tea garden labourers. Enumerate the problems that you would encounter.
15. What special problems are faced in constructing a cost of living index number for a group of people in a time like the present, marked by rapid changes in (i) prices, and (ii) consumption habits.

Section 5.7

16. Discuss the formulae adopted for construction of index numbers.
17. What are weighted aggregates and average of price relatives?
18. Explain (1) Laspeyre's (2) Paasche's price index number formulas both as weighted average and average of price relatives.
19. What are the formulas usually considered for construction of a (1) price (2) quantity index number? Discuss them.
20. What is Fisher's ideal index number.
21. Compare and contrast Laspeyre's and Paasche price index numbers.
22. It is said that Laspeyre's price index tends to overestimate price changes whereas Paasche price index tends to underestimate price changes. Explain.
23. Discuss the method of construction of price index numbers based on price relatives.

Section 5.8

24. What are quantity index numbers? List the important formulas for construction of quantity index numbers.
25. Explain Laspeyre's quantity index number as an average of quantity relatives.
26. Discuss the method of construction of quantity index numbers based on quantity relatives.

Section 5.9

27. What do you mean by tests of an index number? Enumerate the tests.
28. Explain what is meant by factor reversal test. Which of the index numbers satisfy this test?
29. What is Fisher's ideal index number? Why is it so called?
30. Show that Fisher's ideal index number satisfies the time and factor reversal tests.
31. Define Fisher's ideal index number. How far is it ideal?
32. Explain the important criteria or tests that a good index number should satisfy.
33. What are factor and time reversal tests? Explain their significance.
34. Discuss Marshall-Edgeworth's price index number in the light of tests for index numbers.

Sections 5.11-5.12

35. Enumerate the main sources of error in the construction of an index number.
36. Explain formula error, heterogeneity error and sampling error in relation to an index number.
37. Discuss some measures for correcting heterogeneity error.
38. Explain the limitations of index numbers.
39. Do you agree with this statement –

    "It is really questionable whether we should be any worse off if we scrapped all index numbers–so many of them are so ancient and so out-of date, so out of touch with reality, so devoid of practical value that regular calculation must be regarded as nonsense, only serving to mislead the house wife, the businessman and many other"?

    Explain your arguments for and against the above view.
40. Describe a method of constructing a wholesale price index number.
    Discuss briefly the nature of errors that may crop up in this process.

Section 5.10

41. The following table shows the price quantity of the commodities consumed in 1970 and 1978.

	1970		1978	
Commodity	Price (Rs.)	Quantity (kg)	Price (Rs.)	Quantity (kg)
I	4.00	35	4.60	35
II	8.00	10	10.00	10
III	3.00	15	3.50	12
IV	40.00	10	50.00	8
V	2.50	20	3.00	20

Calculate the following price index numbers
(a) Laspeyres
(b) Paasche
(c) Fisher
by using (i) aggregative method, (ii) average of relatives method.

42. Calculate the Marshall-Edgeworth price and quantity index numbers for the data given in Exercise 41.
43. Calculate (a) Laspeyre's, (b) Paasche's, (c) Fisher's quantity index numbers for the data given in Exercise 41.
44. Find Laspeyre's and Paasche's price index numbers from the data given below:

		Price		Quantity	
Commodity	Unit	Base year	Given year	Base year	Given year
A (Rice)	kg	2.50	3.00	40	50
B (Bread)	unit	.80	1.00	60	75
C (Milk)	litre	3.25	4.00	60	60
D (Veg)	kg	2.50	3.00	30	25
E (Fruit)	kg	8.00	10.00	20	25

45. Calculate a suitable price index number to measure change in price of construction from the data given below.

			Price (Rs.)		
	Material Required	Unit	1969	1974	Quantity Required
I	Brick	2,500	600.00	900.00	50,000
II	Timber	Cubic metre	15.00	35.00	1,000
III	Steel rods	Quintal	105.00	315.00	60
IV	Cement	Bag	12.00	60.00	1000

46. Taking 1970 as the base year, compute a suitable quantity index number from the data given below (use both the methods).

Item	Price (1970)	Quantity Produced		
		1970	1975	1980
A	4.50	450	500	300
B	8.00	235	282	329
C	12.00	150	165	180
D	21.00	80	96	100

47. Calculate the cost of living index number from the following data

Groups	Group weights	Price Relative
Food	22.5	1.05
Housing	27.9	1.25
Fuel & Light	5.2	1.00
Clothing	10.6	0.85
Transportation	13.9	1.10
Miscellaneous	19.9	1.20
	100.0	

Section 5.13

48. Are the WPI and CPI strictly comparable? If not explain why they are not strictly comparable.
49. How would you compute annual inflation rate on the basis of change in WPI?
50. Explain the reasons' behind construction of a New Series (of Index Numbers) to replace an Old Series? What does conversion factor from the New to the Old series indicate?

APPENDIX

HUMAN DEVELOPMENT INDEX (HDI)

Human Development Index (HDI) attempts to measure a country's (or an area's) achievements in enhancing human capabilities. It is a measure of empowerment. United Nations Development Programme (UNDP) has been computing HDI every year since 1990 and publishing them in their annual Human Development Reports. While admitting 'HDI, imperfect though it may be,' it states 'it is a viable alternative to Gross National Product (GNP) per capita'. This index is increasingly being used to monitor the progress of the nations. 'Human Development is the end, economic growth a means'. The reports also make policy recommendations. The UNDP Report of 1999 incorporates recommendations, among others, such as job creating growth, strengthening of public services, tapping of new sources of government revenue, and so on. "Cultural liberty in today's diverse world" is emphasized in the latest UNDP Report of 2004.

The computation of HDI has undergone some modifications from year to year since 1990. It is now based on three indicators:

1. Health Status and Longevity, as measured by expectation of life at birth.
2. Educational Attainment to represent level of knowledge and skills, as measured by adult literacy (2/3 weights) and combined primary, secondary and tertiary enrolment ratios (1/3 weight) and
3. Standard of Living, as measured by real Gross Domestic Product per capita (in purchasing power parity dollars).

The above three parameters are considered as essential though not exhaustive. HDI indicates that if people have these basic choices, other opportunities may be easily accessible.

The index for each of the above three indicators is constructed on the basis of fixed minimum and maximum values, in such a way that each index lies between 0 and 1 and the HDI, the simple average of the three indices also lies between 0 and 1. The index of each of the above three indicators is constructed on the basis of fixed minimum, and maximum values as indicated below (maximum; minimum)

1. Life expectancy at birth (85;25).
2. Adult Literacy rate (100;0); combined gross enrolment ratio % (100;0)
3. Gross Domestic Product per capita (PPP US $) (40,000; 100)

There are three indices corresponding to the above (health, education and income). For example, health (corresponding to (1) above) index of a country equals.

$$\frac{e-25}{85-25},$$

where e is the average life expectancy at birth of the country. For India, e = 63.7, the index = 0.645 and so on. Each of the three indexes lies between 0 and 1 and their arithmetic mean (average) which also lies between 0 and 1 constitutes the HDI of the country.

The 2004 UNDP report contains computed values (as of 2002) of 177 countries. While Norway, with HDI 0.956 ranks first, India with HDI 0.595 ranks 127th year. The HDIs and ranks (HDI; rank) of some countries are as follows: USA (0.939;8), JAPAN (0.938;9), UK (0.936;12), CHINA (0.745;94), SRI LANKA (0.740;96), BANGLADESH (0.509;138) and PAKISTAN (0.497;142). INDIA SHOWED AN INCREASING GROWTH; its HDI grew from 0.514 in 1990, to 0.548 in 1995, to 0.579 in 2000 and to 0.595 in 2002. See *UNDP Human Development Report*, 2004, OUP.

UNDP also construct other indexes, such as Human and Income Poverty Index (HPI 1 & HPI 2) and Gender related Development Index (GDI) and many other interesting indicators.

Population Foundation of India compute HDI for India and the States. Amongst States, HDI for Kerala for 1993 was the highest with value 0.628, though it ranked ninth in terms of per capita income (NSDP). ***India Forwards Population and Development Goals***, 1997, OUP.

PART II

Probability and Mathematical Statistics

"As far as the propositions of mathematics refer to reality, they are not certain and as far as they are certain they do not refer to reality."

Albert Einstein (1879-1955)

"Scientific knowledge is a body of statements of varying degrees of certainty–some most unsure, some nearly sure but not absolutely certain".

Richard P. Feynman, NL in 'The Meaning of it All'

"It is remarkable that a science which began with the consideration of games of chance should have become the most important object of human knowledge."

Pierre Simon Laplace (1749-1827)

Chapter 6

Elements of Probability Theory

'Statistics without theory has no roots and statistics without applications bears no fruit.'

6.1. INTRODUCTION

So far we have confined our attention to collection, and to descriptive analysis of data. The methods of summarisation, by way of calculation of a few characteristics like the mean, standard deviation etc., provide us with a good description of the large volume of data. Such data are generally of a population or of a sample of the population when the population is large. It is not possible to deal with the complete data of a large population nor it is possible, even when available, to analyse the data relating to the whole population.

When dealing with a sample from a population, the objective of statistical analysis is to go beyond the stage of mere description or summarisation of data of a sample, and to draw valid and logical conclusions about the population from which the sample data are taken. The statistical methods employed to achieve these objectives are based on the theory of probability.

6.2. DEFINITIONS OF PROBABILITY: DIFFERENT APPROACHES

Random Experiment

Consider an "experiment", say, for example, of tossing a coin. This experiment can be performed under more or less identical conditions, any number of times. When we consider the result of an individual experiment, we are certain that one of the faces (head *or* tail) will show, but we cannot predict whether head will show up or tail will show up. The result will depend on what we call *chance.* There are various factors which influence the result of a toss. For example, the initial position of the coin and the force with which the player tosses the coin; then again, if these conditions are completely known, there are several complexities involved in the resulting motion which influence the final result. As all these factors cannot be fully known and examined, the result cannot be predicted. Such an experiment is known as a *random experiment.* Now since the result of an individual experiment cannot be predicted, what one can possibly do is to find out the 'chance' of a particular result coming, say head. Again instead of a qualitative statement that the chance is very high or very low, a quantitative measure of the chance is a more helpful formulation. Probability gives a quantitative measure of the chance of throwing a head in an experiment.

Another example may be considered: The experiment of throwing a die with its six faces marked 1, 2, 3, 4, 5, 6. We are certain that one of the faces will show up, but how much is a particular face, say 6, likely to show up? Or how likely is a face other than 6 (i.e. 1, 2, 3, 4, or 5) to come up? Intuitively we can say that it is less likely for a face to show 6 *than* to show other than 6. A quantitative measure of the chance will be much more useful, as can be easily imagined.

6.2.1. Classical Definition of Probability

Consider a random experiment and let us refer to the possible results as "cases". Let us assume that each of the possible results or "cases" is *equally likely* and also that the results are *mutually exclusive*, i.e. the happening of one result precludes the happening of other results. Let us define an event as a collection of results. An event may consist of a single result or a group of results taken together. We are now in a position to state the classical definition of probability.

Classical definition of probability

If, consistent with the conditions of an experiment, there are n exhaustive, mutually exclusive and equally likely cases and of them m are favourable to (the occurrence of) an event A, then the probability p of A is defined by $p = \frac{m}{n}$.

This is denoted by $p = P(A) = \frac{m}{n}$.

Here $0 \leq m \leq n$ and n is *finite*; so p lies between 0 and 1.

Two quantities n and m are involved in the definition of probability of an event A. These are to be calculated to find $P(A)$.

For example, a coin is tossed: here there are 2 cases, the results, head and tail. These are:

(i) *exhaustive* so that no other result is possible under the conditions of the experiment. (we exclude the possibility that a coin may stand on its edge neither falling head or tail).

(ii) *mutually exclusive* – that if one result occurs or happens (say head) the other (tail) cannot happen at the same time or in the same throw.

(iii) *equally likely* – for a fair (or unbiased) coin, each face is equally likely to occur. Of the 2 cases, only one is favourable to the event 'head' and only one is favourable to the event 'tail'. So

$$P\,(\text{head}) = \frac{1}{2}$$

$$P\,(\text{tail}) = \frac{1}{2}$$

A die is tossed. There are 6 cases according as it shows 1, 2, 3, 4, 5 or 6. These are exhaustive, mutually exclusive and equally likely. Of these only 1 is favourable to the event A that the die shows 6. So $m = 1, n = 6$ and

$$P\,(A) = \frac{1}{6}.$$

If B is the event that the die shows an even number (a multiple of 2), then the number of cases favourable to B is 3 (die shows 2, 4, or 6), i.e., $m = 3$ while $n = 6$. Thus $P(B) = \frac{3}{6} = \frac{1}{2}$.

Note: $p = P(A) = 1$ implies that the event A is *certain* and $p = 0$ implies that it is *impossible*.

Example 2 (a). Two coins are tossed: to find the probability of getting (i) exactly two heads, (ii) one head and a tail, and (iii) at least one head.

Let us denote head by H and tail by T and let the result (H, T) denote head in the first coin and tail in the second, and so on. We have the following 4 cases:

$$(H, H), (H, T), (T, H) \text{ and } (T, T)$$

so that $n = 4$ (i) of these only one (H, H) is favourable to the event A of two heads i.e. $m = 1$. Hence $P(A) = \frac{1}{4}$.

(ii) Of the $n = 4$ cases, 2 are favourable to the event B of one head and a tail, i.e. the cases (H, T) and (T, H). So that $m = 2$. Hence $P(B) = \frac{2}{4} = \frac{1}{2}$.

(iii) Of the $n = 4$ cases, 3 are favourable to the event C of getting at least one head – the first three cases, so that $m = 3$. Hence $P(C) = \frac{3}{4}$.

Example 2 (b). A bag contains 5 white and 3 black balls. Balls are otherwise identical. One ball is drawn at random. What is the probability that it is white?

Since balls are identical in shape and one ball is to be drawn (without seeing it), the total number of cases is 8 $(= 5 + 3)$ as any one of the 8 balls may be drawn. There are 5 favourable cases to the event that the ball drawn is white, corresponding to the 5 white balls.

Thus $n = 8$ and $m = 5$. So the required probability is $\frac{5}{8}$.

Example 2 (c). A card is taken from a well-shuffled pack of 52 cards. What is the probability that it is an ace?

Any card may be drawn, so the total number of cases is 52 i.e. $n = 52$. There are 4 aces and so the number of cases favourable to drawing an ace is 4, i.e. $m = 4$. Thus the required probability is $\frac{4}{52} = \frac{1}{13}$.

6.2.2. Relative Frequency and Statistical Regularity: Statistical Definition of Probability

Consider a random experiment which can be repeated under more or less similar conditions any number of times. Even if great care is taken to keep the conditions under which the experiment is performed fixed, our knowledge will not be precise enough to enable us to predict the result of an individual experiment. The results may vary from one observation to another in an uncontrollable way. Now let us consider a series or sequence of experiments. When we turn our attention from an individual to a sequence of experiments, we shall notice an extremely important phenomenon. *Though individual results show an irregular behaviour, the average (typical) results of a long sequence of random experiments show a very striking regularity*: a pattern is observable about a long series of experiments.

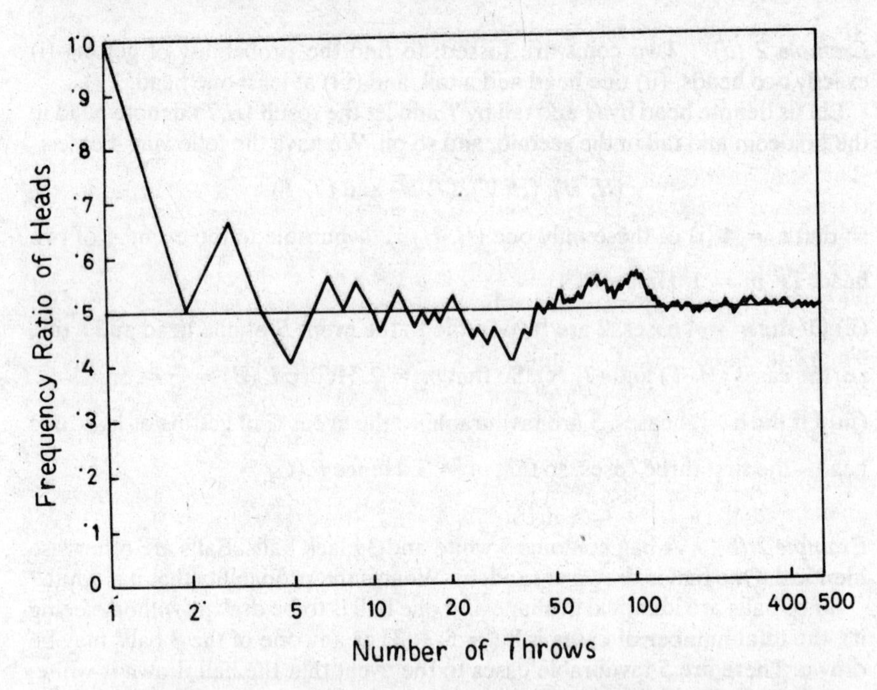

Fig. 6.1. Frequency Ratio of Heads in a coin-tossing experiment (Logarithmic scale for abscissa showing number of throws)

If A is an event and if A occurs r times in n experiments, then the ratio $\frac{r}{n}$ is called *the relative frequency* or the *frequency ratio of the event A*. The ratio will change with the change in the number n. If the frequency ratio $\frac{r}{n}$ for the event A is observed for increasing values of n, it is generally found that there is a tendency for the ratio to be more or less constant for large values of n. This tendency towards regularity is known as *statistical regularity*. This indicates that though the results of an individual experiment are unpredictable, the average results of a long sequence of experiments do show some regularity and are somewhat predictable.

Consider a coin tossing experiment and let A be the event that a throw results in a head. The frequency ratio $\frac{r}{n}$ will fluctuate with small values of n but as n increases, there will be tendency for the ratio to be more or less constant and to be very near to $\frac{1}{2}$.

Thus to any event A connected with a random experiment, it should be possible to ascribe a number p, such that the frequency ratio of A in a long series of experiments tends to p. This is another way of defining probability through relative frequency.

Statistical definition of probability

If $\frac{r}{n}$ is the relative frequency or frequency ratio of an event A connected with a random experiment, then the limiting value of the ratio $\frac{r}{n}$ as n increases indefinitely is called the probability of the event A.

This definition is not suitable from a mathematical point of view. It relies on the actual working and numerical data. From a practical point of view, however, it is useful and this approach is used in many practical situations.

For example, India's population (1991 census) is 843 million out of which 406 million are females. The probability that a randomly chosen person is a female is about 406/843 = 0.482.

The classical definition requires that n is finite and that all cases *are equally likely*. These are very restrictive conditions and cannot cover all situations. If we use the classical definition the probability that a person chosen at random is a female is 0.5, whereas 0.482 is a more reasonable estimate.

In fact, both definitions suffer from certain defects. Another approach has been put forward to overcome them, which we shall now consider.

6.3. AXIOMATIC APPROACH TO PROBABILITY

Sample point and sample space

Consider a random experiment: The results of an individual experiment are unpredictable. The experiment need not be restricted to a laboratory experiment. An experiment is a process of collecting data. We confine to such experiments which can be repeated (observed) under more or less identical conditions and which are such that the result of an individual experiment is unpredictable. For example, an experiment may consist of (i) observing the sex of a newborn baby in a town hospital, (ii) observing the life span of an organism, (iii) observing the number of accidents in a city, (iv) recording the number of telephone calls received per day in an office, (v) observing the wholesale price of a commodity, (vi) observing the daily maximum temperature in a city, and so on. Again there are experiments that may be considered as repeatable. For example, an experiment in which a certain drug is administered to a group of persons may be considered the first of a large number of such experiments and may be considered as repetitive.

Consider the set of results or outcomes of a random experiment, and assume at first that the number of conceivable results is finite. Each *distinct* outcome or result is called a *sample point* or *elementary outcome* or *sample event*. The set or aggregate of all the distinct outcomes will be called *sample space*. A collection of one or more sample point is known as an *event*. An event may be simple (if it comprises of one sample point only) or composite (if it comprises of more than one sample point).

For example, consider again the familiar dice-play experiment and let the observation relate to the number shown in the face of the die. The distinct results of a throw are 1, 2, 3, 4, 5 and 6, and thus these six are the six sample points of the experiment. The set $S = \{1, 2, 3, 4, 5, 6\}$ is the sample space. That the die throws a 6 is an event and as it consists of a single point it is a simple event; that it throws an even number is also an event (which is the collection of 3 elementary outcomes or sample points 2, 4, 6) and it is a composite event; that it throws more than 4 is a composite event (which is the collection of 2 sample points 5 and 6). For the experiment of throwing two coins simultaneously, the sample points are (H, H), (H, T), (T, H) and (T, T) and the sample space is the set $S = \{(H, H), (H, T), (T, H), (T, T)\}$, containing 4 points. The event of throwing at least one head is the set which comprises of the 3 sample points (H, H), (H, T) and (T, H). The event of getting a head in the first coin is the set comprising of two sample points (H, H) and (H, T).

In the terminology of set theory, the sample space is the *space* or the *master* or *universal set S* and an event is a subset of S. Such a subset may contain only one sample point or a group of more than one sample points. Elementary algebra of sets will be needed in our treatment of probability of event. (This will be discussed later). After the sample points (or elementary outcomes) of a random experiment have been identified and enumerated, it would be of interest to know how likely it is for an elementary outcome to happen or for a sample point or simple event to occur and also how likely it is for a composite event to happen.

6.4. PROBABILITY OF A SIMPLE EVENT

It is of interest to construct a mathematical model of the experiment. As a first step towards this, it is required to *assign* a probability to each sample point.

The probability of a simple event is a measure of how likely for the event to happen or the proportion of times it is expected to happen. If a die is thrown for a large number of times, one would expect the relative proportion of times or the relative frequency for a 6 to occur to approach $\frac{1}{6}$. Performance of actual experiments of this kind would show that such an expectation is justified. In an experiment each of the 6 sample points 1, 2, 3, 4, 5 and 6 is *assigned* the probability $\frac{1}{6}$. The above experiment can be actually repeated any number of times. Even without performing the experiment, it can be visualised that of the 6 sample points, the proportion of times each one is expected to occur is $\frac{1}{6}$. This is from symmetry. Logical derivation is possible.

However, in many situations it may not be possible to perform exactly similar experiments, nor is logical derivation possible. In that case other guides or methods are to be used to determine the long run relative frequency. For example, if the experiment relates to finding the probability of a particular person dying during the year, there is no logical way of doing this nor even a way of finding the relative frequency. However, the experience of the insurance companies provides us with a guide as to what relative frequency is for such an event to occur (i.e. a person is expected to die). The probability of a person of a particular age dying during the year can thus be assigned.

The two examples cited above refer to two different types of experiments. In one case (die throwing experiment) it is possible to determine the relative frequency with which an event is expected to occur in the long run by performing the actual experiment a large number of times or even better by logical arguments. In the other case, one has to determine the same from records otherwise available. The basic point is that it is necessary to attach or assign probability to an event: the statistician has the freedom to do so, but then he must use logical arguments for an experiment where this is possible or else when this is not possible, must base on his experience or records of actual situation. In a realistic world, a professional statistician will face such situations quite often.

We confine ourselves here to experiments of the first kind, where it is possible to determine the long-run relative frequency of an event or its probability by logical arguments. Experiments of games of chance are of this kind.

It is to be noted that probability of an event is a number between 0 and 1, since it is long-run relative frequency. Further that the sum of the probabilities of all the sample points of an experiment is 1. As an example of *equally likely* simple *events* (sample points) consider again a die throwing experiment. Suppose that the die is perfect and regular in shape. Then it is plausible to think that each face is equally likely to show up and we can give equal probability to each. In other words the probability of each face showing up is $\frac{1}{6}$. If there are k equally likely sample points in a sample space, then each simple event or sample point will have probability $\frac{1}{k}$.

When nothing is said to the contrary, it will be assumed that each sample point is equally likely.

Example 4 (a). What is the probability of getting 3 heads in tossing 3 coins?

We assume that the coins are distinguishable and if there is only one coin, then this coin is tossed three times. Also that each sample point is equally likely. The sample points are

$$HHH, HHT, HTH, HTT, THH, THT, TTH, TTT.$$

Each of these 8 sample points is equally likely and so the probability of each sample point is $\frac{1}{8}$, and that of getting 3 heads is also $\frac{1}{8}$.

Example 4 (b). Two dice are thrown. Describe the sample space.

What is the probability of getting 6 in both the dice?

We denote the sample points by a pair of numbers, corresponding to the numbers shown on the faces of the two dice. The sample space consists of the following sample points:

1, 1	2, 1	3, 1	4, 1	5, 1	6, 1
1, 2	2, 2	3, 2	4, 2	5, 2	6, 2
1, 3	2, 3	3, 3	4, 3	5, 3	6, 3
1, 4	2, 4	3, 4	4, 4	5, 4	6, 4
1, 5	2, 5	3, 5	4, 5	5, 5	6, 5
1, 6	2, 6	3, 6	4, 6	5, 6	6, 6

There are 36 sample points. The third one in the second column (2, 3) indicates that the first die shows 2 and the second die shows 3, and so on. Each of these 36 sample points is assumed to be equally likely and has the probability $\frac{1}{36}$. Hence the probability of getting 6 in both the dice is $\frac{1}{36}$.

6.5. PROBABILITY OF A COMPOSITE EVENT

A composite event comprises of more than one simple event or sample point. The probability of a composite event is the relative frequency with which it is expected to happen in the long run.

Definition: The probability of a composite event is the sum of the probabilities of the simple events of which it is composed.

Example 5 (a). What is the probability of getting (i) exactly two heads (event A) (ii) at least two heads (event B) in tossing 3 coins?

(i) The event A of exactly two heads comprises of the sample points (HHT), (HTH), (THH). Hence

$$P(A) = \frac{1}{8} + \frac{1}{8} + \frac{1}{8} = \frac{3}{8}$$

(ii) The event B of at least two heads comprises of the sample points (HHH), (HHT), (HTH), (THH). Hence

$$P(B) = \frac{1}{8} + \frac{1}{8} + \frac{1}{8} + \frac{1}{8} = \frac{1}{2}$$

Equally likely simple events: When all the sample points or simple events of an experiment are equally likely and so have equal probabilities, then the probability of a composite event is easily obtained. If there are k sample points and a composite event A contains r of them, then $P(A) = \frac{r}{k}$.

Example 5 (b). Find the probability that the sum of the numbers shown in the two faces, when two dice are thrown (i) is 7 and (ii) is 10.

(i) The event A that the sum of the numbers shown on the two faces is seven consists of the sample points:

$$(1, 6), (2, 5), (3, 4), (4, 3), (5, 2), (6, 1).$$

There are 6 sample points. Hence the required probability is $\frac{\text{[illegible]}}{36}$.

(ii) The event that the sum is 10 consists of the following sample points

$$(4, 6), (5, 5), (6, 4).$$

Hence the required probability is $\frac{3}{36} = \frac{1}{12}$.

The probability assigned to an event is the long run relative frequency ratio i.e., the ratio of the proportion of times an event occurs in the long run. Thus the probability of an event is a number lying between 0 and 1. Further the probability assigned to the whole sample space is 1 since one of the simple events must occur in every experiment. Again the probability of a composite event A is the sum of the probabilities of the simple events comprising A. Thus we can summarise as follows:

The probability P of an event A with regard to a sample space S of an experiment satisfies the following:

(i) $0 \leq P(A) \leq 1$

(ii) $P(A) = P(E_1) + P(E_2) + \ldots + P(E_r)$,

where $E_1, \ldots, E_r$ are the sample points comprising the event A

(iii) $P(S) = 1$.

For the development of the theory of probability it is assumed that the probabilities are *assigned* to all the sample points of the space. In other words probabilities of the sample points are assumed-*given or known* and further, the above are taken as *conventions* or as *axioms*. Then by assigning probabilities to sample points and with the above axioms, the theory of probability is built up. This is the *axiomatic approach* to the theory of probability due to Russian probabilist A.N. Kolmogorov, who advanced this axiomatic theory in 1933. For our purpose, it may be remembered how probabilities are assigned: the probability of an event is the ratio with which it is expected to occur in a long series of experiments. We shall now proceed to derive the laws of probabilities of combinations of events.

An event is a collection of sample points and is a subset of the sample space S. We shall now consider combinations of events. Let A, B, C be three events.

Definition: The event $A \cap B$ or AB (A intersection B) is the set or collection of sample points which belong to *both* A and B.

Events (or sets) can be represented by means of diagrams known as Venn diagrams. Let dots represent sample points. The points lying inside the region marked by the letter A, represent the sample points contained in the corresponding event A, and so on for other events. The sample points lying in the region common to A and B are common to both the events A and B, and the event AB is the collection of these sample points.

Definition: The event $A \cup B$ comprises of the sample points in the whole region bounded by A and B. The event $A \cup B$ is the set of sample points which belong to A or B or both A and B.

$A \cup B$ is also denoted by $A + B$.

When the two events A and B are such that the happening of one prevents the happening of the other, they are called *mutually exclusive* or disjoint: in that case A and B do not have any common region and do not have a common sample point; AB is a null set denoted by the *symbol* 0(also denoted by ϕ). A^c or $\bar{A}$ represents the event not A and is composed of all sample points of the sample space not belonging to A. Consider a die throwing experiment. If the event that the face shown is an even number (2, 4 or 6) is denoted by A and the event that the face shown is a number greater than 4 (i.e. 5 or 6) is denoted by B, then AB is the event that the die shows 6 and $A \cup B$ is the event that the die shows 2, 4, 5 or 6. Note that in $A \cup B$ the number 6 which is common to both A and B is taken only once. The events A and B are *not* mutually exclusive: they do have a common element 6. If the event that the die shows an odd number (1, 3 or 5) is denoted by C, then AC does not have any common element; thus A and C are mutually exclusive or disjoint. A^c or $\bar{A}$ is the event that the die shows 1, 3 or 5, i.e. all the sample points of the sample space excluding those belonging to A. It is clear that A and A^c are mutually exclusive, i.e. $AA^c = 0$ with no sample point in common. The operations of union and intersection may be performed on any number of events and not just two events.

Venn diagrams of union, intersection of two sets and complement of a set are shown in Fig. 6.2.

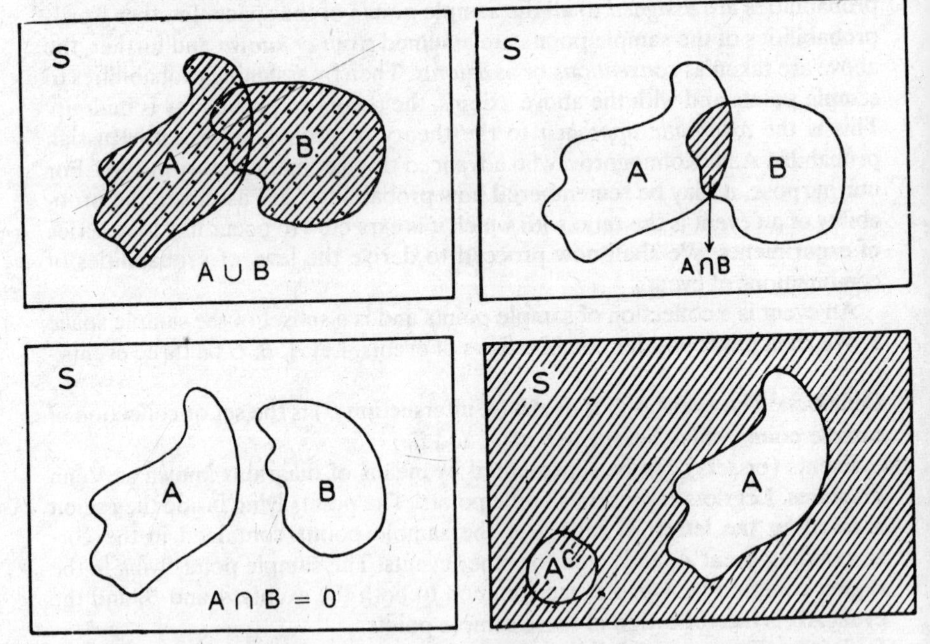

Fig. 6.2. Venn Diagrams of union, intersection of two sets and complement of a set.

6.6. ADDITION RULE

For any two events A and B

$$P(A \cup B) = P(A) + P(B) - P(AB). \qquad (6.1)$$

Proof: Let the number of sample points in the sample space S be n, the number comprising the event A be r, the number comprising the event B be m and the number comprising the event AB be k, (i.e. sample points common to both A and B are k and $0 \le k \le \min(r, m)$) the event $A \cup B$ comprises of sample points that belong to A or B or both; the event will have $r + m - k$ sample points. Thus

$P(A \cup B)$ = sum of the probabilities of the $r + m - k$ sample points belonging to $(A \cup B)$,
$P(A)$ = sum of the probabilities of the r sample points belonging to A,
$P(B)$ = sum of the probabilities of the m sample points belonging to B,
$P(AB)$ = sum of the probabilities of the k sample points belonging to AB.

It follows that

$$P(A \cup B) = P(A) + P(B) - P(AB).$$

Corollary (1) When A and B are mutually exclusive, then $AB = 0$ (i.e. $k = 0$)

and $$P(A \cup B) = P(A) + P(B). \tag{6.2}$$

(2) We have $AA^c = 0$ and $A \cup A^c = S$. Thus

$$P(A) + P(A^c) = P(S).$$
$$= 1$$

or $$P(A) = 1 - P(A^c)\ . \tag{6.3}$$

Example 6 (a). A die is thrown. Find the probability of getting (i) either an even number or a number greater than 4 or both.

The event A that the die shows an even number has probability $P(A) = \frac{3}{6} = \frac{1}{2}$; the event B that the die shows a number greater than 4 is $\frac{2}{6} = \frac{1}{3}$. A and B have one common sample point 6, so that $P(AB) = \frac{1}{6}$. The event that the die shows an even number or a number greater than 4 or both is $A \cup B$. Thus

$$P(A \cup B) = P(A) + P(B) - P(AB)$$
$$= \frac{1}{2} + \frac{1}{3} - \frac{1}{6} = \frac{4}{6} = \frac{2}{3}.$$

Example 6 (b). Two coins are tossed simultaneously. Find the probability of getting (i) exactly two heads or head in the first coin or both; (ii) exactly two heads or exactly two tails or both; (iii) getting head in the first coin or getting head in the second coin or both.

The sample space comprises of 4 sample points

$$HH, HT, TH, TT.$$

(i) Let A be the event of getting exactly 2 heads and B be the event of head in first coin. Then the event of getting exactly two heads or head in first coin or both is $A \cup B$.

We have $$P(A) = \frac{1}{4},\ P(B) = \frac{2}{4} = \frac{1}{2},\ P(AB) = \frac{1}{4}$$

Thus $$P(A \cup B) = \frac{1}{4} + \frac{1}{2} - \frac{1}{4} = \frac{1}{2}.$$

(ii) Let A be the event of getting exactly two heads and B the event of getting exactly two tails. Then A and B are mutually exclusive. Thus the required probability is

$$P(A \cup B) = P(A) + P(B)$$
$$= \frac{1}{4} + \frac{1}{4} = \frac{1}{2}.$$

(iii) Let A be the event of getting head in the first coin and B be the event of getting head in the second coin. We have

$$P(A) = \frac{2}{4} = \frac{1}{2},\ P(B) = \frac{1}{2},\ P(AB) = \frac{1}{4}$$

The required probability is

$$P(A \cup B) = \frac{1}{2} + \frac{1}{2} - \frac{1}{4} = \frac{3}{4}.$$

Example 6 (c). A card is taken from a well-shuffled pack of 52 cards*. What is the probability of getting (i) either a black card or an ace or both, (ii) either a diamond card or an ace or both, (iii) either an ace of diamond or an ace of hearts?

(i) Let A be the event of getting a black card and B be the event of getting an ace. The sample space has 52 sample points since any one of the 52 cards may be obtained in a draw. We have

$$P(A) = \frac{26}{52} = \frac{1}{2},\ P(B) = \frac{4}{52} = \frac{1}{13},\ P(AB) = \frac{2}{52} = \frac{1}{26}$$

Hence the required probability is

$$P(A \cup B) = \frac{1}{2} + \frac{1}{13} - \frac{1}{26} = \frac{13 + 2 - 1}{26} = \frac{7}{13}$$

(ii) Let A be the event of getting a diamond card and B be the event of getting an ace. We get

*A pack of playing cards consists of 4 × 13 = 52 cards of 4 different suits called spades, clubs, diamonds and hearts, each of 13 faces values (2, 3,..., 10, *J*, *Q*, *K*, *A*). The cards. of suites of spades and clubs, numbering 2 × 13 = 26 are black, while the cards of suites of diamonds and hearts are red. Cards of the same face value are called of the same kind. The card with face value *A* is called an ace: there 4 aces—ace of spades, clubs, diamonds and hearts.

$$P(A) = \frac{13}{52} = \frac{1}{4},\ P(B) = \frac{4}{52} = \frac{1}{13},\ P(AB) = \frac{1}{52}$$

Thus the required probability is given by

$$P(A \cup B) = \frac{1}{4} + \frac{1}{13} - \frac{1}{52} = \frac{13 + 4 - 1}{52} = \frac{16}{52} = \frac{4}{13}$$

(iii) Let A be the event of drawing an ace of diamond and B be the event of drawing an ace of hearts. A and B are mutually exclusive. Thus the required probability is given by

$$P(A \cup B) = P(A) + P(B) = \frac{1}{52} + \frac{1}{52} = \frac{2}{52} = \frac{1}{26}.$$

6.7. MULTIPLICATION RULE: CONDITIONAL PROBABILITY

Suppose that two dice are thrown. Then there are 36 sample points and the event A that the first die shows a five consists of 6 sample points (5, 1), (5, 2), (5, 3), (5, 4), (5, 5) and (5, 6). Thus $P(A) = \frac{6}{36} = \frac{1}{6}$. The event B that the sum (total) of numbers in the two dice is 9 consists of the sample points (3, 6), (4, 5), (5, 4), (6, 3) and $P(B) = \frac{1}{9}$. Now suppose that we are given the information that the first die shows a five, then the event that the sum of the numbers shown on the faces of the two dice is nine is a *conditional event*, and this conditional event is denoted by $(B \mid A)$. The probability of the conditional event is called *conditional probability*. For a conditional event, instead of the whole sample space we have only the sample points comprising of the event A, i.e. the 6 sample points (5, 1), (5, 2), (5, 3), (5, 4), (5, 5) and (5, 6), and the conditional probability of each of these is $\frac{1}{6}$. Conditioned by the event A, i.e. that the first die shows a five, the (conditional) event that the sum is nine comprises of only one sample point (5, 4) i.e. the sample point common to both A and B. Here the conditional probability of getting a sum of nine given that the first die shows a five is $\frac{1}{6}$, or in symbols $P(B \mid A) = \frac{1}{6}$.

General formula: Consider a conditional event $(B \mid A)$ i.e. the event B given that A has actually happend. Then for the happening of the event $(B \mid A)$, the sample space is restricted to the sample points comprising the event A. The conditional probability $P(B \mid A)$ is given by

$$P(B \mid A) = \frac{i}{j}, \text{ where}$$

i = number of sample points common to both A and B,
j = number of sample points comprising A,
n = total number of points in the whole (unrestricted) sample space S.

Thus, $$P(AB) = \frac{i}{n},\ P(A) = \frac{i}{n} \text{ and } P(B|A) = \frac{i}{j}$$

Dividing both the numerator and the denominator of $P(B|A) = \frac{i}{j}$ by n, the total number of sample points in the sample space S, we get

$$P(B|A) = \frac{i}{j} = \frac{i}{n} \Big/ \frac{j}{n}$$

$$= \frac{P(AB)}{P(A)} \tag{6.4}$$

It follows that

$$P(AB) = P(B|A)P(A). \tag{6.5}$$

Note : (1) These two results hold only if $P(A) > 0$.

(2) The results are proved under the tacit assumption that n is finite and that each sample point has equal probability $\frac{1}{n}$. It can be shown that the results hold in the general case (without these restrictions).

Example 7 (a). Two dice are thrown. Find the probability that the sum of the numbers in the two dice is 10, given that the first die shows *six*.

Let A be the event that the sum of numbers in two dice is 10,

B be the event that the first die shows 6.

Then AB is the event that the sum is 10 and the first die shows 6 or which is equivalent to the event that first die shows 6 and the second 4.

We have $$P(B) = \frac{1}{6},\ P(AB) = \frac{1}{36}$$

Thus the required probability $= P(A \mid B) = \frac{P(AB)}{P(B)} = \frac{1}{6}$. Thus the conditional probability of getting a sum of 10, given that the first shows 6 is $\frac{1}{6}$. The unconditional probability of getting a sum of 10 is $\frac{1}{12}$ (Example 5 (b)).

Example 7 (b). Two coins are tossed. What is the conditional probability of getting two heads (event B) given that at least one coin shows a head (event A)?

Event A comprises of 3 sample points (HH), (HT), (TH) so that $P(A) = \frac{3}{4}$; the event AB comprises of only one point (HH), so that $P(AB) = \frac{1}{4}$. Thus the conditional probability is

$$P(B|A) = \frac{P(AB)}{P(A)} = \frac{\frac{1}{4}}{\frac{3}{4}} = \frac{1}{3}.$$

Example 7 (c). A box contains 5 black and 4 white balls. Two balls are drawn one by one *without replacement*, i.e. the first ball drawn is not returned to the box. Given that the first ball drawn is black, what is the probability that both the balls drawn will be black?

Before the first draw the sample space consists of 9 points each with probability $\frac{1}{9}$. After the first draw the number of sample points reduces to 8 (as one ball is already out of the box) and the probability of each sample point is $\frac{1}{8}$.

Let A be the event that the first ball drawn is black then $P(A) = \frac{5}{9}$, since there are 5 black balls.

Let B be the event that the second ball drawn is black. Then the conditional event $(B \mid A)$ implies drawing a black ball from the box which contains 4 black and 4 white balls.

Thus
$$P(B \mid A) = \frac{4}{8} = \frac{1}{2}.$$

The event implies that both the balls drawn are black.
We are required to find $P(AB)$. We have

$$P(AB) = P(B \mid A) . P(A)$$
$$= \frac{5}{9} . \frac{1}{2} = \frac{5}{18}.$$

In the above example, consider that the ball drawn in the first draw is returned to the box, so that the composition of the box remains the same before each drawing. Here it is drawing *with replacement.* Now the conditional event $(B \mid A)$, that the second ball is black is not affected by the event that the first draw resulted in a black ball since the ball drawn was returned. Thus the event B and the event A are *independent* and the knowledge of one does not affect the other. Then $P(B \mid A) = P(B)$, independent of $P(A)$. We have then

$$P(B) = \frac{P(AB)}{P(A)}$$

Or,
$$P(AB) = P(A) P(B) \tag{6.6}$$

which gives the *multiplication rule for independent events*, A and B.

Example 7 (d). What is the probability that in 2 throws of a die, six appears in both the dice?

Let A and B be the event that 6 appears in the first and the second throws respectively; these events are independent. Then the event AB implies that six appears in both the dice.

$$P(AB) = P(A)P(B) = \frac{1}{6} . \frac{1}{6} = \frac{1}{36}.$$

Example 7 (e). The probability of getting *HHT* in tossing 3 coins is thus $\frac{1}{2}\cdot\frac{1}{2}\cdot\frac{1}{2}=\frac{1}{8}$, since the three events Head in first throw, Head in second and Tail in third are independent.

Example 7 (f). Find the probability that in a family of 2 children (i) both are boys, (ii) both are of the same sex, assuming that the probability of a child being a boy or a girl is equal $\left(\text{equal to } \frac{1}{2}\right)$

(i) Let A_1, A_2 be the events respectively that the first and the second child is a boy. Then since A_1, A_2 are independent,

$$P(A_1A_2) = P(A_1)P(A_2) = \frac{1}{2}\cdot\frac{1}{2} = \frac{1}{4}$$

Similarly if B_1, B_2 are the events respectively that the first and the second child is a girl, then B_1, B_2 are independent and

$$P(B_1B_2) = P(B_1)P(B_2) = \frac{1}{2}\cdot\frac{1}{2} = \frac{1}{4}$$

(ii) Now the event that both children are of the same sex is equivalent to the event that both are either boys or girls and these are mutually exclusive events. Thus the required probability $= \frac{1}{4}+\frac{1}{4}=\frac{1}{2}$.

Mutually exclusive events and independent events

These ideas are not equivalent ideas. We discuss them to bring out the difference between the two.

When the happening of one event precludes the happening of the other event, the two events are mutually exclusive (or disjoint). For two mutually exclusive events A and B

$$P(AB) = 0.$$

When the happening of one event has no effect on the probability of occurrence or happening of the other event, the two events are independent. For two independent events A and B,

$$P(AB) = P(A)P(B).$$

Two events can be mutually exclusive and not independent. Again two events can be independent and not mutually exclusive.

Suppose two coins are tossed. The events $\{H, H\} \equiv A$ (head on both coins) and the event $\{T, T\} \equiv B$ are mutually exclusive (because if A happens B cannot happen) and

$$P(AB) = 0$$

But $P(A) = \frac{1}{4}$, $P(B) = \frac{1}{4}$ and so $P(AB) = 0 \neq P(A)P(B)$ and hence A and B are not independent.

Consider the events A, B that 6 appears in the first and the second die respectively in throwing 2 dice together (example 7 (d)). The events A and B are independent, as

$$P(AB) = \frac{1}{36} = P(A)\,P(B) = \frac{1}{6} \times \frac{1}{6}.$$

Here $P(AB) \neq 0$ and hence the events are not mutually exclusive.

The only way that two mutually exclusive events A, B can be independent is when both

$$P(AB) = 0 \text{ and } P(AB) = P(A)\,P(B)$$

hold simultaneously. Both of the above can hold simultaneously if at least one of $P(A)$ or $P(B)$ is zero. In case at least one of $P(A)$ or $P(B)$ is zero then the two events A and B are mutually exclusive events as well as independent events.

6.7.1. Number of sample points in a combination of events or sets

Let $N(A)$ denote the number of points in the set A.

Then, using Venn diagrams, it can be easily seen that

$$N(A \cup B) = N(A) + N(B) - N(AB). \tag{6.7}$$

Example 7(g). Suppose that students in an Institution can enrol for one, two or none of the language courses, French (A), German (B). If 30% are enrolled for French, 20% for German and 10% for both French and German, then the number (in percentage) enrolled for at least one of the courses is given by

$$N(A \cup B) = N(A) + N(B) - N(AB)$$

$$= 30 + 20 - 10$$

$$= 40$$

and the percentage not enrolled for any of the courses is

$$100 - 40 = 60$$

Thus the probability that a student selected at random is (i) enrolled for at least one of the courses is 0.4 and (ii) not enrolled for any of the courses is $1 - 0.4 = 0.6$. Given that a student is enrolled for at least one of the courses, the (conditional) probability that he is enrolled for French is

$$\frac{30}{40} = 0.75$$

6.7.2. Discrete Sample Space

So far we considered cases where the sample space contains a finite number of points. We consider the following example where this is not the case.

Example 7 (h). A coin is tossed until a head appears. Describe the sample space. Find the probability that the coin will be tossed (a) exactly 4 times (b) at the most, 4 times. (c) What is the probability that head will appear if the coin is tossed an infinite number of times?

The head may appear at the

(i) very first throw (H)

(ii) second throw, the first toss resulting in a tail (TH)

(iii) third throw, the first two tosses resulting in tails (TTH)

(iv) fourth throw, the first three tosses resulting in tails ($TTTH$)

and so on: an infinite number of throws may be needed to get a head. The sample space consists of an *infinite* number of the *sample* points

$$H, TH, TTH, TTTH, TTTTH, ...$$

The trials are independent. Assume that the coin is fair (unbiased).

The probability of the event $H = \frac{1}{2}$

The probability of the event $TH = \frac{1}{2} \cdot \frac{1}{2} = \left(\frac{1}{2}\right)^2$

The probability of the event $TTH = \left(\frac{1}{2}\right)^2 \cdot \frac{1}{2} = \left(\frac{1}{2}\right)^3$

The probability of the event $TTTH = \left(\frac{1}{2}\right)^3 \cdot \frac{1}{2} = \left(\frac{1}{4}\right)^4$

and so on.

(a) The probability that to get a head, the coin will be tossed exactly 4 times is $\left(\frac{1}{2}\right)^4$.

(b) The event that the coin will be tossed at most 4 times is a compositive event comprising of the 4 sample events H, TH, TTH and $TTTH$. Thus the required probability

$$= \frac{1}{2} + \left(\frac{1}{2}\right)^2 + \left(\frac{1}{2}\right)^3 + \left(\frac{1}{2}\right)^4 = \frac{15}{16}.$$

(c) Suppose that the coin is tossed as many times as is necessary to get a head. The required probability is

$$\left(\frac{1}{2}\right) + \left(\frac{1}{2}\right)^2 + \left(\frac{1}{2}\right)^3 + ...$$

$$= \frac{\frac{1}{2}}{1 - \frac{1}{2}} = 1$$

Thus it is certain that a head will ultimately appear if the coin is tossed indefinitely. The result holds even if the coin is biased.

Incidentally, it is verified that $P(S) = 1$.

Note: In this example we find that though the number of points are infinite, these can be arranged according to the sequence of natural numbers (such that there is one-one correspondence between the natural numbers and the sample points); such an infinity of numbers is called *denumerable infinity* (or *countable infinity*) of numbers.

Discrete Sample space

A sample space that consists of a finite number of sample points or a denumerably infinite number of sample points is called a *discrete sample space.*

6.8. BAYES' FORMULA OR THEOREM

We shall examine an application of the multiplication rule with more involved calculation. Consider the following situation.

An event A happens in conjunction with only one of the events $B_1, B_2, ...B_k$, where $B_1, B_2, ...$ are mutually exclusive and they together comprise the whole sample space S, i.e. $B_1 + B_2 + ... B_k = S$. The following figure (Fig. 6.3) illustrates the situation ($k = 3$). The events B_1, B_2, B_3 are disjoint and they give a partition of the sample space such that

$$AB_1 + AB_2 + AB_3 = A(B_1 + B_2 + B_3)$$
$$= AS = A.$$

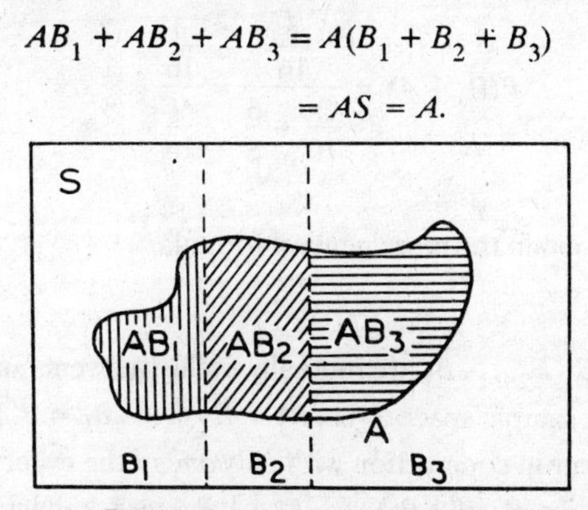

Fig. 6.3. Partition of Sample space

The probabilities $P(B_1)$, $P(B_2)$, ... as also the probabilities $P(A \mid B_1)$, $P(A \mid B_2)$, ... are known. It is of interest to find the probabilities $P(B_1 \mid A)$, $P(B_2 \mid A)$, ... These are given by a formula known as Bayes' formula.

As an illustration let us consider this. There are two boxes. Box-I contains 5 white and 3 black balls and Box II contains 2 white and 6 black balls. One of the boxes is selected at random, the probability of each box being selected being equal to $\frac{1}{2}$. From the box selected one ball is drawn. It is found to be black. What is the probability that it was taken from Box I? Here the event A that the ball drawn is black happens in conjunction with only one of the two events B_1 (that Box I is chosen) or B_2 (that Box II is chosen). The event $(A \mid B_1)$ is the event of drawing a black ball given that it is taken from Box I. Thus

$$P(A \mid B_1) = \frac{3}{8}; P(A \mid B_2) = \frac{6}{8}$$

$P(B_1 \mid A)$ is the probability of the event that given that the ball drawn is black, it was taken from Box I (or Box I was chosen).

Now $$P(B_1 \mid A) = \frac{P(B_1 A)}{P(A)} = \frac{P(AB_1)}{P(A)}$$

$$P(AB_1) = P(B_1)P(A \mid B_1) = \frac{1}{2} \cdot \frac{3}{8} = \frac{3}{16}; P(AB_2) = \frac{1}{2} \cdot \frac{6}{8} = \frac{3}{8}$$

$$P(A) = P(AB_1 + AB_2) = P(AB_1) + P(AB_2)$$

(since AB_1, AB_2 are disjoint events) so that

$$P(A) = \frac{3}{16} + \frac{3}{8} = \frac{9}{16}$$

Thus $$P(B_1 \mid A) = \frac{\frac{3}{16}}{\frac{3}{16} + \frac{3}{8}} = \frac{\frac{3}{16}}{\frac{9}{16}} = \frac{1}{3}$$

We shall now obtain the more general formula.

Bayes' theorem

Suppose that $B_1, B_2, \ldots B_k$ are mutually exclusive events such that they form a partition of the sample space S, i.e. $B_1 + B_2 + \ldots B_k = S$. Let A be an event so that A can occur in conjunction with only *one* of the events $B_1, B_2, \ldots B_k$.

The probabilities $P(A \mid B_1), \ldots, P(A \mid B_k)$ are known; known also are $P(B_1), \ldots, P(B_k)$. Then the probability

$$P(B_1|A) = \frac{P(B_1)P(A \mid B_1)}{\sum_{r=1}^{k} P(B_r)P(A \mid B_r)} \tag{8.1}$$

Proof: We have

$$P(B_1|A) = \frac{P(B_1 A)}{P(A)} = \frac{P(AB_1)}{P(A)} \tag{8.2}$$

Also $$P(AB_1) = P(B_1 A) = P(B_1)P(A|B_1) \tag{8.3}$$

Since A occurs in conjunction with only one of the events $B_1, \ldots B_k$,

$A = AB_1 + AB_2 + \ldots + AB_k$, and the events AB_i are mutually exclusive. Thus

$$\begin{aligned} P(A) &= P(AB_1) + P(AB_2) + \ldots + P(AB_k) \\ &= P(B_1)P(A|B_1) + P(B_2)P(A|B_2) + \ldots + P(B_k)P(A|B_k) \\ &= \sum_{r=1}^{k} P(B_r)P(A|B_r) \end{aligned} \tag{8.4}$$

Substituting in (8.2) the expression of $P(A)$ obtained in (8.4) and the expression of $P(AB_1)$ obtained in (8.3), we get

$$P(B_1|A) = \frac{P(B_1)P(A|B_1)}{\sum_{r=1}^{k} P(B_r)P(A|B_r)}$$

Thus we get Bayes' theorem.

Example 8 (a). The probability that a high school student being male is $\frac{1}{3}$ and that being female is $\frac{2}{3}$. The probability that a male student completes the course successfully is $\frac{7}{10}$ and that a female student does it is $\frac{4}{5}$. A student selected at random is found to have completed the course. What is the probability that the student is (i) male (ii) female? Here A is the event that a student completes the course, B_1, B_2 the events that a student selected is male and female respectively. We have

$$P(B_1) = \frac{1}{3}, P(B_2) = \frac{2}{3}, P(A|B_1) = \frac{7}{10}, P(A|B_2) = \frac{4}{5}$$

The required probabilities are

$$P(B_1|A) = \frac{\frac{1}{3}\cdot\frac{7}{10}}{\frac{1}{3}\cdot\frac{7}{10} + \frac{2}{3}\cdot\frac{4}{5}} = \frac{\frac{7}{30}}{\frac{23}{30}} = \frac{7}{23}$$

$$P(B_2|A) = \frac{\frac{2}{3}\cdot\frac{4}{5}}{\frac{1}{3}\cdot\frac{7}{10} + \frac{2}{3}\cdot\frac{4}{5}} = \frac{16}{23}$$

Example 8 (b). Three suppliers X, Y, Z supply items to an establishment in the proportion of $\frac{1}{2}, \frac{1}{3}$ and $\frac{1}{6}$ respectively. Of the items supplied by X, Y, Z, 5%, 6% and 8% respectively are found to be defective. An item taken at random from the lot of all items supplied is found to be defective. What is the probability that it is supplied by the three suppliers X, Y and Z?

The probabilities that an item is supplied by X, Y and Z are $\frac{1}{2}, \frac{1}{3}$ and $\frac{1}{6}$ respectively. The probabilities that the item is defective given that it is supplied by X, Y, Z is $\frac{5}{100}, \frac{6}{100}$ and $\frac{8}{100}$ respectively. Thus the probability that the item is supplied by X given that it is defective is equal to

$$\frac{\frac{1}{2} \cdot \frac{5}{100}}{\frac{1}{2} \cdot \frac{5}{100} + \frac{1}{3} \cdot \frac{6}{100} + \frac{1}{6} \cdot \frac{8}{100}}$$

$$= \frac{15}{35}$$

The probability that it is supplied by Y given that it is defective equals

$$\frac{\frac{1}{3} \cdot \frac{6}{100}}{\frac{1}{2} \cdot \frac{5}{100} + \frac{1}{3} \cdot \frac{6}{100} + \frac{1}{6} \cdot \frac{8}{100}}$$

$$= \frac{12}{35}$$

The probability that it is supplied by Z given that it is defective equals

$$\frac{\frac{1}{6} \cdot \frac{8}{100}}{\frac{1}{2} \cdot \frac{5}{100} + \frac{1}{3} \cdot \frac{6}{100} + \frac{1}{6} \cdot \frac{8}{100}}$$

$$= \frac{8}{35}$$

The sum of the three probabilities

$$= \frac{15}{35} + \frac{12}{35} + \frac{8}{35} = 1$$

which shows that it is certain that the defective item is supplied by one of the suppliers (as should be clear).

EXERCISES - 6

Sections 6.3 & 6.4

1. A coin is tossed 4 times. Describe the sample space. What is the probability of getting (i) exactly 3 heads (ii) at least 3 heads, (iii) no heads at all.
2. A bag contains three similar balls-1 white, 1 black and 1 red. Two balls are drawn one after another (i) by replacing the first ball drawn before the second draw (ii) without replacing the first ball drawn.
   Describe the sample space in each case. How many points will each of these sample spaces have?

3. Exercise No. 2 (i). What is the probability of getting (a) white balls in the two drawings (b) one white and one red ball (c) at least one white ball?.
4. Find the probability of getting the balls as indicated in Exercise 3 under condition (ii) Exercise 2.
5. Two dice are thrown. Find the probability of getting (i) six in the first die, (ii) even numbers in both the dice, (iii) equal numbers in both the dice, (iv) six in both the dice.

Section 6.5

6. One die is thrown twice. What is the probability that the sum of the numbers shown in the two faces is (i) an odd number, (ii) an even number (iii) greater than 10 (iv) equal to 10?
7. A card is drawn from a full pack. What is the probability of getting (i) the ace of clubs (ii) a hearts card (iii) an ace?
8. Two cards are drawn from a full pack. The card drawn first is returned to the pack (which is then reshuffled) and the second card is drawn from the reshuffled pack of cards. Find the probability of getting (i) two aces (any two), (ii) ace of clubs in the two drawings. (iii) the same ace in both draws.
9. Examine exercise 8, with the difference that the first card drawn is not returned to the pack: what is the probability of getting two aces in the two draws?
10. A die is thrown. Consider the following pairs of events. Find which of them are mutually exclusive.
    Drawing (a) 2, 6
    (b) an even number, an odd number
    (c) an even number, a number greater than 3
    (d) a number greater than 4, a multiple of 3.

Section 6.6

11. If $P(A) = \frac{1}{2}, P(B) = \frac{1}{4}, P(A \cup B) = \frac{3}{8}$, what is the value of $P(AB)$?
12. Show that $P(A \cup B) \leq P(A) + P(B)$. When does the equality hold ?
13. A card is drawn from a full pack. What is the probability that it is either a black card or a face card (*J, Q, K*) or both?
14. A card is drawn from a full pack. Find the probability that it is either an ace or a diamond or both.
15. Three coins are tossed. What is the probability of getting either two heads or two tails or both? Are the two events mutually exclusive?
16. Two dice are thrown. Find the probability of getting the event *A* (even numbers on both) or the event *B* (the same pair of numbers in both the dice) or both *A* and *B*. Are *A* and *B* mutually exclusive?
17. A die is thrown twice. What is the probability of getting at least one six?
18. Two dice are thrown. Let *A* be the event that the sum of the two faces is eight and *B* be the event that the same pair of numbers occur in both. Find the probability of $P(AB)$ and $P(A \cup B)$.
19. Suppose that a sample space is divided into three mutually exclusive regions *A, B, C* with $P(A) = \frac{1}{3}, P(B) = \frac{1}{4}$. Find
    (i) $P(A \text{ or } B)$, (ii) $P(A \text{ or } C)$, (iii) $P(B \text{ or } C)$.
20. Show that for three events *A, B, C*

$$P(A + B + C) = P(A) + P(B) + P(C) - P(AB) - P(AC) - P(BC) + P(ABC)$$

Section 6.7

21. Two balls are drawn, one after another, from a box containing 3 white and 2 black balls. What is the probability that the first ball is black and the second is white? Assume that the ball drawn is replaced before the second drawing (drawing is with replacement).
22. Solve Exercise 21 but with the assumption that the drawing is without replacement.
23. Two balls are drawn from a box containing 5 black and 3 red balls. What is the probability that the balls drawn are of the same colour, when draw is *with* replacement?

24. Solve Exercise 23, but with the assumption that the drawing is *without* replacement.
25. Suppose that A and B are independent and B and C mutually exclusive. Are A and C necessarily mutually exclusive? If $P(A) = 0.2.$, $P(B) = 0.5$, $P(C) = 0.3$ find the probability that (i) either A or B or both occurs (ii) all the three events occur (iii) B occurs given that A occurs, (iv) B occurs given that C occurs.
26. A family has n (> 1) children. Let A be the event that the family has at most one girl and B be the event that not every child is of the same sex. Determine the value of n for which A and B are independent events.

Section 6.8

27. Three urns I, II, and III contain respectively 2 white and 3 black balls. 3 white and 4 black balls, and 4 white and 5 blackballs. One of the urns is chosen at random, the probability of each urn chosen being equal. Then one ball is drawn at random from the urn chosen and found to be black. What is the probability that urn I, II or III was chosen?
28. Repeat exercise 27 but with the difference that urn III contains 4 white and 5 *red* balls.
29. Of the eggs supplied to a cooperative, 30%, 20%, 35% and 15% come from the poultry farms A, B, C, D respectively. Rotten eggs account for 2%, 1%, 2.5% and 1% of the supplies by A, B, C, D respectively. An egg is taken at random and found to be defective. What is the probability that it was supplied by A, B, C or D?
30. Suppose that a blood test is capable of detecting the presence of a disease in 95% of all the cases of people suffering from it. When applied to people not suffering from it. The test reports *incorrectly* that there is suffering amongst 10% of the people not actually infected. Suppose that of the total number of people going for the blood test. 80% are people who actually are infected and 20% are not. An individual is tested and the test indicates that he suffers from the disease. What is the probability that the test has given the (i) correct result (ii) wrong result?

Miscellaneous

31. In a survey of newspaper reading habits of students in a college, it is found that 80% read local English language daily (A), 30% read local regional language daily (B) and 40% read national English daily (C). Further, 25% read both (A) and (B). 30% read both (A) and (C), and 20% read both (B) and (C). Find (i) the percentage of students who read none of the three papers, (ii) percentage of those who read (C) amongst those who read at least one of the three papers.
    Find also the number of students who read at least two papers if the total number of students covered is 300.
    What is the conditional probability that given that a student reads at least two newspapers that he reads all the three?
32. If the sample space of an experiment contains n sample points, show that the total number of events that can be associated with the experiment is 2^n.
33. The number of females per 1000 males of population according to the last 10 Censuses (1901, 1911,......,1991) of India are as follows:
    972, 964, 955, 950, 945, 946, 941, 930, 934, 929,
    Can recording the sex of a person be considered as a random experiment? Discuss.
34. Continuation of Exercise 33.
    Can you use the statistical definition to find the probability that a person selected at random in India is a (i) female, (ii) male?
    What result would you get if you use classical definition? Comment on the results.
35. Give some examples of random experiments from studies relating to social sciences and business.

Rev. Thomas Bayes (1701/2 – 1761). The statistical approach that this British Presbyterian Minister put forward about 240 years ago is now finding wide and useful applications in several areas of modern studies ranging from astronomy to genomics and from pharmaceutical research to legal matters. See 'Bayes Offers a New Way to Make Sense of Numbers' by D. Malakoff in SCIENCE, Vol. 286, 19 Nov. 1999.

Chapter 7

Random Variable and Probability Distribution

7.1. DISCRETE RANDOM VARIABLE

Consider an experiment of tossing two coins. The sample points or outcomes of the experiment are *HH, HT, TH, TT*. The outcomes are qualitative and are described by their attributes, though the probabilities attached to the outcomes are quantitative. We are often interested in some functions of the outcomes or sample points. We wish to find out a number associated with each sample point or in other words find a numerical valued function of the sample points. Suppose that in this example, we consider the number of heads shown and let us denote this by X. Corresponding to the sample point *HH*, we get $X = 2$; to the sample point *HT*, we get $X = 1$, so also for the sample point *TH*, to the sample point *TT* we get $X = 0$. Thus X is a variable defined on the sample space $S = \{HH, HT, TH, TT\}$ of the experiment. X assumes values 2, 1 and 0. We also get

$$P(X = 2) = P(\text{event } HH) = \frac{1}{4}$$

$$P(X = 1) = P(\text{event } HT) + P(\text{event } TH) = \frac{1}{4} + \frac{1}{4} = \frac{1}{2}$$

$$P(X = 0) = P(\text{event } TT) = \frac{1}{4}$$

Thus, in the experiment of tossing two coins, the number of heads shown is a variable X which assumes the values 2, 1, 0 with respective probabilities $\frac{1}{4}, \frac{1}{2}, \frac{1}{4}$. To each sample point there corresponds a value of X. Note that X can assume no other value and that the sum of the probabilities that X assumes these values is

$$P(X = 2) + P(X = 1) + P(X = 0) = \frac{1}{4} + \frac{1}{2} + \frac{1}{4} = 1.$$

Note that though to each sample point there is a value of X, more than one sample point may give the same value. The variable thus associated with a sample space is known as a *random variable.*

Definition: A random variable is a numerical valued function defined on a sample space.

Consider an experiment of throwing two dice, which has 36 sample points. Now let X be the sum of the numbers shown in the two dice. Then for each sample point, we get a numerical value of X. For example, for the sample point (6, 6) we get the value 12 and also the probability of $X = 12$. Similarly for the sample point (5, 3) we get the value $X = 8$; but there are other sample points (4, 4) and (3, 5) corresponding to which also we get $X = 8$. Thus for each sample point we get a numerical value, though there may be two or more sample points to which the same value of X corresponds. X can assume the values 2, 3,..., 12 and the probability that X assumes a particular value can be easily found.

For example

$$P(X = 2) = P\{\text{event } (1, 1)\} = \frac{1}{36}$$

$$P(X = 3) = P\{\text{event } (1, 2) \text{ or event } (2, 1)\} = \frac{2}{36}$$

$$P(X = 4) = P\{\text{event } (1, 3), \text{event } (2, 2) \text{ or event } (3, 1)\} = \frac{3}{36}$$

$$P(X = 12) = P\{\text{event } (6, 6)\} = \frac{1}{36}.$$

Thus X assumes values 2, 3, 4, with the respective probabilities $\frac{1}{36}, \frac{2}{36}, \frac{3}{36}$.

The random variable X assumes in all 11 values from 2 to 12. It can be seen that

$$P(X = 2) + P(X = 3) + \ldots + P(X = 12) = 1.$$

We discuss here mainly those random variables which assume integral values. Such a random variable is known as *discrete* random variable.

Definition: The distinct values of a discrete random variable X together with their associated probabilities define the *probability distribution* or simply the *distribution* of the random variable X.

Consider the experiment of tossing two coins. The probability distribution of the number of heads X is as follows:

$$P(X = 0) = \frac{1}{4}$$

$$P(X = 1) = \frac{1}{2}$$

$$P(X = 2) = \frac{1}{4}$$

Note that $$P(X = 0) + (X = 1) + P(X = 2) = 1$$

In general, if a random variable X assumes values $x_1, x_2, \ldots x_k$ with associated probabilities $p_1, p_2, \ldots p_k$, then the distribution of X is as follows:

$$P(X = x_1) = p_1$$

$$P(X = x_2) = p_2$$

$$\ldots$$

$$P(X = x_k) = p_k$$

This can be written simply as

$$P(X = x_i) = p_i, i = 1, 2, \ldots k.$$

It is to be noted that

$$\sum_{i=1}^{k} P(X = x_i) = 1.$$

Example 1 (a). Three coins are tossed. Find the probability distribution of the number of heads.

The sample space consists of the 8 simple events or sample points

HHH, HHT, HTH, HTT, THH, THT, TTH, TTT

and each of these sample points has the probability 1/8. Let X be the number of heads. Then X can assume the values 3, 2, 1, 0. The probability distribution is as follows:

$$P(X = 3) = P\{\text{event } (HHH)\} = 1/8$$

$$P(X = 2) = P\{\text{events } (HHT), (HTH), \text{ or } (THH)\} = 3/8$$

$$P(X = 1) = P\{\text{events } (HTT), (THT), \text{ or } (TTH)\} = 3/8$$

$$P(X = 0) = P\{\text{event } (TTT)\} = 1/8$$

We can represent the distribution of X by a line diagram as shown below:

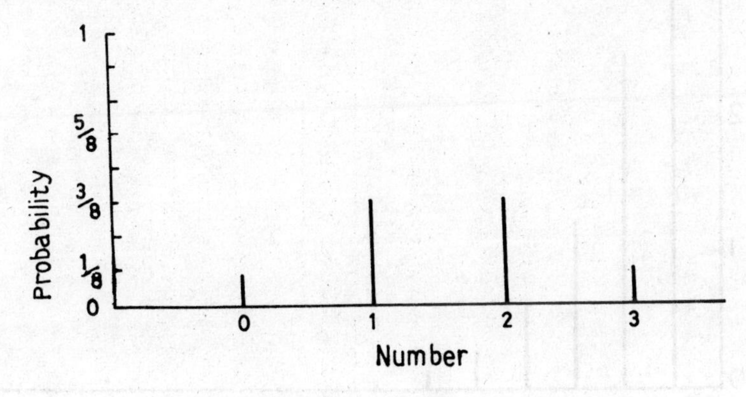

Fig. 7.1 Line Diagram of the Distribution of Ex. 1(a)

A more general probability distribution associated with coin tossing will be discussed later.

Example 1 (b). Consider the experiment of tossing a coin till a head appears for the first time. Let X be the number of tosses required. Find the distribution of X.

X assumes values 1, 2, 3, . . . according as the number of tosses required is 1, 2, 3, . . . etc.

$$P(X = 1) = P(\text{event } H) = \frac{1}{2}$$

$$P(X = 2) = P(\text{event } TH) = \left(\frac{1}{2}\right)^2$$

$$\cdots$$

$$P(X = k) = P(\text{event } \underbrace{TT\ldots T}_{k-1}H) = \left(\frac{1}{2}\right)^k$$

$$\cdots$$

The probability distribution of X is given by

$$P(X = k) = \left(\frac{1}{2}\right)^k, k = 1, 2, 3, \ldots$$

Further $$\sum_{k=1}^{\infty} P(X = k) = 1.$$

This is indicated by the line diagram below:

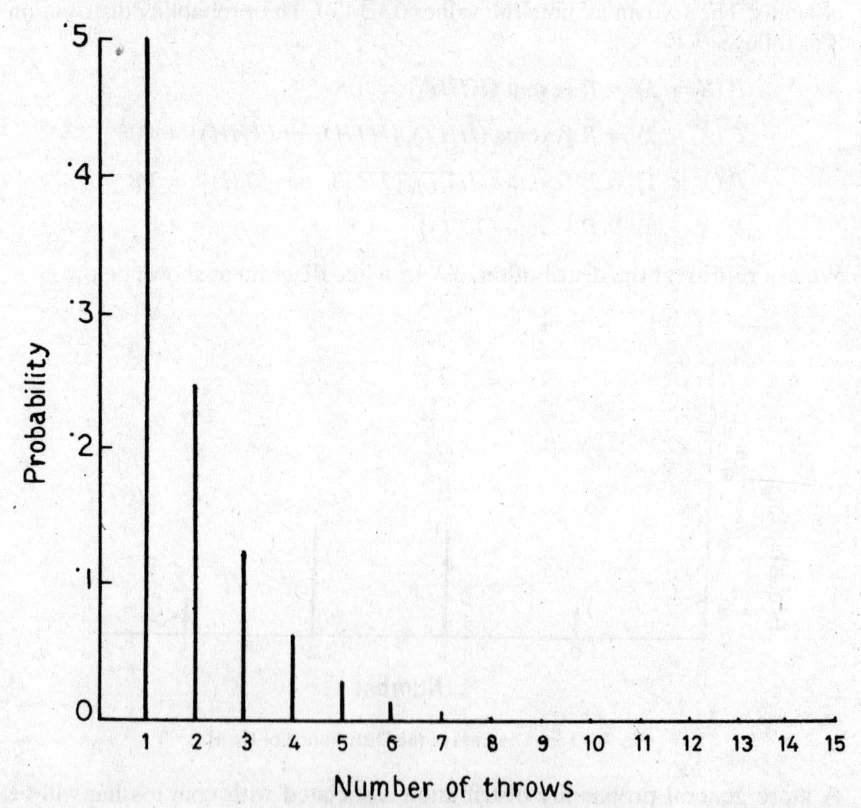

Fig. 7.2. Line diagram of the distribution

We have earlier discussed frequency distribution. A frequency distribution can be constructed from a set of *empirical data* obtained from an experiment or by observing a phenomenon several times. A probability distribution, on the other hand, is a distribution built on a theoretical model and has no relevance to actual data. For example, we have an exact probability distribution for the number of heads in 3 tosses of a coin as given in example above. Now a person may actually take three coins and go on tossing the three coins successively say, a 100 times. In these 100 tosses, he may get 3 heads, say f_1 times; 2 heads, say f_2 times, 1 head, say f_3 times, 0 head, say f_4 times, where $f_1 + f_2 + f_3 + f_4 = 100$; a frequency distribution can be constructed as follows:

No. of heads	3	2	1	0
Frequency	f_1	f_2	f_3	f_4

As we can see, this frequency distribution is quite different from the probability distribution. You can yourself perform the above experiment and construct a frequency distribution of the number of heads.

As in the case of frequency distributions, we may construct a probability histogram, and also take measures to facilitate comparisons of probability distributions; like measures for central tendency, etc.

Probability histogram: We have seen how histograms are constructed for representation of frequency distributions. In a similar manner, we can draw histograms for probability distributions. The values assumed by X are taken on x-axis; a rectangle is drawn with an area equal to the probability for that value of X. The value of X is taken at the centre and the width of the rectangle is taken as 1 unit so that the height of the rectangle corresponds to the probability.

The probability histogram of the distribution considered in Example 1(a) is shown below:

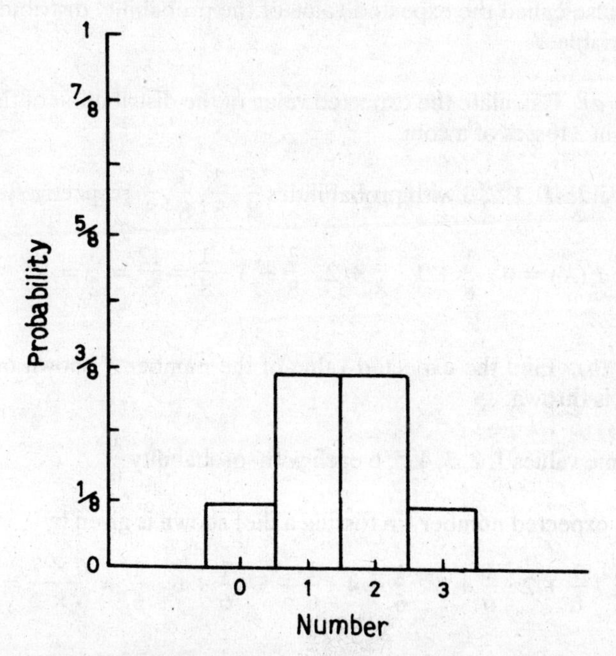

Fig. 7.3 Probability Histogram of the Distribution of Ex. 1 (a)

A *probability frequency polygon* is obtained by joining the mid points of the upper side of the adjoining rectangles. The end points of the polygon are joined to the middle points of the two classes, one added at each extreme. Probability histograms may have different shapes as can be seen from the above example. We can construct measures of central tendency and dispersion in case of frequency distributions. It is also desirable to assign numerical measures of some specific characteristics of the probability distribution, in order to be able to make some sort of comparison amongst probability distributions. There are two aspects, which are to be considered: where does the centre of the distribution lie and how dispersed is the distribution. The measure used for the centre is the expected value and that used for the spread is the variance. We discuss these below.

7.2. EXPECTED VALUE OF A RANDOM VARIABLE

Definition: Let $x_1, x_2, \ldots x_k$ be the values assumed by a random variable X with associated probabilities $p_i = P(X = x_i)$. Then the expected value (or expectation or mean) of the random variable X is defined to be the quantity

$$E(X) = \sum_{i=1}^{k} x_i \; P(X = x_i)$$

$$= \sum_{i=1}^{k} x_i \, p_i$$

The expected value $E(X)$ of X is thus obtained by multiplying each value of X by its associated probability and then taking the sum of these products.

$E(X)$ is also called the expected value of the probability distribution of the random variable X.

Example 2 (a). Calculate the expected value of the distribution of the number of heads X in 3 tosses of a coin.

X takes values 0, 1, 2, 3 with probabilities $\frac{1}{8}, \frac{3}{8}, \frac{3}{8}, \frac{1}{8}$ respectively. So

$$E(X) = 0 \cdot \frac{1}{8} + 1 \cdot \frac{3}{8} + 2 \cdot \frac{3}{8} + 3 \cdot \frac{1}{8} = \frac{12}{8} = \frac{3}{2} = 1.5$$

Example 1 (b). Find the expected value of the number X shown on the face when a die is thrown.

X can assume values 1, 2, 3, 4, 5, 6 each with probability $\frac{1}{6}$.

Hence, the expected number (in tossing a die) shown is given by

$$E(X) = 1 \cdot \frac{1}{6} + 2 \cdot \frac{1}{6} + 3 \cdot \frac{1}{6} + 4 \cdot \frac{1}{6} + 5 \cdot \frac{1}{6} + 6 \cdot \frac{1}{6} = \frac{6 \times 7}{6 \times 2} = \frac{7}{2} = 3.5$$

Note :

(1) The expected value may *not* be any of the values of X. In the first example, the expected number is 1.5 which is different from the values 0, 1, 2, 3 assumed by the variable.

(2) While $E(X)$ will exist whenever X assumes a finite number of values, $E(X)$ may or may not exist when X takes an infinite number of values.

Suppose that some monetary consideration is attached to a dice-play experiment: A person gets as many units of money (as a prize)as the number shown on the face of the die thrown, i.e., he gets Re. 1 (if the die shows 1, Rs. 2 if it shows 2 and so on). Then if a large number of such throws are made, the person is expected to have his average equal to Rs. 3.5 .Thus this has some relevance to the expected value.

Another important fact is the physical interpretation of the expected value as a centre of gravity of a mass having the shape of the probability histogram; a metallic block cut in the shape of the probability distribution of a random variable will have its centre of gravity at the expected value of the random variable. This is why the expected value is used as a measure of the centre of the probability distribution.

The expected value $E(X)$, when it exits, is also called the *Population mean* of the probability distribution . This is usually denoted by $\mu = E(X)$.

Here it is necessary to refer again to the frequency distribution. We have defined mean $\bar{X}$ for a frequency distribution. This is the arithmetic average obtained from the data of some observations. As the data are taken from a sample of the population, the mean $\bar{X}$ is called *sample mean*. For example, if 3 coins are tossed 100 times, and the results show 3 heads for f_1 times, 2 heads for f_2 times, 1 head for f_3 times and 0 head for f_4 times, $f_1 + f_2 + f_3 + f_4 = 100$ then the mean or average number of head equals

$$\frac{3 \times f_1 + 2 \times f_2 + 1 \times f_3 + 0 \times f_4}{100}$$

would give the sample mean. The sample mean will vary from sample to sample i.e. if another person does the same experiment, he might get somewhat different values of f_1, f_2, f_3, f_4 and a different sample mean.

7.2.1. Expected Value of a Function of the Random Variable X

Very often we have to consider the functions of the random variable X and calculate the expectation such functions. For example, functions like $Y = X^2, Z = (X - a)^2$ etc.; such functions of the random variable X are also random variables. Let $g(X)$ be a function of X.

Consider again the dice-play experiment. Suppose that the throw shows face X, then the person gets $g(X) = X^2$ units of money, i.e. Re. 1 if 1 is shown, Rs. 4 if 2 is shown and so on. His expected gain would be

$$E(X^2) = 1 \cdot \frac{1}{6} + (2)^2 \cdot \frac{1}{6} + (3)^2 \cdot \frac{1}{6} + (4)^2 \cdot \frac{1}{6} + (5)^2 \cdot \frac{1}{6} + (6)^2 \cdot \frac{1}{6} = \frac{91}{6}.$$

Definition: The expectation of a function $g(X)$ of a random variable is defined by the quantity

$$E(g(X)) = \sum_{i=1}^{k} g(x_i)P(X = x_i)$$

$$= \sum_{i=1}^{k} g(x_i)p_i$$

where $x_1, x_2, \ldots, x_k$ are the values assumed by the random variable X and $P(X = x_i) = p_i$.

Example 2 (c). In tossing two coins, let X be the number of heads shown. Find $E(X)$ and $E(X-1)^2$.

The sample space is $S = \{(HH), (HT), (TH), (TT)\}$; X assumes values 2, 1, 0 with probabilities $\frac{1}{4}, \frac{1}{2}, \frac{1}{4}$ respectively.

Thus

$$E(X) = 2 \cdot \frac{1}{4} + 1 \cdot \frac{1}{2} + 0 \cdot \frac{1}{4} = 1$$

$$E[(X-1)^2] = (2-1)^2 \cdot \frac{1}{4} + (1-1)^2 \cdot \frac{1}{2} + (0-1)^2 \cdot \frac{1}{4}$$

$$= \frac{1}{4} + 0 + \frac{1}{4} = \frac{1}{2}$$

In fact, we have found $E[(X-1)^2] = E[(X-\mu)^2]$ where $\mu = 1 = E(X)$ is the expected value of X.

Properties of Expectation

Let X be a random variable and a, b and c be constants and X assume values $x_1, \ldots, x_k$ with probability $P(X = x_i) = p_i$. Let $g(X)$ be a function of X. Then from the definition $E(g(X))$, we get

$$E(aX) = \sum_{i=1}^{k} (ax_i)P(X = x_i)$$

$$= a \sum_{i=1}^{k} x_i P(X = x_i) = aE(X).$$

The simple properties of expectation are given below:

(i) $E(a) = a$
(ii) $E(aX) = aE(X)$
(iii) $E(aX + b) = aE(X) + b$
(iv) $E(g(X) + c) = E(g(X)) + c$
(v) $E(cg(X)) = cE(g(X))$.

These properties may be easily verified with the help of the formula for $g(X)$.

If X is the number shown on the face when a die is thrown, then we have seen that $E(X) = \frac{7}{2}$. We get

$$E(2X-5) = 2E(X) - 5 = 2 \cdot \frac{7}{2} - 5 = 2.$$

Variance

Let X be a random variable and $E(X) = \mu$ be its expected value.

Definition: The variance of X denoted by σ^2 or var (X) is defined to be

$$\sigma^2 = \text{var}(X) = E[(X-\mu)^2].$$

We have, using the properties of expectation,

$$\begin{aligned} E[(X-\mu)^2] &= E(X^2) - 2\mu E(X) + \mu^2 \\ &= E(X^2) - 2\mu . \mu + \mu^2. \\ &= E(X^2) - \mu^2. \end{aligned}$$

Thus

$$\sigma^2 = \text{var}(X) = E[(X-\mu)^2] = E(X^2) - \mu^2.$$

It is convenient to use $E(X^2) - \mu^2$ in computing variance of a random variable.

If X assumes values $x_i, \ldots x_k$ with $P(X = x_i) = p_i$, $i = 1, \ldots, k$ then

$$E(X) = \sum_{i=1}^{k} x_i p_i = \mu$$

and

$$\begin{aligned} \text{var}(X) &= E(X^2) - \mu^2 \\ &= \sum_{i=1}^{k} x_i^2 p_i - \mu^2. \end{aligned}$$

Note: While μ may have any value positive or negative, the variance is always positive. Variance is zero when the variable assumes only one value with probability 1. Thus var $(X) \geq 0$.

Example 2 (d). A coin is tossed 3 times. Find the variance of the number of heads shown.

The number of heads X is random variable, which assumes values 3, 2, 1, 0 with probabilities $\frac{1}{8}, \frac{3}{8}, \frac{3}{8}, \frac{1}{8}$ respectively. We write in tabular form:

Values of X	3	2	1	0
Probability	$\frac{1}{8}$	$\frac{3}{8}$	$\frac{3}{8}$	$\frac{1}{8}$
Values of X^2	9	4	1	0

Thus $\quad E(X^2) = 9 \cdot \frac{1}{8} + 4 \cdot \frac{3}{8} + 1 \cdot \frac{3}{8} + 0 \cdot \frac{1}{8} = \frac{24}{8} = 3$

and $\quad \text{var}(X) = E(X^2) - \mu^2 = 3 - \left(\frac{3}{2}\right)^2 = \frac{3}{4}.$

Example 2 (e). For the dice-play experiment find the variance of the number shown.

We have

Values of X	1	2	3	4	5	6
Probability	$\frac{1}{6}$	$\frac{1}{6}$	$\frac{1}{6}$	$\frac{1}{6}$	$\frac{1}{6}$	$\frac{1}{6}$
Values of X^2	1	4	9	16	25	36

Thus $\quad E(X) = \frac{7}{2}$

and

$$E(X^2) = 1 \cdot \frac{1}{6} + 4 \cdot \frac{1}{6} + 9 \cdot \frac{1}{6} + 16 \cdot \frac{1}{6} + 25 \cdot \frac{1}{6} + 36 \cdot \frac{1}{6} = \frac{91}{6}$$

so that $\quad \sigma^2 = \text{var}(X) = \frac{91}{6} - \left(\frac{7}{2}\right)^2 = \frac{91}{6} - \frac{49}{4} = \frac{35}{12}.$

Example 2 (f). Suppose that a die is thrown, and suppose that when 6 occurs, we call it a success (denoted by S), whereas if one of the other faces (1, 2, 3, 4 or 5) occurs we call it a failure (denoted by F). Find the mean and variance of the number of successes when a die is thrown.

Let X be the number of successes-then we can have only 1 success or 0 success, when it is thrown once; failure is a composite event and consists of 5 simple events. In other words X assumes values 1 and 0. We have $P(X = 1) = \frac{1}{6}$ and $P(X = 0) = \frac{5}{6}$.

Thus

Values X	0	1
Probabilities	$\frac{5}{6}$	$\frac{1}{6}$
Values of X^2	0	1

Thus $\quad E(X) = 0 \cdot \frac{5}{6} + 1 \cdot \frac{1}{6} = \frac{1}{6}$

$$E(X^2) = 0 \cdot \frac{5}{6} + 1 \cdot \frac{1}{6} = \frac{1}{6}$$

$$\text{var}(X) = E(X^2) - [E(X)]^2 = \frac{1}{6} - \frac{1}{36} = \frac{5}{36}.$$

Variance is considered to provide a measure of the spread of the distribution, in terms of the squares of the deviations of the random variable X about its expected value $\mu = E(X)$.

The variance is an expectation: expectation of $(X-\mu)^2$. We get the following properties of variance.

(i) var (X) cannot be negative
(ii) var $(X + a)$ = var (X)
(iii) var $(bX) = b^2$ var (X)
(iv) var $(a + bX) = b^2$ var (X).

The *standard deviation* (or simply s.d.) of a random variable X is defined as the positive square root of the variance of X. One advantage of using s.d. is that it gives a measure of spread in the same unit in which X (and also $E(X)$) is expressed.

We have earlier discussed variance and standard deviation of a frequency distribution. Here we have defined variance and standard deviation of a *probability distribution.* The variance and standard deviation of a probability distribution are known as *population variance* and *population standard deviation* respectively, whereas those calculated from a frequency distribution based on sample observations are called *sample variance* and *sample standard deviation.* The mean and variance of a probability distribution are generally denoted by μ and σ^2 respectively, whereas those calculated from a frequency distribution are denoted by $\overline{X}$ and s^2 respectively.

So far we have discussed random variables X which can assume discrete and distinct values. Now we shall consider variables which can assume any value in an interval.

7.3. CONTINUOUS RANDOM VARIABLE

We have already come across examples of continuous variables. Suppose that heights of students are measured: Though we may record the measurements to a nearest unit, it can be easily seen that it could be any measurement within a certain length of a continuous scale. If the measurement is done with the help of a long scale, the height of a student may correspond to any point on the scale. Examples of continuous variables are provided by measurements of heights, weights of individuals, of the amount of rainfall in a city, of the temperature at a certain place, of the amount of time required to perform a task etc. The basic distinction between a discrete and a continuous variable is that the former involves counting while the latter involves measuring. A discrete variable assumes a set of values which are countable; a continuous (random) variable assumes a set of values which are uncountable. We shall discuss here the probability distribution of a continuous random variable.

Consider that a large number of observations of a continuous variable is available. We have seen (in section 3.3) how these data can be grouped, and the histogram and frequency polygon can be drawn. Suppose that a histogram of 100 observations of heights of students is drawn. If another histogram and frequency polygon of say 300 observations of the heights of students are drawn on the same graph sheet, the resulting figures will be of a much bigger height than the earlier ones and the shapes will be quite different. This makes comparison difficult. In order to overcome this difficulty, a standardisation is sought to be done, by requiring that the *area* of the frequency polygon approach 1 as the sample size increases. This is the situation with regard to samples of observation from a continuous variable. The distribution of *continuous* random variable can be represented by a curve such that the total area between the curve and the x-axis is equal to 1. Suppose that the area between the curve (given above) and the x-axis is equal to 1. This curve may correspond to the probability distribution of a continuous random variable X. This curve corresponds to the line diagram of a discrete random variable. The area between, say 1 and 1.5 is then the probability that X assumes a value between 1 and 1.5: this area is the relative frequency that the value of X lies between 1 and 1.5. Thus

$$P(1 \leq X \leq 1.5) = \text{area of the shaded portion (Fig. 7.4).}$$

In the case of the continuous random variable, we get the probability that the variable assumes a value between two values in case of discrete random variables, we get the probability that the variable assumes a *distinct* value. It is considered that the probability of assuming a distinct value by a continuous random variable is the area of the portion, the width of which is only a point, thus the area is infinitesimally small. There is no meaning in considering the probability that a continuous random variable assumes a distinct value, as in the case of a discrete random variable, but instead, one considers the probability that a continuous random variable assumes value in the small neighbourhood of the distinct value. For example, when we think of a distinct value, say $\frac{1}{2}$, it is meaningful to find the probability that the continuous random variable assumes value in the small neighbourhood of $\frac{1}{2}$, say, in $\left(\frac{1}{2} - h, \frac{1}{2} + h\right)$ where h is very, very small.

The curve representing the distribution of the continuous random variable X is known as the frequency curve or the probability density curve. If the equation of the curve is $y = f(x)$, then $f(x)$ is known as the *probability density function* (p.d.f.) or frequency function of the continuous random variable X.

$f(x)\,dx$ is the probability that X assumes a value in $(x, x + dx)$, where dx is small.

We can summarise the properties of $f(x)$ as follows:

The probability density function $f(x)$ of a continuous random variable (or of the distribution of a continuous random variable) is such that

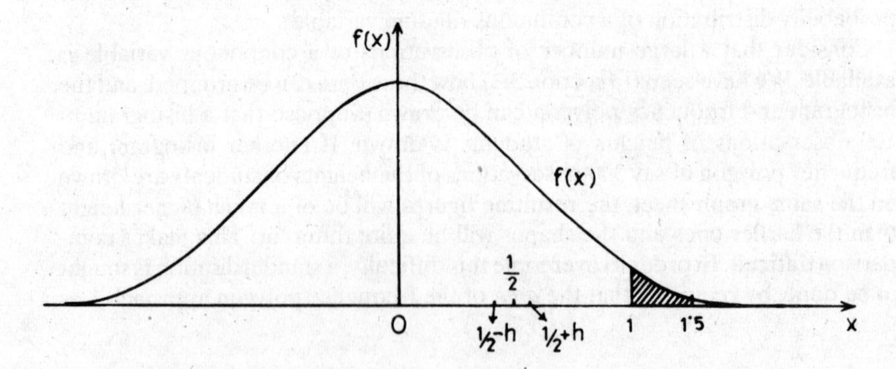

Fig. 7.4 Probability density curve of a continuous random variable.

(i) $f(x)$ is positive or 0 ($f(x) \geq 0$)
(ii) the total area of the region between the curve $y = f(x)$ and the x-axis is 1
(iii) $P(a < X < b)$ = area between the curve $y = f(x)$, and the x-axis between the points a and b
(iv) $P(x < X < x + dx) = P$ (X assumes a value in the small neighbourhood of x), where x is a real number.

Note the essential difference with the probability distribution of a discrete random variable X.

If $x_1 \ldots x_k$ are the values assumed by X, then we get $P(X = x_1) = p_i$ such that

(i) p_i is positive or 0

(ii) $\sum_{i=1}^{k} p_i = 1$

(iii) $P(X = x_i \text{ or } x_j) = P(X = x_i) + P(X = x_j)$

(iv) $P(X = x_i) = p_i$ at a distinct point x_i.

The simplest continuous r.v. is the one where distribution is constant over an interval and zero outside. The p.d.f. is then a straight line $y = f(x) = c$. Such a r. v. is said to have *uniform* or *rectangular* distribution. In Fig. 7.5 suppose that $OA = c$ and $AB = 1/c$; X has a uniform distribution in the interval OA. One can easily obtain $P(a < x < b)$ as the area of a rectangle.

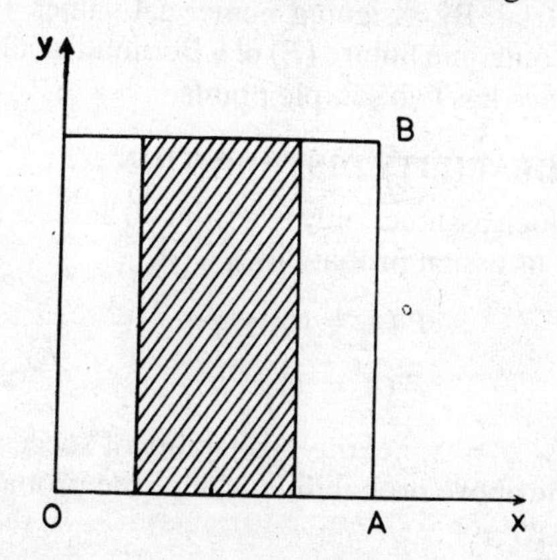

Fig. 7.5 Probability density function of a uniform random variable.

In general, $y = f(x)$ is not a straight line. Calculations are then more involved, but the principles are the same. We shall discuss an important continuous probability distribution later on.

7.4. STANDARD PROBABILITY DISTRIBUTIONS

We have seen that a discrete random variable X assumes a set of values $x_1, x_2, \ldots x_k$ with probabilities $p_i = P(X = x_i)$, $i = 1, 2, \ldots, k$ such that (i) $0 < p_i < 1$ and (ii) $\sum_{i=1}^{k} p_i = 1$. In other words the set p_i, satisfying

$P(X = x_i) = p_i, \Sigma p_i = 1$ define the probability distribution of a random variable. For example, if $P(X = 5) = \frac{1}{2}, P(X = 4) = \frac{1}{4}, P(X = 3) = \frac{1}{4}$, then these probabilities satisfy the properties (i) and (ii) and thus define the probability distribution of X. In fact there could be an infinite number of such probability distributions. But we would be interested only in such distributions which could serve as the probability model of real phenomena. Such a model would be useful in solving problems of statistical inference. We shall consider some such standard probability distributions.

First we develop a few basic models which could be used for representing the distribution of the number of occurrences of an event in repeated trials of an experiment. By *trial* is meant each repetition of an experiment. For example, if a coin is tossed a number of times, each repetition is a trial in coin tossing. In such a case, there are *two* possible outcomes, head and tail, in each trial. We can consider an experiment of a somewhat more general kind than coin-tossing where each trial results in two possible outcomes. Let us call these two outcomes success (S) and failure (F). For example, the outcomes of each trial in a die throwing experiment can be broadly grouped into two possible outcomes: success (S), when a face with an even number appears and failure (F) when a face with an odd number appears. A trial with two possible outcomes and which can be repeated any number of times is known as a *Bernoulli Trial.* The two possible outcomes are generally termed as success (S) and Failure (F). Trials are *independent* so that the outcome of one trial does not in any way affect the outcome of any other trial. By assigning numerical values 1 to the outcome success (S) and 0 to the outcome failure (F) of a Bernoulli trial we get a random variable; the sample space has two sample points.

7.5. BERNOULLI PROBABILITY DISTRIBUTION

A random variable X which assumes only two values 0 and 1 is called a *Bernoulli random variable X.* Let us assign probabilities

$$P(X = 0) = q$$

$$P(X = 1) = p$$

where $0 < p < 1$, $p + q = 1$, i.e. the probability of success is p and that of failure is $q = 1 - p$. The above probabilities define the probability distribution of the random variable X.

We have

$$E(X) = 0 \,.\, q + 1 \,.\, p = p$$

$$E(X^2) = 0^2 \,.\, q + 1^2 \,.\, p = p$$

and

$$\text{var}(X) = E(X^2) - E(X)^2 = p - p^2 = p(1-p) = pq.$$

Example: In tossing a fair coin, we can regard head as a success and tail as a failure and can take $p = q = \frac{1}{2}$. Then the expected number of head(s) in a trial is $p = \frac{1}{2}$.

Example: In throwing a die, we can regard the appearance of an even number as success and appearance of an odd number as failure and take $p = q = \frac{1}{2}$.

If we regard the appearance of 6 as success and the appearance of a face with any of the other numbers (1, 2, 3, 4, or 5) as failure, we shall be justified to take $p = \frac{1}{6}$ and $q = \frac{5}{6}$ $(p + q = 1)$.

Thus the two outcomes need not necessarily imply that they are equally likely and each has a probability $\frac{1}{2}$. They may have unequal probabilities p and q such that $p + q = 1$.

Suppose now that a fair coin (with equal probability for each face) is tossed twice. The number of heads (or successes, when a head is called a success) is a random variable X which assumes values 2, 1, 0 with probabilities (or probability distribution)

$$P(X = 2) = P(\text{event } HH) = \left(\frac{1}{2}\right) \cdot \left(\frac{1}{2}\right) = \left(\frac{1}{2}\right)^2$$

$$P(X = 1) = P(\text{event } HT) + P(\text{event } TH) = 2 \cdot \left(\frac{1}{2}\right)^2$$

$$P(X = 0) = P(\text{event } TT) = \left(\frac{1}{2}\right)^2$$

Since trials are independent $P(\text{event } HH) = P(H) \times P(H) = \left(\frac{1}{2}\right)^2$ and so on. Note that

$$P(X = 2) + P(X = 1) + P(X = 0)$$

$$= \left(\frac{1}{2}\right)^2 + 2\left(\frac{1}{2}\right)^2 + \left(\frac{1}{2}\right)^2 = 1$$

X is the number of successes when a Bernoulli trial with $p = \frac{1}{2}$ is repeated 2 times.

The number of heads in 3 tosses of a coin is a random variable X which assumes values 3, 2, 1, 0 with probabilities

$$P(X=3) = P(\text{event } HHH) = \frac{1}{2} \cdot \frac{1}{2} \cdot \frac{1}{2} = \left(\frac{1}{2}\right)^3$$

$$P(X=2) = P(\text{event } HHT) + P(\text{event } HTH) + (\text{event } THH) = 3 \cdot \left(\frac{1}{2}\right)^3$$

$$P(X=1) = P(\text{event } HTT) + P(\text{event } THT) + P(\text{event } TTH) = 3. \left(\frac{1}{2}\right)^3$$

$$P(X=0) = P(\text{event } TTT) = \left(\frac{1}{2}\right)^3$$

Note that $P(X=3)+P(X=2)+P(X=1)+P(X=0)=1$.

Here X is the number of successes when a Bernoulli trial with $p=\frac{1}{2}$ is repeated 3 times. Consider again the die throwing experiment with the appearance of 6 considered as success P (event S) $=\frac{1}{6}$ and P (event F) $=\frac{5}{6}$. The number of successes in 2 throws is a random variable X which assumes values 2, 1, 0 with probabilities.

$$P(X=2) = P(\text{event } SS) = \frac{1}{6}\cdot\frac{1}{6} = \left(\frac{1}{6}\right)^2$$

$$P(X=1) = P(\text{event } SF) + P(\text{event } FS) = \frac{1}{6}\cdot\frac{5}{6}+\frac{5}{6}\cdot\frac{1}{6} = 2\cdot\frac{1}{6}\cdot\frac{5}{6}$$

$$P(X=0) = P(\text{event } FF) = \frac{5}{6}\cdot\frac{5}{6} = \left(\frac{5}{6}\right)^2$$

Note that $P(X=2)+P(X=1)+P(X=0)=1$.

Here X is the number of successes when a Bernoulli trial with $p=\frac{1}{6}$ is repeated 2 times.

Consider that 3 dice are thrown and as before 6 is counted as success (S) and other than 6 is counted as failure (F). Then the number of successes is a random variable X which assumes values 3, 2, 1, 0 with probabilities

$$P(X=3) = P(\text{event } SSS) = \left(\frac{1}{6}\right)^3$$

$$P(X=2) = P(\text{event } SSF) + P(\text{event } SFS) + P\text{ event }(FSS) = 3\cdot\left(\frac{1}{6}\right)^2\cdot\frac{5}{6}$$

$$P(X=1) = P(\text{event } SFF) + P(\text{event } FSF) + P(\text{event } FFS) = 3\cdot\frac{1}{6}\cdot\left(\frac{5}{6}\right)^2$$

$$P(X=0) = P(\text{event } FFF) = \left(\frac{5}{6}\right)^3$$

Note that $P(X=3)+P(X=2)+P(X=1)+P(X=0)=1$.

Here X is the number of successes when a Bernoulli trial with $p=\frac{1}{6}$ is repeated 3 times.

Consider now that a Bernoulli trial with probability of success p is repeated 4 times. The number of successes is a random variable X which assumes values 4, 3, 2, 1, 0 with probabilities

$P(X = 4) = P(\text{event } SSSS) = p.p.p.p. \quad = p^4 = \binom{4}{0}p^4 = \binom{4}{4}p^4$

$P(X = 3) = P(\text{event } SSSF)$

$+ P(\text{event } SSFS)$

$+ P(\text{event } SFSS)$

$+ P(\text{event } FSSS) \quad = 4p^3q = \binom{4}{1}p^3q = \binom{4}{3}p^3q$

$P(X = 2) = P(\text{event } SSFF)$

$+ P(\text{event } SFFS)$

$+ P(\text{event } FFSS)$

$+ P(\text{event } FSSF)$

$+ P(\text{event } SFSF)$

$+ P(\text{event } FSFS) \quad = 6p^2q^2 = \binom{4}{2}p^2q^2$

$P(X = 1) = P(\text{event } SFFF)$

$+ P(\text{event } FSFF)$

$+ P(\text{event } FFSF)$

$+ P(\text{event } FFFS) \quad = 4pq^3 = \binom{4}{3}pq^3 = \binom{4}{1}pq^3$

$P(X = 0) = P(\text{event } FFFF) \quad = q^4 = \binom{4}{4}q^4 = \binom{4}{0}q^4$

Since trials are independent $P(SSSF) = P(S)\,P(S)\,P(S)\,P(F) = p.p.p.q = p^3q$ and so on. To find $P(X = 3)$ we consider all possible sequences of outcomes that have 3 S's and 1 F, i.e. 3 S's out of 4 or 1 F out of 4. There are $\binom{4}{3} = \binom{4}{1} = 4$ such sequences. For $(X = 2)$ we have to consider all possible sequences that have 2 S's and 2 F's; there are $\binom{4}{2} = 6$ such sequences. In general $\binom{n}{r} = \binom{n}{n-r}$.

Thus in a sequence of 4 Bernoulli trials with probability of success p at each trial and with probability of failure q $(p + q = 1)$, the number of successes X is a random variable with probability distribution

$$P(X = 4) = \binom{4}{0}p^4 = \binom{4}{4}p^4$$

$$P(X = 3) = \binom{4}{1}p^3q = \binom{4}{3}p^3q$$

$$P(X = 2) = \binom{4}{2} p^2 q^2$$

$$P(X = 1) = \binom{4}{3} pq^3 = \binom{4}{1} pq^3$$

$$P(X = 0) = \binom{4}{4} q^4 = \binom{4}{0} q^4$$

Note that

$$P(X = 4) + P(X = 3) + P(X = 2) + P(X = 1) + P(X = 0)$$

$$= \binom{4}{0} p^4 + \binom{4}{1} p^3 q + \binom{4}{2} p^2 q^2 + \binom{4}{3} pq^3 + \binom{4}{4} q^4$$

$$= (p + q)^4$$

$$= 1, \text{ since } p + q = 1.$$

7.6. BINOMIAL DISTRIBUTION

We can now extend the above arguments to the case of a more general number n of Bernoulli trials with probability of success p and that of failure q ($p + q = 1$) at each trial. The number of successes X is a random variable which assumes values $n, n-1, \ldots, 2, 1, 0$. The random variable X is said to be a binomial random variable with parameters n and p.

$X = n$ when all the n trials result in successes and no failure in any of the n trials, or n successes in n trials. $P(X = n) = P(SS\ldots S) = P(S) \ldots P(S) = \binom{n}{0} p^n = \binom{n}{n} p^n$

$X = n - 1$, when $(n - 1)$ of the n trials result in successes and 1 out of n trials results in a failure, the probability of $(n - 1)$ successes and 1 failure is $p^{n-1} q$ and since 1 failure out of n can occur in $\binom{n}{1}$ ways

$$P(X = n-1) = \binom{n}{1} p^{n-1} q = \binom{n}{n-1} p^{n-1} q$$

$$\ldots \qquad \ldots \qquad \ldots$$

$$P(X = 0) = \binom{n}{0} q^n = \binom{n}{n} q^n$$

In general:

$X = k$, when (k) trials result in successes and $n-k$ trials out of n result in failures and

$$P(X = k) = \binom{n}{k} p^k q^{n-k} = \binom{n}{n-k} p^k q^{n-k}$$

where $k = 0, 1, 2, \ldots, n$.

Definition: The random variable X is said to have a *binomial distribution with parameters n and p,* when X assumes values $0, 1, \ldots, n$ with probability mass function (pmf)

$$P(X = k) = \binom{n}{k} p^k q^{n-k}, \quad k = 0, 1, 2, \ldots, n$$

$$p + q = 1, p > 0, \; q > 0$$

Note that

$$\sum_{k=0}^{n} P(X = k) = (p+q)^n = 1$$

and that $P(X = k), k = 0, 1, 2, \ldots, n$ correspond to the terms of the binomial expansion of $(p + q)^n$.

The probabilities are obtained by binomial expansion of $(p + q)^n$, and the distribution derives its name from this fact in algebra called *binomial theorem.* Binomial distribution was first enumerated by James Bernoulli in 1713.

Probability Histogram :

Plotting the values of X along the x-axis and with each value of x as a centre,

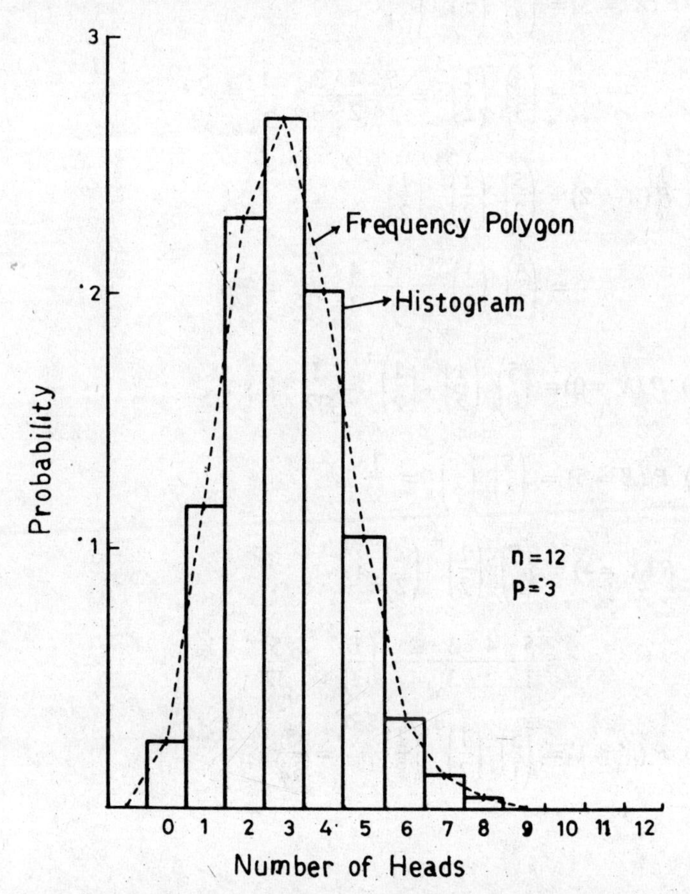

Fig. 7.6. Probability Histogram and Probability Frequency polygon (dotted line) of Binomial distribution for $n = 12, p = 0.3$

drawing a rectangle with area equal to the probability $P(X = x)$ we get a histogram which is known as a probability histogram. A probability frequency polygon is obtained by joining the mid-points of the upper side of the adjoining rectangles.

Probability Histogram for Binomial Distribution with $n = 12, p = 0.3$ is shown in Fig. 7.6. The figure with the dotted line shows the corresponding probability frequency polygon.

Example 6 (a). Five coins are tossed. What is the probability of obtaining (i) 3 heads and 2 tails, (ii) 2 heads, 3 tails, (iii) 5 tails, (iv) 5 heads (v) 4 heads and 1 tail (vi) 1 head and 4 tails (vii) at least 3 heads? We assume that head and tail are equally likely, i.e. $p = q = \frac{1}{2}$.

The number of heads X has a binomial distribution with parameters $n = 5$ and $p = \frac{1}{2}$.

The event of getting exactly 3 heads and 2 tails is equivalent to getting exactly 3 heads when naturally there will be 5 – 3 = 2 tails.

Thus

$$\text{(i) } P(X = 3) = \binom{5}{3}\left(\frac{1}{2}\right)^3\left(\frac{1}{2}\right)^{5-3}$$

$$= \binom{5}{3}\left(\frac{1}{2}\right)^5 = \frac{5 \cdot 4 \cdot 3}{1 \cdot 2 \cdot 3} \cdot \frac{1}{2^5} = \frac{5}{16}$$

$$\text{(ii) } P(X = 2) = \binom{5}{2}\left(\frac{1}{2}\right)^2\left(\frac{1}{2}\right)^{5-2}$$

$$= \binom{5}{2}\left(\frac{1}{2}\right)^5 = \frac{5 \cdot 4}{1 \cdot 2} \cdot \frac{1}{2^5} = \frac{5}{16}$$

$$\text{(iii) } P(X = 0) = \binom{5}{0}\left(\frac{1}{2}\right)^0\left(\frac{1}{2}\right)^5 = \frac{1}{32}$$

$$\text{(iv) } P(X = 5) = \binom{5}{5}\left(\frac{1}{2}\right)^5 = \frac{1}{32}$$

$$\text{(v) } P(X = 4) = \binom{5}{4}\left(\frac{1}{2}\right)^4\left(\frac{1}{2}\right)^{5-4}$$

$$= \frac{5 \cdot 4 \cdot 3 \cdot 2}{1 \cdot 2 \cdot 3 \cdot 4} \cdot \left(\frac{1}{2}\right)^5 = \frac{5}{32}$$

$$\text{(vi) } P(X = 1) = \binom{5}{1}\left(\frac{1}{2}\right)^1\left(\frac{1}{2}\right)^{5-1} = \frac{5}{32}.$$

(vii) The event of getting at least 3 heads is equivalent to three mutually exclusive events (a) 3 heads 2 tails (b) 4 heads 1 tail (c) 5 heads 0 tail. Thus the required probability is

$$P(X = 3) + P(X = 4) + P(X = 5) = \frac{5}{16} + \frac{5}{32} + \frac{1}{32} = \frac{16}{32} = \frac{1}{2}.$$

Note that X can assume only the values 5, 4, 3, 2, 1, 0 and that

$$P(X=5) + P(X=4) + P(X=3) + P(X=2) + P(X=1) + P(X=0) = 1.$$

Example 6 (b). A die is tossed 4 times. Find the probability of getting (i) exactly one 6, (ii) no 6 at all (iii) four 6's.

Here getting a 6 is taken as success so that $p = \frac{1}{6}$.

As $n = 4$, we get

$$P(X = 1) = \binom{4}{1}\left(\frac{1}{6}\right)^1\left(\frac{5}{6}\right)^{4-1} = \frac{4 \times 5^3}{6^4}$$

$$P(X = 0) = \binom{4}{0}\left(\frac{1}{6}\right)^0\left(\frac{5}{6}\right)^4 = \left(\frac{5}{6}\right)^4$$

$$P(X = 4) = \binom{4}{4}\left(\frac{1}{6}\right)^4\left(\frac{5}{6}\right)^0 = \left(\frac{1}{6}\right)^4$$

Example 6 (c). From past experience it is found that some screws manufactured by a machine are defective and that the probability that a screw is defective is $\frac{1}{100}$ independently of each other. A lot of 6 screws are taken at random. Find the probability that (i) there is only one defective screw (ii) no defective screw, (iii) at most one defective screw. (iv) exactly 4 defective screws in the lot.

Getting a defective screw is considered as success, with $p = \frac{1}{100}$. As $n = 6$, X is the number of defective screws.

(i) $P(X = 1) = \binom{6}{1}\left(\frac{1}{100}\right)^1\left(\frac{99}{100}\right)^{6-1} = \frac{6 \cdot (99)^5}{(100)^6}$

(ii) $P(X = 0) = \binom{6}{0}\left(\frac{1}{100}\right)^0\left(\frac{99}{100}\right)^6 = \frac{(99)^6}{(100)^6}.$

(iii) The event of getting at most one defective is equivalent to (a) exactly one defective, and (b) no defective. The required probability is

$$P(X = 1) + P(X = 0) = 6 \cdot \frac{(99)^5}{(100)^6} + \frac{(99)^6}{(100)^6}$$

(iv) $P(X = 4) = \binom{6}{4}\left(\frac{1}{100}\right)^2\left(\frac{99}{100}\right)^4 = 15 \cdot \frac{(99)^4}{(100)^6}$.

Example 6 (d). **In a family of 6 children, what is the probability of having (i) all daughters, (ii) all sons, (iii) at least one son, (iv) at most one son?**

Here the probability of success (that of a child being a male) is $\frac{1}{2}$, i.e. $p = \frac{1}{2}$; $n = 6$ and X is the number of male children.

(i) $P(X = 0) = \binom{6}{0}\left(\frac{1}{2}\right)^0\left(\frac{1}{2}\right)^6 = \frac{1}{2^6} = \frac{1}{64}$

(ii) $P(X = 6) = \binom{6}{6}\left(\frac{1}{2}\right)^6\left(\frac{1}{2}\right)^0 = \frac{1}{2^6} = \frac{1}{64}$

(iii) $P(X \geq 1) = 1 - P(X = 0) = 1 - \frac{1}{64} = \frac{63}{64}$

(iv) $P(X = 0) + P(X = 1) = \frac{1}{64} + \frac{6}{64} = \frac{7}{64}$.

Recurrence Relation for Probabilities

We have

$$P(X = k + 1) = \binom{n}{k+1} p^{k+1} q^{n-k-1}$$

$$= \frac{n-k}{k+1} \cdot \left\{\frac{n!}{k!(n-k)!} \cdot p^k q^{n-k}\right\} \cdot \frac{p}{q}$$

$$= \frac{n-k}{k+1} \cdot \frac{p}{q} P(X = k),$$

$$k = 0, 1, 2, \ldots, n - 1.$$

The relation enables one to find the probabilities $P(X = k)$, $k = 1, \ldots, n$ recursively from $P(X = 0)$; also to find k for which $P(X = k)$ is maximum.

7.6.1. The Mean and Variance of Binomial Distribution

Consider that X is a binomial random variable with $n = 3, p = \frac{1}{2}$.

We have

$$E(X) = \sum_{k=0}^{3} kP(X = k) = 0 \cdot \frac{1}{8} + 1 \cdot \frac{3}{8} + 2 \cdot \frac{3}{8} + 3 \cdot \frac{1}{8}$$

$$= \frac{12}{8} = \frac{3}{2}$$

and $$E(X^2) = \sum_{k=0}^{3} k^2 P(X = k) = 0^2 \cdot \frac{1}{8} + 1 \cdot \frac{3}{8} + 4 \cdot \frac{3}{8} + 9 \cdot \frac{1}{8}$$

$$= \frac{24}{8} = 3$$

$$\text{Var}(X) = 3 - \left(\frac{3}{2}\right)^2 = \frac{3}{4}$$

We note that $E(X) = \frac{3}{2} = 3 \cdot \frac{1}{2} = np$

and $\text{var}(X) = \frac{3}{4} = 3 \cdot \frac{1}{2} \cdot \frac{1}{2} = npq.$

It can be easily seen that the random variable X can be expressed as the sum of 3 independent random variables:

$$X = X_1 + X_2 + X_3$$

where X_1 is the number of successes in the first trial, X_2 in the second trial and X_3 in the third trial. X_1 assumes two values 0 and 1 each with probability $\frac{1}{2}$; so is the case with X_2 and X_3 independently of each other. Thus X_1, X_2, X_3 are independent Bernoulli variables each with expected value $\frac{1}{2}$ and variance $\frac{1}{4}$.

Thus

$$E(X) = E(X_1) + E(X_2) + E(X_3) = \frac{1}{2} + \frac{1}{2} + \frac{1}{2} = \frac{3}{2}$$

$$\text{var}(X) = \text{var}(X_1) + \text{var}(X_2) + \text{var}(X_3) = \frac{1}{4} + \frac{1}{4} + \frac{1}{4} = \frac{3}{4}.$$

We can now generalise this result for any number n of Bernoulli variables with probability p of success at each trial:

Theorem 7.1. Let X be a binomial random variable with parameters n and p. Then

$$E(X) = np$$

and $$\text{var}(X) = npq, \quad p + q = 1$$

Proof: X can be expressed as the sum of n independent Bernoulli variables $X_1, \ldots, X_n$, i.e.

$$X = X_1 + X_2 + \ldots + X_n$$

In other words, the total number of successes in n Bernoulli trials can be considered as the sum of the number of successes in the n trials, X_1 being the number of successes in the 1st trial, X_2 that in the 2nd trial and so on.

Now for a Bernoulli variable X_1, $E(X_1) = p$, var $(X_1) = pq$

Thus

$$E(X) = E(X_1) + E(X_2) + \ldots + E(X_n)$$

$$= p + p + \ldots + p$$

$$= np$$

$$\text{var}(X) = \text{var}(X_1) + \text{var}(X_2) + \ldots\ldots + \text{var}(X_n)$$

$$= pq + pq + \ldots + pq$$

$$= npq$$

This completes the proof.

There is a direct way of getting these results, but this involves application of binomial theorem and is more tedious.

Example 6 (e). Let X be the number of 6's obtained in 10 throws of a die. Find $E(X)$ and var (X).

X is a binomial variable with $p = \frac{1}{6}$ (since getting a 6 is considered as success) and $n = 10$. Thus

$$E(X) = 10 \cdot \frac{1}{6} = \frac{5}{3}$$

$$\text{var}(X) = 10 \cdot \frac{1}{6} \cdot \frac{5}{6} = \frac{25}{18}$$

7.6.2. Skewness and Kurtosis of Binomial Distribution

For binomial random variable X with parameters n and p, $\mu = E(X) = np$, $\sigma^2 = \text{var}(X) = npq$; the third and four moments about the mean μ are

$$\mu_3 = E\{(X-\mu)\}^3 = npq(q-p)$$

$$\mu_4 = E\{(X-\mu)^4\} = 3n^2p^2q^2 + npq(1-6pq)$$

so that the coefficient of skewness $\gamma_1 = \frac{\mu_3}{\sigma^3} = \frac{q-p}{\sqrt{npq}} = \frac{1-2p}{\sqrt{npq}}$

and the coefficient of kurtosis is $\gamma_2 = \frac{\mu_4}{\sigma^4} - 3 = \frac{1-6pq}{npq}$

The skewness is positive if $p < \frac{1}{2}$, negative for $p > \frac{1}{2}$, zero for $p = \frac{1}{2}$. For large n, γ_1 and γ_2 approach zero.

7.7. THE HYPERGEOMETRIC DISTRIBUTION

Suppose that a box contains 3 red and 5 black balls. One ball is drawn at random. The probability of getting a red ball is $\frac{3}{8}$. Now suppose that we are to make a second drawing. This can be done in two ways: either (i) by replacing the ball drawn, or (ii) without replacing the ball drawn.

Consider the first case, where drawing is done *with replacement*, the number of red and black balls remains the same, their total number being 8, and the result of the first trial has no effect whatsoever on the second drawing (we assume that balls are thoroughly mixed before each drawing). If we confine to the colour of the ball drawn, and consider drawing a red ball as a success, then each trial is a Bernoulli trial with probability of success $p = \frac{3}{8}$.

If 4 such drawings with replacement are made and X is the number of red balls drawn, then X is a binomial random variable with $n = 4, p = \frac{3}{8}$. X assumes values 0, 1, 2, 3, 4. Let k be a specific value of X, then $\{X = k\}$ is the event that out of a total of 4 drawings, in k drawings red balls are drawn and this implies that in the other $(4-k)$ drawing black balls are drawn.

$$\begin{aligned} P(X = k) &= P(k \text{ red balls}, n-k \text{ black balls}) \\ &= \binom{n}{k} p^k q^{n-k}, \ k = 0, 1, 2, 3, 4, \ n = 4 \end{aligned}$$

Consider now the second case, where drawing is done *without replacement*. Here the ball drawn at the first drawing is *not* replaced or put back into the box, so that for the second drawing there will be 7 balls, for the 3rd drawing there are 6 balls and so on. After 8 drawings, the box will be empty. Here the result of the 1st drawing influences or affects the second drawing. Trials are *not independent*. To analyse such an experiment, the concept of Bernoulli trials cannot be applied. The number of red balls drawn will be a random variable different from a binomial. Such a random variable will be called *hypergeometric random variable*. Here if two drawings are made, we may get two red balls, one red and one black balls or two black balls. If X is the number of red balls, then X assumes value 2, 1 or 0. Take the case of $X = 1$, then out of 2 balls drawn, one is red and the other black: the red ball must come from the 3 red balls and the black ball from the 5 black balls. Two balls can be taken from $3 + 5 = 8$ balls in $\binom{8}{2}$ ways; 1 red ball can be taken from 3 red balls in $\binom{3}{1}$ ways, 1 black ball can be taken from 5 black balls in $\binom{5}{1}$ ways, and each way of drawing a black and a red ball may be combined, so that the number of ways in which 1 red ball and a black ball can be taken is $\binom{3}{1} \cdot \binom{5}{1}$.

Thus

$$P(X = 1) = P(\text{one red and one black balls})$$

$$= \frac{\binom{3}{1}\binom{5}{1}}{\binom{8}{2}}$$

We consider a more general form below:

Suppose that there is a finite population of N of two kinds of things, of which K are of the first kind and $N-K$ of the second kind. A sample of n is drawn at random *without replacement* from the sample. Let X be the number of things of the 1st kind drawn. Then $X = k$ implies that of the n drawn, k are of the 1st kind and $n-k$ of the second kind. We have

$$P(X = k) = P(k \text{ of 1st kind and } n-k \text{ of 2nd kind})$$

$$= \frac{\binom{K}{k}\binom{N-K}{n-k}}{\binom{N}{n}}, \quad k \leq \min(n, K) \tag{7.1}$$

Now consider the range of values of X.

If $n < N-K$, then X assumes values from $k = 0$ to $k = \min(n, K)$.

If $n > N-K$, and $r = n-(N-K)$, then X assumes values from $k = r\,(>0)$ to $k = \min(n, K)$.

For example, if $N = 15$, $K = 10$ and $n = 4$, then X assumes values from $k = 0$ to $k = \min(4, 10) = 4$, i.e. $k = 0, 1, 2, 3, 4$; if $N = 15$, $K = 10$ and $n = 8$, then $n > N-K$ and $r = 8-(15-10) = 3\,(>0)$ and then X assumes values from $k = 3$ to $k = \min(n, K) = \min(8, 10) = 8$, i.e. $k = 3, 4, \ldots, 8$.

The above distribution is known as *hypergeometric distribution*. It holds for $\max(0, n-(N-K)) \leq k \leq \min(n, K)$.

Example 7 (a). A box contains 10 screws out of which 3 are defective. Two screws are taken at random from the box. Find the distribution of the number of defective screws drawn.

Here $N = 10$, $K = 3$, $N-K = 7$, $n = 2$. Thus $k = 0, 1, 2$

If X is the number of defective screws drawn then

$$P(X = 0) = \frac{\binom{3}{0}\binom{7}{2}}{\binom{10}{2}} = \frac{\frac{7.6}{1.2}}{\frac{10.9}{1.2}} = \frac{7 \times 6}{10 \times 9} = \frac{7}{15}$$

$$P(X = 1) = \frac{\binom{3}{1}\binom{7}{1}}{\binom{10}{2}} = \frac{3.7}{\frac{10.9}{1.2}} = \frac{3 \times 7}{5 \times 9} = \frac{7}{15}$$

$$P(X=2)=\frac{\binom{3}{2}\binom{7}{0}}{\binom{10}{2}}=\frac{3.1}{\frac{10.9}{1.2}}=\frac{3}{5\times 9}=\frac{1}{15}$$

X assumes values $k = 0, 1, 2$ with probabilities $\frac{7}{15}, \frac{7}{15}, \frac{1}{15}$ respectively.

Note that values assumed by X cannot exceed $K = 3$ nor $n = 2$. Further,

$$P(X=0)+P(X=1)+P(X=2)=\frac{7}{15}+\frac{7}{15}+\frac{1}{15}=1.$$

Example 7 (b). Four cards are taken from a well shuffled pack of 52 cards. What is the probability that (i) 2 are black and 2 are red (ii) there is no black card (iii) there is exactly one ace?

Here $N = 52, n = 4$

For (i) and (ii) cards are considered to be as of two kinds, black and red: $K = 26$ black and $N-K = 26$ red cards

If X is the number of black cards drawn, then X assumes the value k, where $k = 0, 1, 2, 3, 4$. Thus

$$\text{(i)}\quad P(X=2)=\frac{\binom{26}{2}\binom{26}{2}}{\binom{52}{4}}$$

$$\text{(ii)}\quad P(X=0)=\frac{\binom{26}{0}\binom{26}{4}}{\binom{52}{4}}$$

For (iii) cards are considered to be of two kinds ace ($K = 4$) and non ace ($N-K = 52-4 = 48$).

$$P(X=1)=\frac{\binom{4}{1}\binom{48}{3}}{\binom{52}{4}}$$

The total population N is divided into two kinds, K of one kind and $N-K$ of another. We can mark each of the (K) similar elements of the first kind by the number 1 and each of the ($N-K$) similar elements of the second kind by the number 0, so that out of the total N elements, there are K number of 1's and $N-K$ number of 0's. Suppose that n drawings are made at random, and k of the first kind is found the event is equivalent to finding k number of 1's in n drawings. The event $X = k$ (i.e. k are of the first kind) is equivalent to getting k number of 1's and $n-k$ number of 0's out of n drawn.

Thus

$$X = X_1 + X_2 + ... + X_n$$

where each of the X_i's takes two values 1 and 0 and the event $X = k$ is equivalent to saying that k of the X_i's take the value 1 and the remaining $n - k$ of the X_i's take the value 0. We may also associate "success" with 1 and "failure" with 0. However, the probability that X_1 takes the value 1 (or 0) does not remain the same for all i in hypergeometric case, as the number of elements go on decreasing as further drawings are made, as the drawing is without *replacement*. In case of Binomial distribution, the probability p remains the same for each drawing, as the drawing is *with replacement*; while trials are independent in case of drawing with replacement, they are not independent in case of drawing without replacement.

7.7.1. The Mean and Variance of Hypergeometric Distribution

Without going into the algebraic details, we write down the expressions for the mean (or expected value) and variance.

$$\text{Expected value} = E(X) = n\left(\frac{K}{N}\right)$$

$$\text{variance} = \text{var}(X) = n\left(\frac{K}{N}\right)\left(1-\frac{K}{N}\right)\left(\frac{N-n}{N-1}\right).$$

Here $p = \dfrac{K}{N}$ is the proportion of elements of the first kind in the original population (or $\dfrac{K}{N}$ is the proportion of the defectives before any drawing is made).

Writing $p = \dfrac{K}{N}$ and $q = 1 - p = 1 - \dfrac{K}{N}$, we have

$$P(X = k) = \frac{\binom{K}{k}\binom{N-K}{n-k}}{\binom{N}{n}}$$

$$E(X) = np$$

and

$$\text{var}(X) = npq\left(\frac{N-n}{N-1}\right)$$

7.7.2. Hypergeometric Distribution for Large N

When N is very large compared to n (i.e. a small sample is taken from a large population), the value of $\dfrac{n}{N}$ is quite small. It can then be seen that

$$\frac{N-n}{N-1} = \frac{1-\frac{n}{N}}{1-\frac{1}{N}} \text{ (dividing both the numeration and denomination by } N)$$

As $\frac{n}{N}$ and $\frac{1}{N}$ are small, the quantity $\frac{N-n}{N-1}$ is very near to 1, so that var $(X) = npq\left(\frac{N-n}{N-1}\right)$ approaches npq. Thus, for large $N, E(X) = np$, var $(X) = npq$ and the mean and variance of hypergeometric distribution are the same as those of binomial distribution. It can also be shown that

$$P(X = k) = \frac{\binom{K}{k}\binom{N-K}{n-k}}{\binom{N}{n}} \text{ approaches } = \binom{n}{k} p^k q^{n-k}$$

so that for N large compared to n, hypergeometric distribution can be treated as binomial distribution with parameters $n, \frac{K}{N}$.

Thus, *when a sample of comparatively small size is taken from a large population, then there is practically no difference between sampling with and without replacement. This should be intuitively clear also.*

7.7.3. Extension of Hypergeometric Distribution to More than 2 Categories

Suppose that there are more than 2 categories of elements which comprise the population. For example, let there be 3 categories with K elements of the first kind, M elements of second kind and $N-K-M$ elements of the third kind with a total of $(K) + (M) + (N-K-M) = N$. Suppose that n elements are drawn at random without replacement. The probability that the drawings result in k elements of the first kind, m of the second kind and $n-m-k$ of the third kind is given by

$$\frac{\binom{K}{k}\binom{M}{m}\binom{N-K-M}{n-k-m}}{\binom{N}{n}}$$

since k elements of the first kind in the sample come from the total of K elements of the first kind in the population and so on.

Example 7 (c). A box contains 10 red, 6 black and 9 white balls. 5 balls are drawn at random (without replacement). What is the probability that there are

(i) 3 red, 1 black and 1 white balls
(ii) 1 red, 1 black and 3 white balls
(iii) 1 red, 3 black and 1 white balls

Required probabilities are

(i) $$\frac{\binom{10}{3}\binom{6}{1}\binom{9}{1}}{\binom{25}{5}}$$

(ii) $$\frac{\binom{10}{1}\binom{6}{1}\binom{9}{3}}{\binom{25}{5}}$$

(iii) $$\frac{\binom{10}{1}\binom{6}{3}\binom{9}{1}}{\binom{25}{5}}$$

7.8. GEOMETRIC DISTRIBUTION

Binomial distribution arises in the context of repeated trials. Hypergeometric distribution also arises out of similar but not independent trials. Another important distribution arising in the context of Bernoulli trials, is geometric distribution. In Binomial and hypergeometric distributions, the number of trials n is fixed in advance, while in Binomial distributions p is also fixed. Binomial distribution is concerned about the number of successes in a fixed number n of trials with fixed probability of success p at each trial. Now let us consider the situation that n is not fixed in advance but p is constant and that trials are repeated so long as necessary to get a success, the first success or success for the first time. If X is the number of trials necessary to get the first success, then X is a random variable which assumes the value 1 if the very first trial results in a success; the value 2, if the first trial results in a failure and the second in a success, and so on. Further, the number of trials needed may be infinitely large. Thus

$$\begin{aligned} P(X = 1) &= P(S) = p \\ P(X = 2) &= P(FS) = qp \\ P(X = 3) &= P(FFS) = q^2 p \\ &\dots \quad \dots \quad \dots \\ P(X = k) &= P(\underbrace{FF\ldots F}_{k-1} S) = q^{k-1} p. \end{aligned} \tag{8.1}$$

We can write the distribution of the number of trials needed to get the first success or success for the first time as

$$P(X = k) = q^{k-1} p, \ k = 1, 2, 3, \ldots, \infty \tag{8.2}$$

This is called *geometric distribution.*

Note that $$\sum_{k=1}^{\infty} P(X = k) = \sum_{k=1} q^{k-1} p = p(1 + q + q^2 + \ldots)$$

$$= \frac{p}{1-q} = 1.$$

The series $1 + q + q^2 + \ldots$ is a geometric series from which the distribution gets its name.

The mean and variance of geometric distribution

These can be calculated by algebraic operations.

We give below the expressions:

$$P(X = k) = q^{k-1}p, \quad k = 1, 2, 3, \ldots$$

$$q = 1-p, \quad 0 < p < 1$$

$$\text{Expected value of } X = E(X) = \frac{1}{p} \tag{8.3}$$

$$\text{variance of } X = \text{var}(X) = \frac{q}{p^2} = \frac{1-p}{p^2}. \tag{8.4}$$

Example 8 (a). A die is thrown until a 6 occurs. Find the distribution of the number of throws needed and find its mean and variance.

Here getting a 6 is termed "success", and the probability p of a success is $\frac{1}{6}$.

Thus if X is the number of throws needed

$$P(X = k) = \left(\frac{5}{6}\right)^{k-1}\left(\frac{1}{6}\right), \; k = 1, 2, \ldots$$

$$E(X) = \frac{1}{p} = 6$$

and

$$\text{var}(X) = \frac{q}{p^2} = \frac{\frac{5}{6}}{\left(\frac{1}{6}\right)^2} = 30$$

Example 8 (b). A person tries to get a busy telephone number. The probability that the person gets the number in an attempt is $\frac{1}{20}$ and is the same for all attempts. What is the probability that he would get the first call in the third attempt. What is the expected number of attempts needed to get the first call through?

Here X is the number of calls needed to get the call through and $p = \frac{1}{20}$.

$$P(X = 3) = q^{3-1}p = q^2p = \left(\frac{19}{20}\right)^2 \cdot \frac{1}{20} = \frac{361}{8000}.$$

$$E(X) = \frac{1}{p} = 20$$

The expected value implies that on an average 20 attempts will be needed to get the call through.

7.8.1. Another Way of Defining Geometric Distribution

This is described below :

Let γ denote the number of failures preceding the first success in a sequence of Bernoulli trials with probability of success p. Trials are repeated till success occurs for the first time: if success occurs at the very first trial, then $\gamma = 0$ and the probability for this to happen is p, i.e.

$$P(\gamma = 0) = P(\text{success at first trial}) = p$$

$$P(\gamma = 1) = P(\text{first success at second trial})$$

$$= P(FS) = qp$$

and so on. Thus

$$P(\gamma = k) = q^k p, \quad k = 0, 1, 2, \ldots \tag{8.5}$$

also defines a geometric distribution.

We have

$$\sum_{k=0}^{\infty} P(\gamma = k) = p + qp + q^2 p + \ldots$$

$$= 1$$

Expected value of γ

$$= E(\gamma) = \frac{q}{p} \tag{8.6}$$

variance of γ

$$= \text{var}(\gamma) = \frac{q}{p^2} \tag{8.7}$$

7.9. POISSON DISTRIBUTION

Consider the binomial distribution.

The expression $P(X = k) = \binom{n}{k} p^k q^{n-k}$ enables us to calculate the value of the probability of k successes in n Bernoulli trials with probability of success p at each trial. For example, if $n = 5, p = \frac{1}{2}$, we can readily calculate

$$P(X = 3) = \binom{5}{3}\left(\frac{1}{2}\right)^3\left(\frac{1}{2}\right)^2 = \frac{5.4}{1.2}\cdot\left(\frac{1}{2}\right)^5 = \frac{5}{16}$$

If $n = 1000, p = \frac{1}{100}$, then

$$P(X = 3) = \binom{1000}{3}\left(\frac{1}{100}\right)^3\left(\frac{99}{100}\right)^{1000-3}$$

and it can be calculated, but it would be tedious and time consuming. Of course tables of binomial probabilities are available which give probabilities for certain values of n and p, but no table gives them for all possible values of n and p. Even apart from this aspect of calculation, there is an interesting phenomena: the limiting behaviour of binomial distribution. We have seen that when N is large compared to n, the limiting behaviour of hypergeometric is binomial. It can be shown that the binomial distribution

$$P(X = k) = \binom{n}{k} p^k q^{n-k}, \quad k = 0, 1, 2, ..., n$$

exhibits an interesting limiting behaviour for n large, p small and np of moderate magnitude. The limiting form of binomial distribution is known as *Poisson distribution*. This distribution is named after a famous French mathematician Simeon D. Poisson(1781-1840) who described it in 1837.

It can be mathematically proved that when n is large, p is small, $np = a$ is of moderate magnitude, then

$$\binom{n}{k} p^k q^{n-k} \to \frac{e^{-a} a^k}{k!}$$

for all $k = 0, 1, 2, ...$

This is considered in *Theorem* 7.2 given after Example 9 (*b*).

Definition: A random variable X, taking set of values 0, 1, 2, 3,... is said to have Poisson distribution with parameter a if for a (> 0)

$$P(X = k) = \frac{e^{-a} a^k}{k!}, \quad k = 0, 1, 2, 3, ... \tag{9.1}$$

where, e is a constant (the base of Naperian logarithm). The value of e = 2.78128, correct upto the 5th place of decimals. The random variable X having Poisson distribution is also known as *Poisson random variable.*

The above expression gives the probability that X assumes the value k, or in other words it gives the probability of k occurrences of an event, the probability of one occurrence is *small.* An event with a small probability of occurrence is an improbable event or a rare event (an event which occurs rarely). Because of this, Poisson distribution (or law) is also known as the *distribution (or law) of rare events* or *distribution (law) of improbable events.*

There is a very wide range of phenomena occurring in diverse areas, where Poisson law or distribution could be applied. Further table of powers of e as well as extensive tables of the probabilities

$$\frac{e^{-a} a^k}{k!}$$

are available for a wide range of values of a, facilitating numerical calculation of probabilities in specific cases.

Some values of e^{-a} are given below :

a	0	0.5	1.0	1.5	2.0	3	4	5
e^{-a}	1	0.607	0.368	0.223	0.135	0.050	0.018	0.004

It may be verified that

$$\sum_{k=0}^{\infty} P(X=k) = P(X=0) + P(X=1) + P(X=2) + \ldots$$

$$= 1$$

Note that

$$P(X \geq r) = \sum_{k=r}^{\infty} P(X=k) = 1 - \sum_{k=0}^{r-1} P(X=k) \qquad (9.2)$$

Example 9 (a). Suppose that the number of births resulting in twins during a year has a Poisson distribution with parameter $a = 1$. Calculate the probability that during a year there (i) is no twin birth (ii) exactly one twin birth (iii) less than 2 twin births (iv) greater than 1 twin birth.

Let X be the number of twin births during a year: then,

since $a = 1$

$$P(X=k) = \frac{e^{-1}1^k}{k!} = \frac{e^{-1}}{k!}, k = 0, 1, 2, \ldots$$

(i) $P(X=0) = \frac{e^{-1}}{0!} = e^{-1} = 0.368$

(ii) $P(X=1) = \frac{e^{-1}}{1!} = e^{-1} = 0.368$

(iii) $P(X<2) = P(X=0) + P(X=1) = 0.368 + 0.368 = 0.736$

(iv) $P(X>1) = P(X=2) + P(X=3) + P(X=4) + \ldots$

$= 1 - [P(X=0) + P(X=1)]$

$= 1 - [0.368 + 0.368] = 0.264$

Example 9 (b). Suppose that the number of defective screws produced by a sophisticated machine per day has a Poisson distribution with parameter 2. What is the probability that out of the total produce of a day, there is (i) no defective screw (ii) exactly 2 defective screws (iii) at least one defective screw (iv) less than 3 defective screws?

Let X be the number of defective screws produced in a day.

$$P(X=k) = \frac{e^{-a}a^k}{k!} = \frac{e^{-2}2^k}{k!}, k = 0, 1, 2, \ldots$$

(i) $P(X=0) = \frac{e^{-2}2^0}{0!} = e^{-2} = 0.135$

(ii) $P(X=2) = \frac{e^{-2}2^2}{2!} = \frac{e^{-2}4}{2} = 0.270$

(iii) $P(X \geq 1) = 1 - P(X=0) = 1 - 0.135 = 0.865$

(iv) $P(X<3) = P(X=0) + P(X=1) + P(X=2)$

$= 0.135 + 0.135 + 0.270 = 0.540$

We now prove an important result stated earlier.

Theorem 7.2. If the probability of success in a single trial p approaches 0 while the number n of trials becomes infinite such that the mean $np = a$ is of moderate and fixed magnitude, then the binomial distribution will approach the Poisson distribution with parameter (mean) a.

Proof. Consider a *r.v.* X having binomial distribution with parameters n and p, where n is large and p is small but $np = a$ is of a moderate (fixed) magnitude. We have

$$P(X = k) = \binom{n}{k} p^k q^{n-k}$$

$$= \frac{n(n-1)\ldots(n-k+1)}{k!} p^k (1-p)^{n-k}$$

$$= \frac{\dfrac{n(n-1)\ldots(n-k+1)}{n^k} . n^k p^k (1-p)^{n-k}}{k!}$$

$$= \frac{\left(\dfrac{n}{n}\right)\left(\dfrac{n-1}{n}\right)\ldots\left(\dfrac{n-k+1}{n}\right)(a)^k\left(\dfrac{1-a}{n}\right)^n}{k!(1-p)^k}$$

Now as $n \to \infty$,

$$\left(\frac{n}{n}\right) = 1, \left(\frac{n-1}{n}\right) = \left(1-\frac{1}{n}\right) \to 1, \ldots, \left(\frac{n-k+1}{n}\right) = \left(1-\frac{k-1}{n}\right) \to 1$$

$$(1-p)^k = \left(1-\frac{a}{n}\right)^k \to 1$$

$$\left(1-\frac{a}{n}\right)^n = \left(1-\frac{a}{n}\right)^{\frac{n}{a}.a} = \left[\left(1-\frac{a}{n}\right)^{\frac{n}{a}}\right]^a \to (e^{-1})^a = e^{-a}$$

Thus as $p \to 0$ and $n \to \infty$, we have

$$P(X = k) \to \frac{(1.1 \ldots 1)a^k e^{-a}}{k\,!(1)} = \frac{e^{-a} a^k}{k!}$$

$$k = 0, 1, 2, \ldots$$

which proves the theorem.
An illustration is given below.

Example 9(c). It is known that the probability that a screw produced by a machine is defective is 0.01. A lot of 30 screws are taken at random. Find the probability of the number of defective items in the lot by using binomial distribution as well as by approximation using Poisson distribution.
Here $p = 0.01$, $n = 30$; X is the number of defective items. We use binomial distribution.

$$P(X = k) = \binom{n}{k} p^k q^{n-k} \tag{1}$$

and also Poisson approximation, with $a = np = 30 \times 0.01 = 0.3$

$$P(X = k) = \frac{e^{-a} a^k}{k!} \tag{2}$$

The probabilities are given in the Table below:

k	Binomial Probability (1)	Poisson approximation (2)
0	0.7397	0.7408
1	0.2242	0.2222
2	0.0328	0.0334
3	0.0031	0.0033
4	0.0002	0.0003
.	.	.
.	.	.
.	.	.
	1.0000	1.0000

We can see from the above how close Poisson approximation is to the binomial.

Recurrence Relation for Probabilities

We have

$$P(X = k + 1) = \frac{e^{-a} a^{k+1}}{(k+1)!}$$

$$= \frac{e^{-1} a^k}{k!} \cdot \frac{a}{k+1}$$

$$= \frac{a}{k+1} \cdot P(X = k),$$

$$k = 0, 1, 2, \ldots$$

The relation enables one to find $P(X = k), k = 1, 2, \ldots$ recursively from $P(X = 0)$; also to find k for which $P(X = k)$ is maximum.

7.9.1. Mean and Variance of Poisson Distribution

These can be calculated from those of binomial distribution, using the fact that Poisson distribution is a limiting form of binomial distribution.

For a binomial random variable X, we have

$$E(X) = np$$

and $\quad \text{var}(X) = npq.$

Now binomial distribution tends to Poisson distribution when n is large, p is small and $np = a$ is of moderate magnitude. Again when p is small, $q = 1-p$ is approximately equal to 1. Thus for a Poisson random variable X,

$$E(X) = a \tag{9.3}$$

and

$$\text{var}(X) = a.(q) = a \tag{9.4}$$

Thus the mean and variance of Poisson distribution with parameter a are both equal to a.

These can also be obtained by direct calculation as follows: We have

$$E(X) = \sum_{k=0}^{\infty} \frac{ke^{-a}a^k}{k!} = e^{-a}\left(0 + 1.a + 2.\frac{a^2}{2!} + 3.\frac{a^3}{3!} + ...\right)$$

$$= ae^{-a}\left(1 + a + \frac{a^2}{2!} + ...\right)$$

$$= ae^{-a}e^a = a$$

$$E(X^2) = \sum_{k=0}^{\infty} \frac{k^2e^{-a}a^k}{k!} = e^{-a}\left(0^2 + 1^2.a + \frac{2^2.a^2}{2!} + \frac{3^2.a^3}{3!} + ...\right)$$

$$= ae^{-a}\left(1 + 2a + \frac{3a^2}{2!} + \frac{4a^3}{3!} + ...\right)$$

$$= ae^{-a}\left[\left(1 + a + \frac{a^2}{2!} + \frac{a^3}{3!} + ...\right) + \left(a + 2\frac{a^2}{2!} + \frac{3a^3}{3!} + ...\right)\right]$$

$$= ae^{-a}\left[e^a + a\left(1 + a + \frac{a^2}{2!} + ...\right)\right]$$

$$= ae^{-a}\left[e^a + a.e^a\right] = ae^{-a}(a+1)e^a$$

$$= a(a+1).$$

Hence

$$\text{var}(X) = E(X^2) - [E(X)]^2$$

$$= a(a+1) - a^2$$

$$= a$$

as found above. The standard deviation is $\sqrt{a}$.

Example 9 (c). If the number of accidents occurring in an industrial plant during a day is given by a Poisson random variable with parameter 3, find the (i) probability that no accident occurs on a day (ii) the expected number of accidents per day and also its variance.

Here X is the number of accidents per day and

$$P(X = k) = \frac{e^{-3}3^k}{k!}, \quad k = 0, 1, 2, ...$$

(i) $P(X = 0) = e^{-3} = 0.05$

(ii) $E(X) = 3$ and $\text{var}(X) = 3$.

7.10 NORMAL DISTRIBUTION

So far we have discussed discrete probability distributions. We shall now confine our attention to continuous random variable (or probability distribution). In section 7.3, we have explained the basic concept of a continuous random variable and have indicated the essential difference between the probability distributions of discrete and continuous random variables. In the case of a continuous random variable X, we speak of the *probability density function* (*p.d.f.*) $f(x)$ of X; the basic features of *p.d.f.* were already enumerated there.

We shall now consider an important continuous random variable, known as *normal random variable* and its distribution, *normal distribution*. The *p.d.f.* $f(x)$ of normal distribution (or of normal random variable X) is given by

$$f(x) = \frac{1}{\sigma\sqrt{(2\pi)}} \exp\left(-\frac{(x-\mu)^2}{2\sigma^2} \right), \text{ for } -\infty \leq x \leq \infty$$

Here μ is the mean and σ is the standard deviation of X; $e = 2.78128$ (the base of Naperian logarithm) and $\pi = 3.14159$, the values of e and π being correct upto 5 decimal places. The normal distribution has two parameters, the mean (μ) and the *s.d* (σ).

The actual shape of the frequency curve $y = f(x)$ is bell shaped and is as shown in Fig. 7.7.

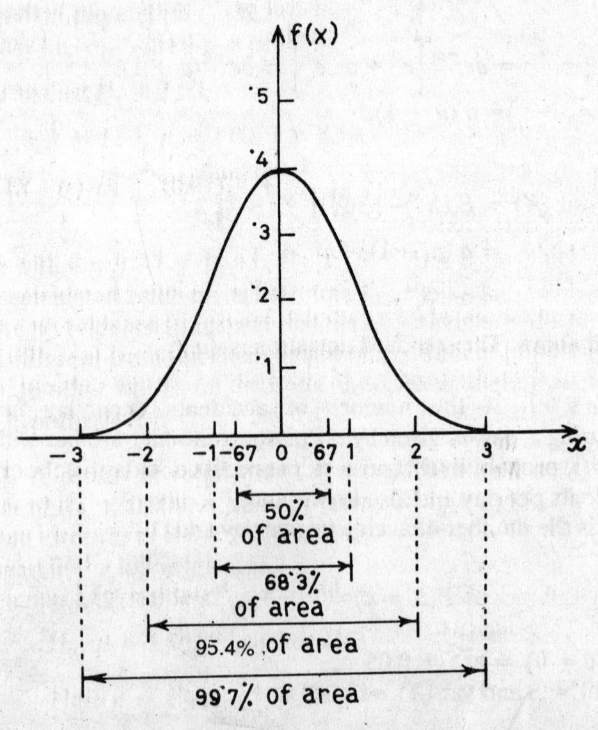

Fig 7.7 Probability density curve of Normal Distribution with mean 0 and s.d. 1.

The curve is symmetrical about the point $x = \mu$. The total area between the curve $y = f(x)$ and the x-axis from $x = -\infty$ to $x = +\infty$ is 1. The frequency curve of normal distribution is also known as *normal curve.*

The normal distribution is also known as *Gaussian* distribution (after Carl Gauss, 1777-1855) and also as *Laplacian* distribution (after Pierre Laplace, 1749-1827). It was Abraham De Moivre (1667-1754) who first obtained normal distribution in 1733.

Using results of calculus, we see that

$$\int_{-\infty}^{\infty} f(x)dx = \int_{-\infty}^{\infty} \frac{1}{\sigma\sqrt{2\pi}} \exp\left(\frac{-(x-\mu)^2}{2\sigma^2}\right) dx = 1$$

Figure 7.7 is the frequency curve for a pair of specific values of μ and σ. We get a frequency curve for every pair of values of μ and σ. We have two such curves:

(1) $\mu = 0,\ \sigma = 1$ and $\mu = 1,\ \sigma = 1$ (Fig. 7.10)

(2) $\mu = 0,\ \sigma = 1$ and $\mu = 0,\ \sigma = 3$ (Fig. 7.11)

Consider that $\mu = 0$ and $\sigma = 1$: the curve is given in Fig. 7.7.

We can find the probabililty $P(a < x < b)$ as the area between the curve $y = f(x)$, and the x-axis between the point $x = a$ and $x = b$. We have $P(0 \le x < \infty)$ = area between the curve and the x-axis between 0 and $\infty = \frac{1}{2}$.

This can be easily seen from the fact that the curve is symmetrical about 0 and that the total area is 1. Similarly

$$P(-\infty < x \le 0) = \frac{1}{2}$$

This implies that the area of the portion covered on each side of the mean μ is $\frac{1}{2}$. Each portion is symmetrical about μ.

For example, if we consider $\mu = 1$. We shall have the area of each side μ to be equal to $\frac{1}{2}$.

As already discussed the expected value $\mu = E(X)$ is the centre of the distribution: it is the centre of gravity about which a metallic plate in the shape of the distribution (frequency function) will balance.

This is about the mean: the position of the mean μ with regard to the curve. What is the corresponding position of the standard deviation? This is explained below. We should not bother about how these calculations are made.

The area of the curve between $\mu-\sigma$ and $\mu+\sigma$ is 0.683, i.e. the probability that the variable X assumes a value between $\mu-\sigma$ and $\mu+\sigma$ is 0.683.

The area of the curve between $\mu-2\sigma$ and $\mu+2\sigma$ is 0.954, i.e. the probability that X assumes a value in $(\mu-2\sigma, \mu+2\sigma)$ or between $\mu-2\sigma$ and $\mu+2\sigma$ is 0.954.)

The area of the curve between $\mu-3\sigma$ and $\mu+3\sigma$ is 0.997, i.e. the probability that X assumes a value in $(\mu-3\sigma, \mu+3\sigma)$ or between $\mu-3\sigma$ and $\mu+3\sigma$ is 0.997

For example, take specific cases: (1) $\mu = 0,\ \sigma = 1$. Then the probability that X assumes value in (–1, 1) is 0.683, that it assumes values in (–2, 2) is 0.954 and that it assumes values in (–3, 3) is 0.997.

(2) $\mu = 1, \sigma = 2$.

Then the probability that X assumes values in (–1, 3) is 0.683, that it assumes values in (–3, 5) is 0.954 and that it assumes value in (–5, 7) is 0.997.

The interpretation of μ and σ in terms of the curve should now be clear. Using the notation of calculus, we can write

$$\int_{-\infty}^{\infty} f(x)dx = 1$$

$$\int_{-\infty}^{\mu} f(x)dx = \int_{\mu}^{\infty} f(x)dx = \frac{1}{2}$$

$$\int_{\mu-\sigma}^{\mu+\sigma} f(x)dx = 0.683,$$

$$\int_{\mu-2\sigma}^{\mu+2\sigma} f(x)dx = 0.954$$

$$\int_{\mu-3\sigma}^{\mu+3\sigma} f(x)dx = 0.997$$

We have started with a normal distribution with mean μ and s.d. σ. Normal distribution has two parameters μ and σ. We denote such a distribution by $N(\mu, \sigma)$. The particular normal distribution that has mean 0 and s.d. 1 (i.e. $\mu = 0, \sigma = 1$) is known as *standard normal distribution* and is denoted by $N(0, 1)$. Its frequency curve has already been shown in Fig. 7.7.

We have learnt about probabilities in specific intervals $(\mu - \sigma, \mu + \sigma)$ $(\mu - 2\sigma, \mu + 2\sigma)$ and $(\mu - 3\sigma, \mu + 3\sigma)$. What can we say about probabilities in an arbitrary interval (a, b) ? Tables are available from which we can at once get the probability that X assumes values between (a, b): Tables are called *normal probability tables.* We shall explain how to use such a table for calculating probabilities.

Note : Some authors use the notation $N(\mu, \sigma^2)$. We shall use $N(\mu, \sigma)$: here μ and σ are the location and scale parameters respectively.

7.10.1. Calculation of Probabilities: Use of Tables

It is convenient to deal with standard normal distribution. It is also possible to transform a $N(\mu, \sigma)$ variable to a standard normal variable by using a suitable and simple transformation.

Suppose that X is a $N(\mu, \sigma)$, so that $E(X) = \mu$ and var $(X) = \sigma^2$. Transform X to another random variable Z such that

$$Z = \frac{X - \mu}{\sigma} \quad (\mu, \sigma \text{ are constants});$$

then Z has also a normal distribution.
Using properties of expectations, we get

$$E(Z) = E\left[\frac{1}{\sigma}X - \frac{\mu}{\sigma}\right] = E\left[\frac{X}{\sigma}\right] - E\left[\frac{\mu}{\sigma}\right]$$

$$= \frac{E(X)}{\sigma} - \frac{\mu}{\sigma} = \frac{\mu}{\sigma} - \frac{\mu}{\sigma} = 0$$

and $$\text{var}(Z) = \text{var}\left(\frac{1}{\sigma}X - \frac{\mu}{\sigma}\right)$$

$$= \frac{1}{\sigma^2}\text{var}(X) = \frac{\sigma^2}{\sigma^2} = 1$$

Thus Z is a $N(0, 1)$.

From $$Z = \frac{X-\mu}{\sigma}, \text{ we get } X = \mu + \sigma Z,$$

so that

$$P(a < X < b) = P(a < \mu + \sigma Z < b)$$

$$= P\left(\frac{a-\mu}{\sigma} < Z < \frac{b-\mu}{\sigma}\right)$$

Thus from the tables of probabilities for Z, probabilities for X can be calculated. To illustrate the point, we take X to be $N(100, 10)$.

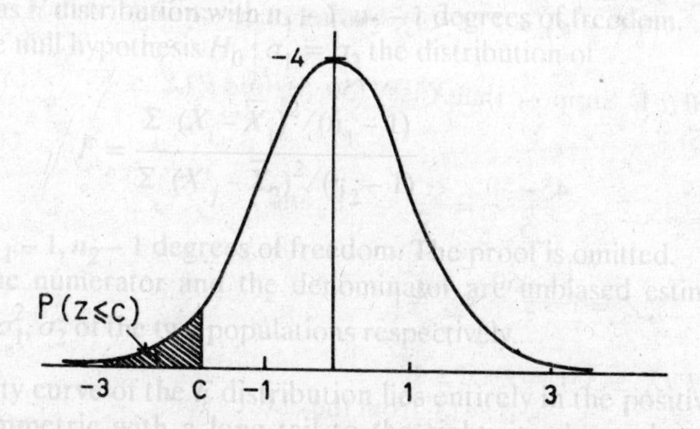

Fig 7.8 P ($Z \leq c$) (Shaded portion) where Z is standard normal variate.

Then $Z = \dfrac{X-100}{10}$ is $N(0, 1)$.

Suppose that we are required to find

(i) $P(X < 130)$, (ii) $P(X > 120)$, (iii) $P(120 < X < 130)$.

(i) When $X = 130$, $Z = \dfrac{130 - 100}{10} = 3$

$$P(X < 130) = P(Z < 3)$$

$$= 0.9987, \text{ from tables.}$$

(ii) When $X = 120, Z = \dfrac{120 - 100}{10} = 2$

$$
\begin{aligned}
P(X > 120) &= P(Z > 2)\\
&= 1 - P(Z < 2)\\
&= 1 - 0.9772\\
&= 0.0228, \text{ from tables}
\end{aligned}
$$

$$
\begin{aligned}
\text{(iii) } P(120 < X < 130) &= P(2 < Z < 3)\\
&= P(Z < 3) - P(Z < 2)\\
&= 0.9987 - 0.9772\\
&= 0.0215
\end{aligned}
$$

Thus from the tables of probabilities $P(Z < c)$, where Z is $N(0, 1)$ we can find the probabilities of X (where X is $N(\mu, \sigma)$) lying in any interval.
We give below an extract of values of $P(Z < c)$ for same values of c.

Values c :	−3.5	−3	−2	−1	0	1	2	3	3.5
$P(Z < c)$:	0.0002	0.0013	0.0228	0.1587	0.5000	0.8413	0.9772	0.9987	0.9998

Values of $P(Z < c)$ for c less than -3.5 is very very small, and is about 0, while for c greater than 3.5, $P(Z > c)$ is about 1. Tables give values of $P(Z < c)$ for a very large range of values of c between $c = -3.5$ to $c = 3.5$.

Note : For a continuous r.v. X, $P(X = a) = 0$ so that $P(X \geq a) = P(X > a)$ and $P(X \leq a) = P(X < a)$, but not necessarily so when X is discrete.

Example 10 (a). Suppose that X is $N(50, 5)$, find $P(X \leq 55)$, $P(X > 45)$ and $P(X < 55)$.

When $X = 45$, $Z = \dfrac{45-50}{5} = \dfrac{-5}{5} = -1$, and

when $X = 55$, $Z = \dfrac{55-50}{5} = \dfrac{5}{5} = 1$.

$$
\begin{aligned}
\text{Thus } \quad P(X \leq 55) &= P(Z \leq 1)\\
&= 0.8413, \text{ from the above table.}\\
P(X \leq 45) &= 1 - P(Z < -1)\\
&= 1 - 0.1587, \text{ from the above table}\\
&= 0.8413
\end{aligned}
$$

$$
\begin{aligned}
P(45 < X \leq 55) &= P(-1 < Z \leq 1)\\
&= P(Z < 1) - P(Z < -1)\\
&= 0.8413 - 0.1587, \text{ from the above table}\\
&= 0.6826
\end{aligned}
$$

$$
\begin{aligned}
\text{Note that } P(Z \leq -1) &= 0.1587\\
\text{and} \quad P(Z > 1) &= 1 - P(Z \leq 1) = 1 - 0.8413\\
&= 0.1587
\end{aligned}
$$

Thus $P(Z \leq -1) = P(Z > 1)$

In other words, if c is positive, then

$$P(Z < -c) = P(Z > c).$$

Example 10 (b). Find $P(1.2 < X < 12.5)$ when X is $N(10, 5)$.

When $X = 12.5, Z = \frac{12.5-10}{5} = 0.5$

and when $X = 1.2, Z = \frac{1.2-10}{5} = -1.76.$

We have

$P(X < 12.5) = P(Z < 0.5)$

$= 0.6915$ from the table given in Appendix

$P(X < 1.2) = P(Z < -1.76)$

$= 0.0392$ from the table given in Appendix

$$\begin{aligned} P(1.2 < X < 12.5) &= P(-1.76 < Z < 0.5) \\ &= P(Z < 0.5) - P(Z < -1.76) \\ &= 0.6915 - 0.0392 \\ &= 0.6523. \end{aligned}$$

Example 10 (c). The height X of students of a large College is found to have normal distribution with mean 162.50 cm and s.d. 6 cm. Find the probability that a student selected at random will have

(i) height greater than 168 cm.

(ii) height less than or equal to 150 cm.

(iii) height between 150 cm and 168 cm.

Here X is $N(162.5, 6)$

When $X = 168$ cm, $Z = \frac{168-162.5}{6} = \frac{5.5}{6} = 0.91;$

when $X = 150$ cm, $Z = \frac{150-162.5}{6} = -\frac{12.5}{6} = -2.08$

$$\begin{aligned} P(X > 168) &= P(Z > 0.91) \\ &= 1 - P(Z < 0.91) \\ &= 1 - 0.8186 \\ &= 0.1814 \end{aligned}$$

$$\begin{aligned} P(X \leq 150) &= P(Z \leq -2.08) \\ &= 0.0188 \end{aligned}$$

$$P(150 < X < 168) = P(-2.08 < Z < 0.91)$$
$$= P(Z < 0.91) - P(Z < -2.08)$$
$$= 0.8186 - 0.0188 = 0.7998$$

Interpreting probability as long run relative frquency, we can conclude as follows:

If the frequency distribution of heights of students approximate a normal probability distribution with mean 162.5 cm and s.d. 6 cm then the percentage of students with heights

(i) greater than 168 cm is 18.14

(ii) less than or equal to 150 cm is 1.88

(iii) between 150 cm and 168 cm is 79.88

Note: The mean and the standard deviation are given in the same unit: centimetre in this case. Therefore the quantity Z is a pure number, for example, when $X = 168$ cm, $Z = \frac{(168-162.5) \text{ cm}}{6 \text{ cm}} = \frac{5.5}{6}$; for simplicity, we have written $Z = \frac{168-162.5}{6}$, omitting the units. This is done in all such examples dealing with units.

Another point to note that while the standard deviation is in the same unit as the mean, its square, the variance is *not* in the same unit as the mean.

Example 10 (d). A machine produces small metallic balls (used for ball bearing); the specification is 3 gms in weight. However, the balls produced do not all have this standard weight, but the weights of the large number of balls produced are found to have a normal distribution with mean 3 gms and s.d. 0.1 gm. Out of the total number of balls produced, what proportion of balls will have weight (i) less than or equal to 3.1 gms., (ii) more than 3.1 gms, (iii) between 2.8 and 3.1 gms?

Here X is the weight of the balls produced:

(i) When $X = 3.1$, $Z = \frac{3.1-3}{0.1} = 1$, so that

$$P(X < 3.1) = P(Z < 1) = 0.8413$$

i.e. 84.13% of the balls will have weight less than 3.1 gm.

(ii) $P(X > 3.1) = P(Z > 1)$

$$= 1 - P(Z < 1) = 1 - 0.8413 = 0.1587$$

i.e. 15.87% of the balls will have weight more than 3.1 gm.

(iii) When $X = 2.8$, $Z = \frac{2.8-3}{0.1} = \frac{-0.2}{0.1} = -2$

$$P(X < 2.8) = P(Z < -2) = 0.0228$$

$$P(2.8 < X < 3.1) = P(-2 < Z < 1) = P(Z < 1) - P(Z < -2)$$
$$= 0.8413 - 0.0228$$
$$= 0.8185$$

i.e., 81.85% of the balls will be between 2.8 gm and 3.1 gm in weight.

7.10.2. Properties of Normal Distribution

We sum up the characteristic properties of normal distribution as follows:

(1) It is a continuous distribution or distribution of a continuous random variable X.

(2) It is bell shaped; the normal frequency (probability density) curve rises gradually to a maximum and then decreases in a similar manner. It is symmetric in shape about its mean.

(3) The curve is *asymptotic*, i.e. it gets closer and closer to X-axis on both sides but never actually touches it.

(4) The probability density function $f(x)$ of X has the form

$$f(x) = \frac{1}{\sigma\sqrt{2\pi}} \exp\left(-\frac{(x-\mu)^2}{2\sigma^2}\right)$$

(5) The distribution involves two parameters μ and σ.
The mean of X is μ and the variance of X is σ^2.
For every pair of values of (μ, σ) we get a normal distribution and a corresponding curve. The shape of every curve, though bell-shaped and symmetrical, will vary with the values of μ and σ.

(6) The probabilities that X lies in the interval around the mean μ are given by

(i) $P(\mu - \sigma < X < \mu + \sigma) = 0.683$

i.e probability that X lies within one time s.d. from each side of mean is 0.683.

(ii) $P(\mu - 2\sigma < X < \mu + 2\sigma) = 0.954$

i.e. probability that X lies within two times s.d. from each side of mean is 0.954.

(iii) $P(\mu - 3\sigma < X < \mu + 3\sigma) = 0.997$

i.e. probability that X lies within three times s.d. from each side of mean is 0.997.

In other words, 68.3% of values of X fall between the mean and plus and minus s.d; 95.6% between the mean and plus and minus two times s.d., and 99.7% between the mean and plus and minus three times s.d. See Fig. 7.9.

(7) The effect in the change in the mean from μ to a higher value μ_1 (with the s.d. remaining the same) is that the curve is shifted from the centre μ to the new centre to the right at μ_1, with no change in the shape of the curve.

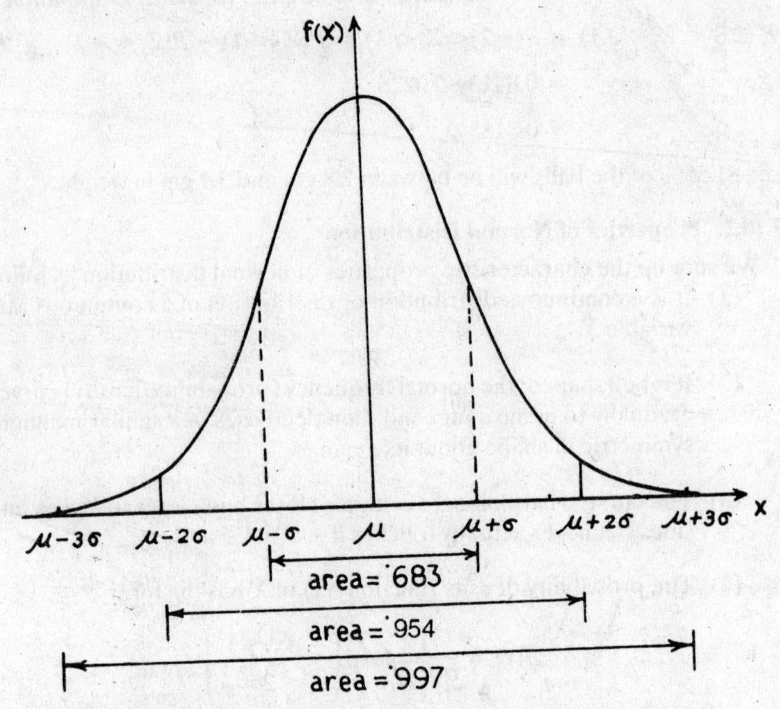

Fig. 7.9. Probability density function of normal distribution with mean μ and s.d. σ

(8) The effect in the change of the s.d. from σ to a higher value σ_1 (with mean remaining the same) is a change in the shape of the curve with centre remaining at the same point (mean). The maximum height decreases as s.d. increases.

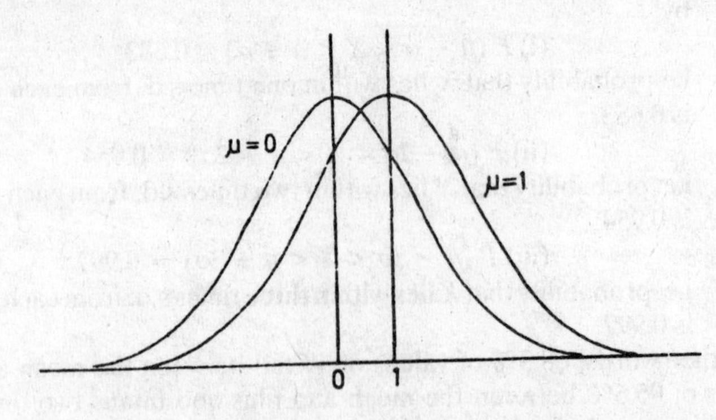

Fig. 7.10. Probability density functions of normal distribution with means 0 and 1 and the same s.d.

(9) The mean, the median and the mode coincide in case of the normal curve (as also for any symmetric curve). Symmetry means that there is

no skewness and so the coefficient of skewness is zero for a normal curve. The coefficient of kurtosis is also zero.

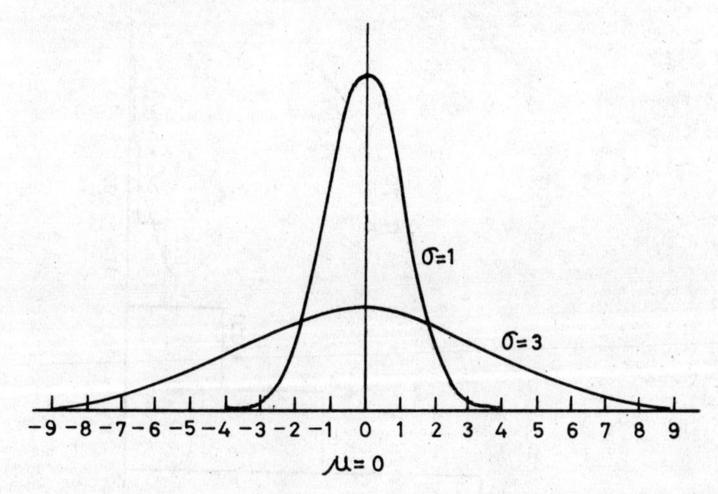

Fig. 7.11. Probability density functions of normal distribution with the same mean 0 and s.d. $\sigma = 1$ and $\sigma = 3$

(10) If X is normally distributed with mean μ and s.d. σ,

then $Z = \dfrac{X-\mu}{\sigma}$ is normally distributed with mean 0 and s.d. 1.

In other words, if X is $N(\mu, \sigma)$ then $Z = \dfrac{X-\mu}{\sigma}$ is standard normal $N(0, 1)$.

7.10.3. Importance of Normal Distribution

The normal distribution was obtained in 1733 by a French mathematician De Moivre, as the limiting form of binomial distribution. Without going to the mathematical derivation of (normal distribution as the limiting form of bino-

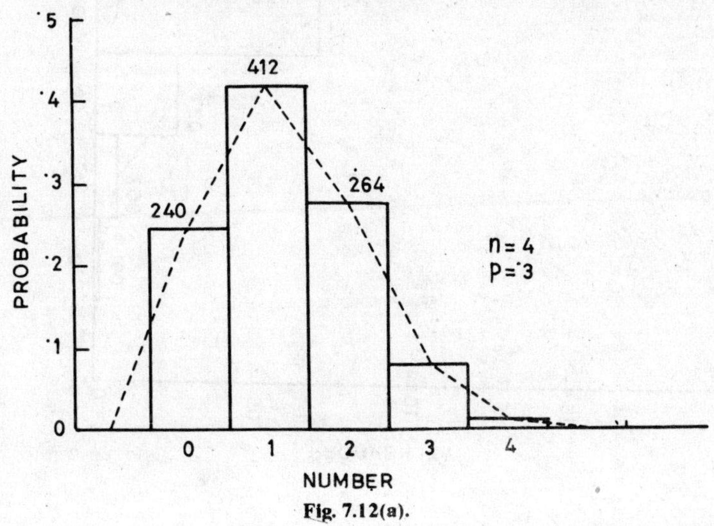

Fig. 7.12(a).

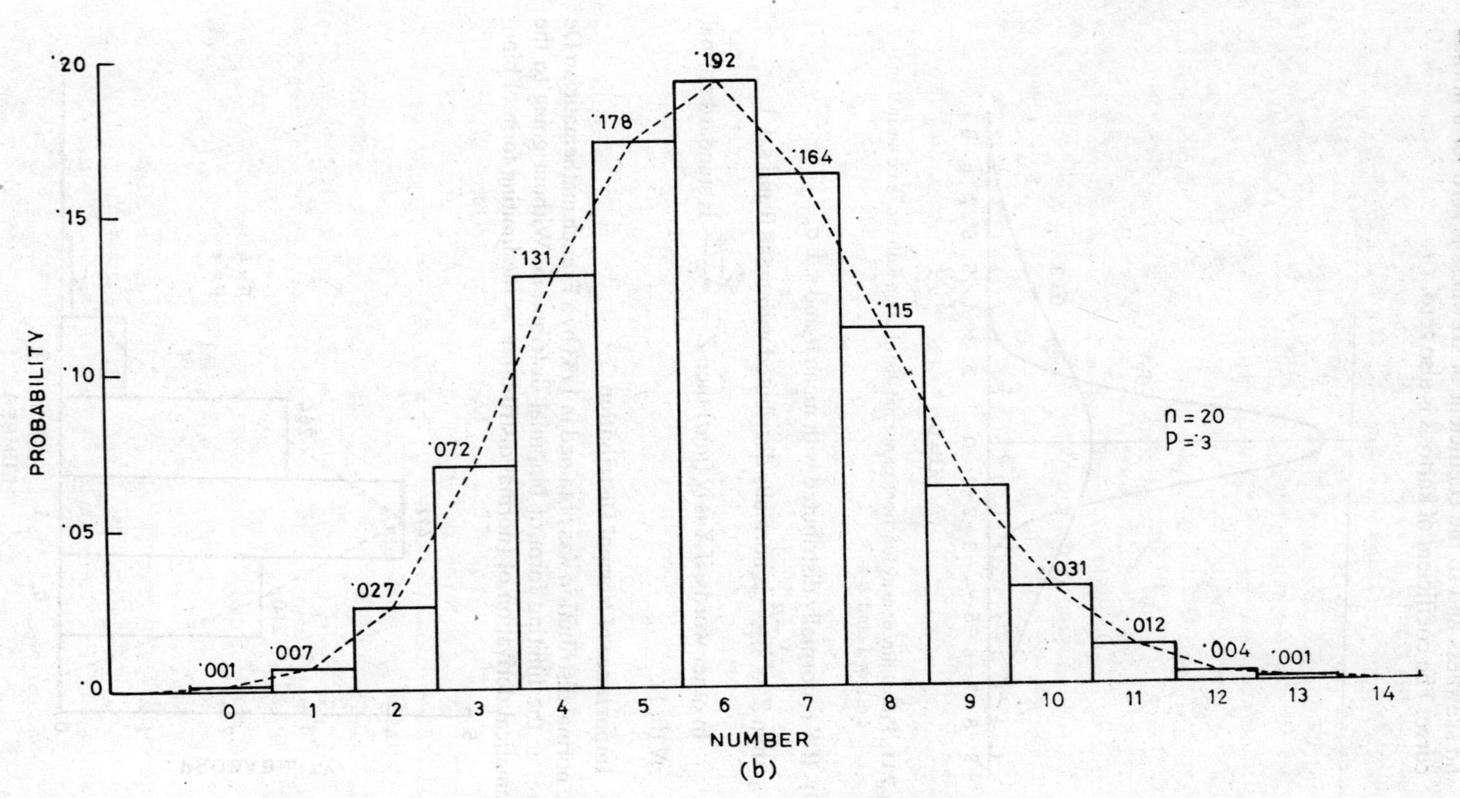

Fig. 7.12. Probability Histogram for Binomial Distribution (a) $n = 4, p = 0.3$ (b) $n = 20, p = 0.3$ [Dotted line Graph for Probability Frequency Polygon]

mial) we can see that as n increases while p is kept fixed, the shape of binomial distribution will resemble that of a normal distribution. The two figures of binomial distribution given below may be observed (see Figs. 7.12 (a) and (b)).

Poisson distribution also tends to normal distribution as λ, the parameter of Poisson distribution increases indefinitely.

De Moivre showed mathematically that as $n - \infty$, p is kept fixed, the probability distribution of binomial will tend towards that of the normal. He obtained normal distribution mathematically. The proof is beyond the scope of this book.

De Moivre's discovery (in 1733) seems to have attracted little attention until long afterwards; in 1809, Carl Gauss and later Laplace both obtained normal distribution while dealing with the theory of errors of measurements: They found that the distribution of errors of physical measurements is normal. When the same or similar thing is measured by an experimenter or a group of experimenters, however, cautious the experimenter might be, there will be errors (difference between observed and actual) of measurements. These errors are found to obey normal distribution.

We can think of the situation like this. A very large number of observations are taken, and errors noted in each case. A frequency distribution is obtained of the data (of errors). The shape of the frequency distribution will have a shape very similar to that of a normal distribution: A mathematical explanation of why a normal distribution is obtained in a situation dealing with errors of measurements is provided by a theoretical result–the Central Limit Theorem: this was first attempted by Laplace.

The point is that the errors (deviations of measurements from the mean or expected) of physical, astronomical or other types of observations were found to obey the normal law, and hence the normal law is also called 'normal law of error.'

Under the influence of Gauss and Laplace, it was, for a fairly long time, regarded as an axiom that statistical (frequency) distributions of *practically* all kinds would approach the normal distribution, if a sufficiently large number of observations were considered. Even biological measurements (e.g., heights of a large number of individuals) were found to approach the normal law. In fact, it was considered that every kind of real-life data would conform to the bell shaped normal curve. Frequency distribution of real life data do conform to the pattern of normal distributions very often. Because of this, the distribution came to be known as *normal distribution*.

Though this belief prevailed for a very long time, it was found later that normal distribution would not give a satisfactory 'fit' in every situation. Still, the normal distribution is found to be reasonably satisfactory in dealing with several kinds of data.

We observe that normal distribution is encountered from two different directions: the distribution or the law can be mathematically derived, as was first done by De Moivre, from purely *theoretical considerations of probability*; it can as well be derived from *experimental facts* (e.g. errors of observations of measurements), as was first done by Gauss. As observed by Lippman "everybody believes in the law of errors (normal law), the experimenters because they think it is a mathematical theorem, the mathematicians because they think it is an experimental fact."

We find that not only binomial, but many other probability distributions have normal distributions as limiting forms. Again it will be seen that a large number of sampling distributions tend to normal distribution as the sample size increases. Thus, in a large number of situations, under certain conditions, the three types of distributions: frequency distribution, probability distribution and sampling distribution tend to normal distribution.

Apart from the above considerations of applicability and derivability of the normal law, there is another aspect of importance of this distribution. This particular distribution plays a central role in statistical analysis. Inference procedures in statistics lean heavily on this distribution. We shall see this later when we study some inference procedures. In view of these facts, it is said that the normal distribution plays a key and important role in statistics: from descriptive statistics to statistical inference. This distribution, in a way forms the backbone of the current methodology of statistical analysis. The distribution has a wide range of applications in various areas of statistics.

EXERCISES-7

Sections 7.1 & 7.2

1. Four coins are tossed. Find the probability distribution of the number of tails. Verify that the total probability is 1. Find also the mean number of heads.
2. A biased coin, having probability equal to 0.6 of showing head is tossed three times. Find the probability distribution of the number of heads, its expected value. Draw the probability histogram.
3. A random variable X has uniform distribution over the integers 1, 2, . . . , 10. i.e. X assumes each of the values 1, 2, . . . , 10, with equal probability. Find the expected values of X and X^2. Hence, find the variance of X.
4. Two dice are thrown. If X is the sum of the numbers shown on the faces of the two dice, find $E(X)$.
5. Consider Exercise 3. If $Y = 2X + 5$ find $E(Y)$ and var (Y).
6. Consider Exercise 4. If $Z = \frac{X}{2}$ find var (Z).
7. The random variable X has the distribution:

Value of X	−1	0	1
Probability	0.5	0.3	0.2

   Find $E(X)$, var (X). If $Z = \frac{X-2}{3}$ find the distribution of Z, $E(Z)$ and var (Z).
8. Two dice are thrown. If getting a number more than 4 is considered as success, obtain the probability distribution of the number of successes and its expected value.
9. Two cards are drawn at random from a well-shuffled pack of 52 cards. Find the probability distribution of the number of spade cards drawn, if the drawings are (i) with replacement, (ii) without replacement.
10. An urn contains 4 white and 6 black balls. Three balls are drawn at random (i) with replacement, (ii) without replacement. Find the distribution of the number of white balls drawn.
11. The probability that there is at least one error in the preparation of a statement of accounts by three accountants A, B, C are 0.1, 0.25 and 0.3 respectively. Find the expected number of correct statements.

Hint: Let X_1, X_2, X_3 be the number of correct statements prepared by A, B, C respectively. X_1 takes two values 0 with probability 0.1 and 1 with probability $(1 - 0.1) = 0.9$, so that

$E(X_1) = 0 \times (0.1) + 1 \times 0.9 = 0.9$

Similarly

$E(X_2) = 0 \times (0.25) + 1 \times (1 - 0.25) = 0.75$

$E(X_3) = 0 \times (0.3) + 1 \times (1 - 0.3) = 0.7$

Thus the required expectation = $E(X_1)+E(X_2)+E(X_3)$
= 0.9 + 0.75 + 0.7 = 2.35 (out of a total of 3, with 1 each).

12. Two dice are thrown. Let X, Y be the numbers shown on the two dice. Find the distribution of the random variable $Z = X-Y$. Find also $E(Z)$ and var (Z).

Section 7.3

13. Let X be a continuous random variable with *p.d.f.* $f(x) = k$ for $1 < x < 3$ Find (*i*) k, (ii) $P(1 < x < 2)$, (iii) $P(X < 3)$, (iv) $P(X > 2)$.
14. The daily demand X for a certain commodity has the p.d.f. $F(x)$ given by
$f(x) = k, 100 \le x \le 200$
Find (i) k, (ii) $P(X < 100)$, (iii) $P(X > 200)$ (iv) $P(100 \le x \le 125)$
15. The p.d.f. $f(x)$ of a continuous random variable is given by

$$f(x) = \frac{1}{2}, \; 0 \le x \le 1$$
$$= \frac{1}{4}, \; 1 < x \le 2$$
$$= \frac{1}{4}, \; 2 < x < 4$$

Draw the frequency curve for X
Find (i) $P(0 < X < 3)$, (ii) $P(0 < X < 4)$, (iii) $P(X > 4)$.

Sections 7.4 & 7.5

16. What do you mean by a Bernoulli trial? An urn contains 2 black and 3 white balls. Two balls are drawn *without replacement.* Could you call the trials (of drawing balls) Bernoulli trials? If not, why not?
17. X is a Bernoulli random variable assuming only two values 0 and 1 and $P(X = 1) = \frac{1}{10}$.
Find $E(X)$ and var (X).
18. (a) Consider a sequence of 4 Bernoulli trials with probability of success $\frac{1}{3}$ at each trial. Find the distribution of the number of successes X.
(b) The probability that a sales programme results in success is 0.6 and the probability of its failure is 0.4. Success brings a profit of Rs. 1,000 and failure a loss of Rs. 300. Find the expectation of gain (profit or loss).

Section 7.6

19. Six fair (unbiased) coins are tossed simultaneously. Find the probability of getting
(i) exactly 6 heads
(ii) no head
(iii) at least 3 heads
(iv) not more than 3 heads.
20. A lot of material contains 10% defective items; 4 items are drawn with replacement. Find the probability of getting (i) exactly 2 defective items (ii) more than 2 defective items (iii) not more than 2 defective items (iv) no defective item. Find the mean and variance of the number of defective items drawn.
21. A die is tossed five times. Getting less than 4 is considered a failure. Find the probability of getting (i) exactly 3 successes, (ii) no success, (iii) more than 3 successes.
22. Let X be a binomial random variable with mean 1 and variance $\frac{3}{4}$. Find (i) $P(X < 5)$ (ii) $P(X < 2)$, (iii) $P(X = 0)$.
23. Can you have a binomial distribution with (i) mean 1 and variance 2 (ii) mean 1 and variance $\frac{8}{9}$? If your answer is no, explain why not. If your answer is yes, find the parameters of the distribution.
24. A binomial distribution has mean 2 and s.d. $\frac{2}{\sqrt{3}}$. Write out the distribution completely. Find its skewness and kurtosis.

Section 7.7

25. An urn contains 4 white and 6 black balls. Three balls are drawn at random without replacement. Find the distribution of the number of black balls drawn. Also find its mean and variance.
26. A lot of 100 contains 5 defective bolts. 3 bolts are drawn at random from the lot without replacement. Find the probability of getting (i) no defective bolt, (ii) two defective bolts (iii) at most one defective bolt.
27. A box contains 6 fuses, of which 2 are blown out. If three fuses are drawn at random without replacement, what is the probability of getting (i) no defective fuse (ii) at least one defective fuse (iii) all the three fuses defective?
    Find the mean number of defective fuses drawn.
28. A key ring contains 5 similar keys, of which exactly one fits the almirah. If the keys are tried successively *without replacement*, what is the probability that the almirah will be opened at (i) the third trial (ii) at most in 3 trials?
29. A box contains 1000 items, of which 10 are defective. Five items are drawn at random without replacement. Using proper approximation find the probability that the number of defective items drawn is at most 1. Find also the expected number of defective items drawn.
30. A box contains 30 items, of which 20 are good, 5 damaged but repairable and 4 not repairable. 5 screws are drawn at random without replacement. Find the probability that there will be

    (i) 2 good, 2 repairable and 1 not repairable screws,
    (ii) at most 3 good screws, and
    (iii) at least 1 not repairable screw.

Section 7.8

31. The probability of failure of an equipment when it is tested is equal to 0.1. Find the probability that the equipment fails for the first time (i) in 4th trial (ii) in less than 4 trials (tests or trials are assumed to be independent, i.e. one test does not affect subsequent tests)
32. A certain programme has the probability of success $\frac{1}{3}$ and the probability of failure $\frac{2}{3}$.
    Programmes assumed to be independent, can be made any number of times.
    The profit associated with success is Rs. 1,000 and the loss associated with failure is Rs. 4,000. Find the probability that first success occurs in the 3rd trial. Find the expected cost of continuing the programmes till the first success occurs.
33. A tries to contact a particular person P on a phone which serves many other persons. The probability that A gets P on any call is 0.2. A goes on trying till he gets P. If A has to pay Re. 1 for each call, what is the probability that A gets P by spending at most Rs. 5 ? What is the expected cost if A goes on calling as many times as necessary to get P?
34 (a) If X has geometric distribution
    (i) with mean 4; find p
    (ii) with variance 20; find p.
    (b) The probability that a person hits a target with a single shot is 0.2. He goes in trying till he hits the target. Find the probability that he hits the target for the first time in the 4th shot. Find the mean and variance of the number of shots needed to hit the target for the first time.

Section 7.9

35. Assume that X has Poisson distribution with mean 5.
    Find (i) $P(X = 2)$, (ii) $P(X < 2)$, (iii) $P(X > 2)$ (iv) $P(1 < X < 3)$.
36. Suppose that X has Poisson distribution such that $P(X = 0) = 0.050$. Find (i) $E(X)$. (ii) var (X), (iii) $P(X > 0)$ (iv) $P(X < 2)$ (v) $P(4 < X < 6)$.
37. Of the total weekly produce of an item, the number of defective items is found to have Poisson distribution with mean 2. Find the probability that the produce for a certain week does (i) not have any defective item, (ii) at most 1 defective item.
38. From long experience it is found that the daily number of absentees in a factory follows Poisson distribution with mean 5. Find the probability that during a particular day. (i) there is no absentee (ii) there are more than 5 absentees.
39. The number of ticketless travellers detected per checking follows a Poisson distribution with mean 10. Find the probability that during a checking, (i) no ticketless traveller is found (ii) 3 ticketless travellers are found.

40. The number of failures of machines per day in a certain firm has Poisson distribution with mean 1.5. Failures of 2 or less can be handled during the course of the shift, whereas more than 3 failures involve overtime work. Find the probability of needing overtime work on a given day.
41. X has Poisson distribution. Given that
$P(X = 3) = P(X = 2)$, find (i) $E(X)$, (ii) $P(X = 2)$.
42. X has binomial disribution with $n = 100$ and $p = 0.01$; using Poisson approximation, find (i) $P(X = 0)$, (ii) $P(X < 2)$, (iii) $E(X)$.

Section 7.10

43. X has normal distribution with mean 10 and s.d. 3. For what values of a, b, we shall have
(i) $P(a < X < b)$, (ii) $P(a < X < b) = 0.954$
(iii) $P(a < X < b) = 0.987$
Find from the table also
(iv) $P(X < 17.5)$
(v) $P(X > 12)$
(vi) $P(12 < X < 18)$
44. The weights X of students are found to have normal distribution with mean 56 kg and s.d. 2 kg. Find the probability that a student selected at random will have weight.
(i) greater than 54 kg
(ii) less than 52 kg
(iii) between 50 and 60 kg.
45. The scores of the performance of students in a certain subject in a large public examination are approximately normal with mean 55 and s.d. 10. Find the percentage of students with scores
(i) greater than 80
(ii) between 45 and 70
(iii) less than 35.
If students getting less than 35 fail, and there are 10,000 students, how many would fail? If those securing 80 or above get distinction in the subject, how many are expected to qualify for distinction?
46. The contents of a certain beverage in a can are stated to be approximately 250 gm; but the distribution of weights of contents is N (250 gm). What percentage of the cans will have contents
(i) greater than 250 gm (ii) less than 250 gm (iii) between 240 gm and 260 gm ?
47. Find the mean and s.d. of the normal variate X if the values of Z (where Z is $N(0, 1)$) corresponding $X = 90$ and $X = 80$ are $Z = -1.5$ and $Z = -2.5$ respectively.
48. Find μ and σ for X which is $N(\mu, \sigma)$ given that $P(X < 10) = 0.0062$ and $P(X > 16) = 0.3085$

Hints: $0.0062 = P(Z < -2.5)$ and $1 - 0.3085 = 0.6915 = P(Z < 0.5)$
Thus $X = 10$ corresponds to $Z = 2.5$ and $X = 16$ corresponds to $Z = 0.5$. We get $\mu = 15, \sigma = 2$

49. The weight of ball bearings produced in a certain factory are normally distributed with mean, 0.50 newtons and s.d. 0.002 newtons. Find the percentage of ball bearings produced with weights
(i) less than 0.498 newtons (ii) more than 0.501 newtons (iii) between 0.4995 and 0.5005 newtons.
50. The heights X of 500 soldiers are found to have normal distribution. Of them 258 are found to be within 2 cm. of the mean height of 170 cm. Find the s.d. of X.

Hints: $P(170 - 2 < X < 170 + 2) = \dfrac{258}{500} = 0.516$

and $P(168 < X < 172) = \dfrac{.516}{2} = 0.258$

$X = 168$ corresponds to $Z = 0$; suppose $X = 172$ corresponds to $Z = a$, then $P(0 < Z < a) = 0.258$. From tables we find $a = 0.7$
Thus $X = 172$ corresponds to $Z = 0.7$

as $Z = \dfrac{X-\mu}{\sigma}$, we get $0.7 = \dfrac{172-170}{\sigma} = \dfrac{2}{\sigma}$

so that $\sigma = \dfrac{2}{0.7} = 2.857$ cm.

Chapter 8

Elements of Sampling Theory

8.1. INTRODUCTION

We have considered in the earlier chapters frequency distribution and (exact) probability distribution.

Let us consider the population of students in a big college. If we take measurements, say, of the height, of 100 students, we can form a frequency distribution, and then calculate from it, the mean, s.d. and other characteristics.

We get *one* sample of 100 students and the mean, s.d. calculated are of this particular sample: these are called *statistics,* while the mean, s.d. of the population are *population mean* and *population s.d.* It should be intuitively clear that a very large sample would approach the actual population, and then sample mean would approach the population mean. If another investigator takes a sample of 100 students (by some valid method) he would not get the same 100 students: from the measurements of heights of these 100 students, a frequency distribution can be constructed and mean, s.d. etc. can be calculated. We do not expect these two samples of 100 each to be exactly the same: as such the two frequency distributions and their mean, s.d. are *not* expected to be the same. Each such sample, of 100 students is generally taken with a view to find out the characteristics of the whole population of students. That is, we wish to *infer*, on the basis of the results obtained from the sample, the corresponding characteristic of the population. For example, we may be interested in knowing what is the mean height of students of the whole population. Then we would try to draw some inference of the *population* mean from the sample mean. It is easily understandable that if the inference about the population is to be done on the basis of the sample, the sample must conform to certain criteria: the sample must be *representative.* If we are interested in the mean age of the whole population of college students (of all the 6 or 7 classes) would a sample of students taken from the first year class be good enough? Our conclusion about the average age of student of the whole population from that obtained from the sample would be suspect, as the sample confined to 1st year class would not be representative.

The question arises as to what is a representative sample and how such a sample can be selected from a population. One method by which a representative sample may be obtained from a population, is by a process called *random sampling.* We shall discuss this process later.

Again, let us come to the question of drawing two samples. The mean and s.d. obtained from *two* different samples of the same population would not be the same, even if both the samples are representative ones and drawn by valid methods. The question arises: whether the difference between the results of the two samples is really *significant* or are they due to *chance variation* of selecting these particular samples.

These questions are sought to be answered through what is known as *sampling theory*.

We shall first discuss what is a representative sample and also explain how this is taken. We shall also try to see how, or in what way, results vary, if two or more samples of the same population are taken.

8.2. SAMPLING WITH AND WITHOUT REPLACEMENT

Consider all possible (representative) samples of size n that can be drawn from a population. We can calculate a statistic (such as the mean) from each sample, and get a number of sample means, from which we get a distribution of this statistic. This distribution will be known as the *sampling distribution* of the *statistic*. If the statistic is the sample mean we get a *sampling distribution* of the mean. Similarly we get sampling distribution of the variance, or of the median, or of proportion etc. etc. See also section 8.6 for a general definition.

Sampling with and without replacement

Suppose that an urn contains similar balls numbered 1, 2, 3, 4. If we draw a ball, a ball with one of these numbers marked will be obtained. If we wish to draw a ball second time, we may do it in two ways: (1) We may replace the ball drawn in the first time and then draw a ball; we would have the same 4 balls in the urn. The drawing is said to be *with replacement. (2) We may not* replace the ball drawn in the first time; we would have 3 balls instead of 4 and one would be missing (which could be any one of them 1, 2, 3, or 4). The drawing is then said to be *without replacement.*

8.3. SAMPLING DISTRIBUTION OF THE SAMPLE MEAN: SAMPLING WITH REPLACEMENT

Consider the following example.

Example 3 (a). A population consists of 3 balls numbered 1, 2, 3. (Our interest is on the number shown on the balls). Find the population mean and the population variance.

List the possible samples of size 2 that can be drawn from the population *with replacement,* find the sample means and construct the sampling distribution of the (sample) means. Also find the mean and variance of the sampling distribution of means (also call them respectively the mean and variance of the sample means).

Solution: Denote the population mean by μ and population variance by σ^2.

We have $$\mu = \frac{1+2+3}{3} = 2$$

$$\sigma^2 = \frac{(1-2)^2 + (2-2)^2 + (3-2)^2}{3} = \frac{1+1}{3} = \frac{2}{3}$$

Samples of size 2 are:

(1, 1)	(1, 2)	(1, 3)
(2, 1)	(2, 2)	(2, 3)
(3, 1)	(3, 2)	(3, 3)

There are in all 3 × 3 = 9 possible samples of size 2; these constitute the population of samples of size 2.
The corresponding sample means are :

1	1.5	2
1.5	2	2.5
2	2.5	3

The distribution of sample means is as given below :

Sample mean	1	1.5	2	2.5	3	Total
Frequency	1	2	3	2	1	9

The mean of the sampling distribution of means (mean of the sample means) is given by

$$\mu_{\bar{x}} = \frac{1 \times 1 + 1.5 \times 2 + 2 \times 3 + 2.5 \times 2 + 3 \times 1}{9}$$

$$= \frac{18}{9} = 2$$

and the variance of the sampling distribution of means is given by

$$\sigma_{\bar{x}}^2 = \frac{(1-2)^2 \cdot 1 + (1.5-2)^2 \cdot 2 + (2-2)^2 \cdot 3 + (2.5-2)^2 \cdot 2 + (3-2)^2 \cdot 1}{9}$$

$$= \frac{3}{9} = \frac{1}{3} = \frac{1}{2}\sigma^2$$

We find that $\mu_{\bar{x}} = \mu$ i.e. the population mean is equal to the mean of the sample means and that $\sigma_{\bar{x}}^2 = \frac{\sigma^2}{2}$, where σ^2 is the population variance. The sampling distribution of means have the same mean as the population but its variance $\frac{\sigma^2}{2}$ is smaller than population variance σ^2; it is the value obtained by dividing the population variance σ^2 by the sample size 2. Because of smaller variance, the sample means will cluster round the mean and will not be as dispersed as the population values.

Example 3 (b). **Sampling *without* replacement.** Consider the above example taking sampling without replacement, i.e. the element drawn in the first draw is taken out and is not replaced.

There can be $3 \times 2 = 6$ samples of size 2.
Samples of size 2 that can be drawn are:

(1, 2) (1, 3) (2,1) (2, 3) (3,1) (3,2)

The corresponding sample means are

1.5 2 1.5 2.5 2 2.5

The mean of the sampling distribution of means (mean of the sample means)

$$\mu_{\overline{X}} = \frac{1.5 + 2 + 1.5 + 2.5 + 2 + 2.5}{6} = 2$$

and the variance of the sampling distributions of means (variance of the sample means) is given by

$$\sigma^2_{\overline{X}} = \frac{2 \times (1.5-2)^2 + 2 \times (2.5-2)^2}{6} = \frac{1}{6}.$$

We find that the two means are equal, i.e. the mean $\mu_{\overline{X}}$ of the sample means equals the population mean μ.

Now let us consider sampling from a probability distribution: Consider that a random variable X is uniformly distributed and assumes values $k = 1, 2, 3$ each with equal probability i.e. $P(X = k) = \frac{1}{3}, k = 1, 2, 3$. Suppose that sample of size 2 is taken from the distribution (*with* replacement). We shall consider the sampling distribution of the sample mean.

When an observation or sample is taken, this observation (call it X_1) could be any one of the numbers 1, 2, 3 and in a random sample each of the numbers is equally likely to occur. That is, the observation or sample can be considered as a random variable having the same distribution as the random variable X, i.e. $P(X_1 = k) = \frac{1}{3}, k = 1, 2, 3.$

Similarly, if a second sample is drawn (with replacement of the first sample), this sample X_2 can also be considered as a random variable with the same distribution as X, i.e. $P(X_2 = k) = \frac{1}{3}, k = 1, 2, 3$. Consider now the sample mean $\overline{X} = \frac{X_1 + X_2}{2}$; the sample mean is the mean of the two random variables X_1 and X_2. Let us obtain the distributions of X, X_1, X_2, $X_1 + X_2$ and of $\frac{X_1 + X_2}{2}$

Distribution of X

Value of X (k)	1	2	3	Total
Probability $P(X = k)$	$\frac{1}{3}$	$\frac{1}{3}$	$\frac{1}{3}$	1

We have
$$E(X) = 1.\frac{1}{3} + 2.\frac{1}{3} + 3.\frac{1}{3} = 2$$
$$E(X^2) = 1^2.\frac{1}{3} + 2^2.\frac{1}{3} + 3^2.\frac{1}{3} = \frac{14}{3}$$
$$\text{var }(X) = E(X^2) - [E(X)]^2 = \frac{14}{3} - 4 = \frac{2}{3}$$

i.e.,
$$\mu = 2, \sigma^2 = \frac{2}{3}$$

Distributions of X_1, X_2 will be the same as above. (X_1, X_2) assume the pairs of values

$$(1, 1),\ (1, 2),\ (1, 3),\ (2, 1),\ (2, 2),\ (2, 3),\ (3, 1),\ (3, 2) \text{ and } (3, 3),$$

so that $X_1 + X_2$ assumes the values

$$(1+1), (1+2), (1+3), (2+1), (2+2), (2+3), (3+1), (3+2), (3+3)$$

i.e. values 2, 3, 4, 5, 6.
Since X_1, X_2 are independent, we have

$$P(X_1 + X_2 = 2) = P(X_1 = 1, X_2 = 1) = P(X_1 = 1)\, P(X_2 = 1)$$
$$= \frac{1}{3}.\frac{1}{3} = \frac{1}{9}$$
$$P(X_1 + X_2 = 3) = P(X_1 = 1, X_2 = 2) + P(X_1 = 2, X_2 = 1)$$
$$= \frac{1}{9} + \frac{1}{9} = \frac{2}{9}$$
$$P(X_1 + X_2 = 4) = P(X_1 = 1, X_2 = 3) + P(X_1 = 2, X_2 = 2) + P(X_1 = 3, X_2 = 1)$$
$$= \frac{1}{9} + \frac{1}{9} + \frac{1}{9} = \frac{3}{9}$$
$$P(X_1 + X_2 = 5) = P(X_1 = 2, X_2 = 3) + P(X_1 = 3, X_2 = 2)$$
$$= \frac{1}{9} + \frac{1}{9} = \frac{2}{9}$$
$$P(X_1 + X_2 = 6) = P(X_1 = 3, X_2 = 3) = \frac{1}{9}.$$

The distribution $X_1 + X_2$ is as given below:

Values	2	3	4	5	6	Total
Probability	$\frac{1}{9}$	$\frac{2}{9}$	$\frac{3}{9}$	$\frac{2}{9}$	$\frac{1}{9}$	1

The random variable $\frac{X_1 + X_2}{2}$ assumes values

$$\frac{1+1}{2}, \frac{1+2}{2}, \frac{1+3}{2}, \frac{2+1}{2}, \frac{2+2}{2}, \frac{2+3}{2}, \frac{3+1}{2}, \frac{3+2}{2}, \frac{3+3}{2}$$

i.e., values 1, 1.5, 2, 2.5, 3. We have

$$P\left(\frac{X_1+X_2}{2}=k\right) = P(X_1+X_2=2k), k = 1, 1.5, 2, 2.5, 3.$$

$$\text{i.e. } P\left(\frac{X_1+X_2}{2}=1\right) = P(X_1+X_2=2) \text{ and so on.}$$

The distribution of $\frac{X_1+X_2}{2} = \bar{X}$ is given below:

Values of $\bar{X}$:	1	1.5	2	2.5	3	Total
Probability :	$\frac{1}{9}$	$\frac{2}{9}$	$\frac{3}{9}$	$\frac{2}{9}$	$\frac{1}{9}$	1

We have

$$\mu_{\bar{X}} = E(\bar{X}) = 1\times\frac{1}{9} + 1.5\times\frac{2}{9} + 2\times\frac{3}{9} + 2.5\times\frac{2}{9} + 3\times\frac{1}{9}$$

$$= \frac{18}{9} = 2$$

$$E\{(\bar{X})^2\} = 1^2\cdot\frac{1}{9} + (1.5)^2\cdot\frac{2}{9} + (2)^2\cdot\frac{3}{9} + (2.5)^2\cdot\frac{2}{9} + (3)^2\cdot\frac{1}{9}$$

$$= \frac{1+\frac{9}{4}.2+4.3+\frac{25}{4}.2+9}{9} = \frac{39}{9}$$

$$\sigma^2_{\bar{X}} = \text{var}(\bar{X}) = E\{(\bar{X})^2\} - [E(\bar{X})]^2 = \frac{39}{9} - 4 = \frac{3}{9} = \frac{1}{3}.$$

Thus we find that the sampling distribution of mean has mean 2 and variance $\frac{1}{3}$. Now

$$E(\bar{X}) = 2 = \mu, \sigma^2_{\bar{X}} = var\ (\bar{X}) = \frac{1}{3} = \frac{2}{3}\cdot\frac{1}{2} = \frac{\sigma^2}{2}$$

Note: We find from the above and also from Example 3 (*a*) that irrespective of the form of the distribution of X, $E(X) = \mu$, and var $(X) = \frac{\sigma^2}{n}$, n being the sample size.

Now instead of taking samples from a uniform distribution, let us consider taking samples from an arbitrary distribution. This is considered in the example below:

Example 3 (c). Suppose that in a city with a large number of households, counts are taken of the number X of radio (and transistor) sets per household, and that

X is found to have the following distribution :

Value of X (k)	0	1	2	Total
Probability $P(X=k)$	0.4	0.5	0.1	1.0

This implies that the proportions of households with number of sets 0, 1, 2 are 4 : 5 : 1, respectively, or in other words 40% of households have no set, 50% have axactly 1 set and 10% have two sets. Here probability is interpreted as long run frequency ratio.

In this preceding example we have assumed that the distribution is uniform, here we consider a non-uniform distribution, i.e. X assumes the three values 0, 1, 2 with different probabilities.
We have

$$\mu = E(X) = 0 \times 0.4 + 1 \times 0.5 + 2 \times 0.1 = 0.7$$

$$E(X^2) = 0^2 \times 0.4 + 1^2 \times 0.5 + 2^2 \times 0.1 = 0.9$$

$$\text{Var } (X) = E(X^2) - [E(X)]^2 = 0.9 - (0.7)^2 = 0.9 - 0.49 = 0.41$$

Suppose that a sample of size 2 is taken from the households. Each unit of the sample X_1, X_2 has the same distribution as the r.v. X. To find the distribution of $\overline{X} = \dfrac{X_1 + X_2}{2}$, we first find the distribution of $X_1 + X_2$. Noting that samples X_1 and X_2 are independent, we get

$$P(X_1 + X_2 = 0) = P(X_1 = 0, X_2 = 0) = P(X_1 = 0) \,.\, P(X_2 = 0)$$
$$= (0.4)\,(0.4) = 0.16$$

$$P(X_1 + X_2 = 1) = P(X_1 = 1, X_2 = 0) + P(X_1 = 0, X_2 = 1)$$
$$= P(X_1 = 1)\, P(X_2 = 0) + P(X_1 = 0)\, P(X_2 = 1)$$
$$= (0.5)\,(0.4) + (0.4)\,(0.5) = 0.4$$

$$P(X_1 + X_2 = 2) = P(X_1 = 2, X_1 = 0) + P(X_1 = 1, X_2 = 1) + P(X_1 = 0, X_2 = 2)$$
$$= (0.1)\,(0.4) + (0.5)\,(0.5) + (0.4)\,(0.1)$$
$$= 0.04 + 0.25 + 0.04 = 0.33$$

$$P(X_1 + X_2 = 3) = P(X_1 = 2, X_1 = 1) + P(X_1 = 1, X_2 = 2)$$
$$= (0.1)\,(0.5) + (0.5)\,(0.1)$$
$$= 0.05 + 0.05 = 0.1$$

$$P(X_1 + X_2 = 4) = P(X_1 = 2, X_2 = 2) = (0.1)\,(0.1) = 0.01$$

The distribution $\overline{X} = \dfrac{X_1 + X_2}{2}$ is as follows :

Values of $\overline{X}$	0	0.5	1	1.5	2	Total
Probability	0.16	0.4	0.33	0.1	0.01	1.0

$$\mu_{\overline{X}} = E(\overline{X}) = 0 \times 0.16 + 0.5 \times 0.4 + 1 \times 0.33 + 1.5 \times 0.1 + 2 \times 0.01$$

$$= 0 + 0.2 + 0.33 + 0.15 + 0.02 = 0.7$$

$$E(\overline{X})^2 = (0)(0.16) + (0.5)^2(0.4) + (1)^2(0.33) + (1.5)^2(0.1) + (2)^2(0.01)$$

$$= (0.25)(0.4) + (1)(0.33) + (2.25)(0.1) + (4)(0.01)$$

$$= 0.1 + 0.33 + 0.225 + 0.04 = 0.695$$

$$\sigma^2_{\overline{X}} = \text{var}(\overline{X}) = E(\overline{X})^2 - [E(\overline{X})]^2 = 0.695 - (0.7)^2$$

$$= 0.695 - 0.49 = 0.205$$

We find that the sampling distribution of means has mean = 0.7 = population mean μ[i.e. $\mu_{\overline{X}} = \mu$]

and

$$\sigma^2_{\overline{X}} = \text{variance} = 0.205 = \frac{0.41}{2} = \frac{\sigma^2}{2} = \frac{\text{population variance}}{\text{sample size}}$$

We now come to the general results concerning the sampling distribution of the sample mean $\overline{X}$.

First we describe (or define) a *random sample*. A random sample of size n taken with replacement from a population (or random variable X) with distribution $p(x) = P(X = x)$, is a collection or set of random variables $X_1, X_2, \ldots, X_n$ each having the same distribution as X. Further, $X_1, X_2, \ldots, X_n$ are independent. In other words, $P(X_1 = x) = p(x), \ldots, P(X_n = x) = p(x)$ for every value x assumed or which is the same thing as $P(X_i = x) = p(x)$ for every x and every i.

Further, if $E(X) = \mu$ and var $(X) = \sigma^2$, then

$$E(X_1) = E(X_2) = \ldots = E(X_n) = \mu$$

and $$\text{var}(X_1) = \text{var}(X_2) = \ldots = \text{var}(X_n) = \sigma^2$$

We now state the general theorem as follows:

Theorem 8.1. If μ and σ^2 are the mean and variance of a population(or of a random variable X) and random samples with replacement of size n are taken, then the sampling distribution of the sample mean $\overline{X}$ has mean μ and variance $\frac{\sigma^2}{n}$.

Proof: Let the sample mean $\overline{X}$ be

$$\overline{X} = \frac{X_1 + \dots + X_n}{n}$$

Then $$E(\overline{X}) = \frac{E(X_1) + E(X_2) + \dots + E(X_n)}{n}$$

$$= \frac{\mu + \mu + \dots + \mu}{n} = \frac{n\mu}{n} = \mu$$

and $$\text{var}(\overline{X}) = \text{var}\left(\frac{X_1 + \dots + X_n}{n}\right)$$

$$= \frac{1}{n^2}\text{var}(X_1 + \dots + X_n)$$

$$= \frac{1}{n^2}(\sigma^2 + \sigma^2 + \dots + \sigma^2),$$

as $X_1, X_2, \dots X_n$ are independent

$$= \frac{1}{n^2} \cdot n\sigma^2 = \frac{\sigma^2}{n}$$

Note :

(1) The result holds irrespective of the form of distribution of the population from which the random sample is drawn. The variance of the distribution of the sample mean $\overline{X}$ involves the variance σ^2 of the population *and* the sample size n.

(2) The theorem also explains why we could get the results in Example 3(a) about the mean and variance of the sampling distribution of the mean, *without* assuming anything about the distribution of the population. Whatever might be the population distribution of the numbers of the ball, the results obtained hold; for example, we may assume that the numbers 1, 2, 3 occur with equal probability, i.e. if X is the number shown in the ball, then we may assume that $P(X=1) = P(X=2) = P(X=3) = \frac{1}{3}$ or we may assume that $P(X = 1) = p, P(X = 2) = q, P(X = 3) = r$, where $p + q + r = 1$.

(3) From the above theorem, we can obtain the mean and variance of the sampling distribution of $\overline{X}$. That would give us some idea of the distribution: the mean about the location of the distribution and the variance about its spread or dispersion around the centre. To get a complete picture or full information of the distribution, we need the *actual* distribution itself, i.e. the distribution of $\overline{X}$. We shall see what can be said about the distribution $\overline{X}$ from the two theorems given below. An outline of proof, without giving details, is also given.

Theorem 8.2. The sample mean $\overline{X}$ from a random sample of size n from a *normal* population with mean μ and variance σ^2 has also *normal* distribution with mean μ and variance $\frac{\sigma^2}{n}$.

Proof: A random sample of size n is a collection of n independent random variables $X_1, X_2, \ldots, X_n$ each having the same distribution as X. The sample mean $\bar{X} = \dfrac{X_1+X_2+\ldots+X_n}{n}$ is the sum of n independent random variables $\left(\dfrac{X_1}{n}\right), \left(\dfrac{X_2}{n}\right), \ldots, \left(\dfrac{X_n}{n}\right)$, each of which is again normal. Now as the sum of a number of independent random variables is normal, $\bar{X}$ will be normal. From Theorem 8.1 we get that $E(\bar{X}) = \mu$ and var $(\bar{X}) = \dfrac{\sigma^2}{n}$.

Fig. 8.1 shows the distribution of the normal population and the distribution of the sample mean of samples of size 25 from N (60, 3).

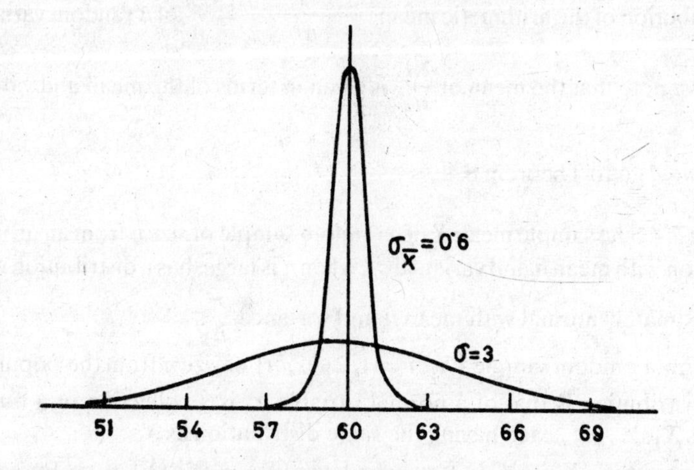

Fig. 8.1. Sampling distribution of the sample means taken from samples of size 25 from normal population with $\mu = 60, \sigma = 3$

Now we shall go to the case where the distribution of the population is other than normal. We get an interesting result here as given in Theorm 8.4 below. But before we go to Theorem 8.4, we state, without proof, another interesting Theorem 8.3.

Theorem 8.3. Central Limit Theorem

Suppose that $X_1, X_2, \ldots, X_n$ are independent random variables all having the same mean $E(X_i) = \mu$ and variance var $(X_i) = \sigma^2$ for $i = 1, 2, \ldots n$. Then, *when n is large*, the distribution of the sum $S_n = X_1 + \ldots + X_n$ is *approximately normal* with

$$\text{mean} = E(S_n) = n\,\mu$$

and $$\text{variance} = \text{var}\,(S_n) = n\,\sigma^2.$$

Then it follows that the distribution of

$$\frac{S_n}{n} = \frac{X_1 + \ldots + X_n}{n} \text{ is also normal}$$

$$\text{with mean } \frac{n\mu}{n} = \mu \text{ and variance } \frac{n\sigma^2}{n^2} = \frac{\sigma^2}{n}.$$

A proof of this important theorem is beyond the scope of the book. Apart from the theoretical proof, an understanding of the result stated in the Theorem, is of much greater importance.We simply note the sum $S_n = X_1 + ... + X_n$ of n independent random variables $X_1,...X_n$ (whatever-might be their distributions) is approximately normal when n is large; so also is the distribution of the arithmetic mean $\frac{X_1+...+X_n}{n} = \frac{S_n}{n}$ of n random variables; further we note that the mean of $\frac{S_n}{n}$ is given in terms of the mean and variance of X_i's.

We can now go to Theorem 8.4.

Theorem 8.4. The sample mean $\overline{X}$ of a random sample of size n from an arbitrary population with mean μ and variance σ^2, when n is large, has a distribution which is approximately normal with mean μ and variance $\frac{\sigma^2}{n}$.

Proof: Now a random sample $\{X_i, i = 1, 2 \ldots, n\}$ of size n from the population whose distribution is that of a normal variable X, is a collection of a random variables $X_1, \ldots, X_n$ each having the same distribution as X.

We thus get $E(X_i) = E(X) = \mu$ and var (X_i) = var $(X) = \sigma^2$ for each of $i = 1, 2, \ldots, n$. Now by the Central limit theorem, irrespective of the form of distribution of X, the sample mean $\overline{X} = \frac{X_1 + \ldots, X_n}{n}$, when n is large, has *approximately* normal distribution. Further the sample mean $\overline{X}$ has mean $E(\overline{X}) = \mu$ and var $(\overline{X}) = \frac{\sigma^2}{n}$.

In other words, for X having an arbitrary distribution with mean μ and variance σ^2, the sampling distribution of mean $\overline{X}$ of a random sample of size n from X, when n is large, is approximately normal with mean μ and variance $\frac{\sigma^2}{n}$.

When n is large, and X is $N(\mu, \sigma)$, then the sample mean $\overline{X}$ is $N\left(\mu, \frac{\sigma}{\sqrt{n}}\right)$

If we write

$$Z = \frac{\overline{X} - \mu}{\sqrt{\left(\frac{\sigma^2}{n}\right)}}$$

then Z is apporoximately $N(0, 1)$.

The above theorems are of considerable importance in statistics and in statistical inference. They also point to the general importance of normal distribution. A sampling distribution, that of $\overline{X}$, for large n, tends to normal distribution.

We saw earlier that one gets normal distribution as a limiting form of some *probability distribution,* like *binomial* and *Poisson* distributions. Here we find normal distribution as the limiting form of a *sampling distribution.* This shows the universality of normal distribution and justifies it being called *normal:* under certain conditions, it is *normal* to expect a normal distribution.

The above theorems also have interesting practical applications. We shall show some applications through some examples.

Example 3(d). Assume that the weights of male students of a large College are normally distributed with mean 55 kg and s.d. 2 kg. A number of samples of 100 students each are taken at random with replacement from the population. What would be the mean and the standard deviation of the sampling distribution of the mean? Find the probability that the sample mean will differ from the population mean by less than 0.4 kg.

Here n = size of sample = 100

$\mu = 55,\ \sigma = 2,\ \sigma^2 = 4.$

$\overline{X}$, the sample mean will also have normal distribution (Theorem 8.3) with mean 55 and s.d. $\dfrac{2}{\sqrt{100}} = \dfrac{2}{10} = 0.2$ kg.

Thus $Z = \dfrac{\overline{X}-55}{0.2}$ is $N(0, 1)$.

Here we are to find the probability that the sample mean $\overline{X}$ will differ from the population mean 55 kg by less than 0.4 kg, i.e. we are to find

$$P(55 - 0.4 < \overline{X} < 55 + 0.4) = P(54.6 < \overline{X} < 55.4)$$

$\overline{X} = 55.4$ corresponds to $Z = \dfrac{55.4 - 55}{0.2} = \dfrac{0.4}{0.2} = 2$

and $\overline{X} = 54.6$ corresponds to $Z = \dfrac{54.6-55}{0.2} = \dfrac{-0.4}{0.2} = -2$

$$\begin{aligned}\text{Thus } P(54.6 < \overline{X} < 55.4) &= P(-2 < Z < 2)\\ &= P(Z<2) - P(Z<-2)\\ &= 0.9772 - 0.0228\\ &= 0.9544\end{aligned}$$

Example 3(e). It is given that the diameters of small metallic balls manufactured in a factory are normally distributed with mean 12 mm and s.d. 0.1 mm. A sample of 25 balls is taken from balls produced each day during a month of 26 working days. Find the mean and s.d. of the distribution of sample means.

Find the probability that the sample mean of balls will (1) exceed 12.01 mm (2) be less than 11.98 mm (3) lie between 11.98 and 12.1 mm.

Here μ = 12 mm, s.d. σ = 0.1 mm

As $n = 25$, the distribution of X is normal with mean 12 mm and s.d. $= \frac{0.1}{\sqrt{25}} = 0.02$ mm.

Thus $\quad Z = \frac{X - 12}{0.02}$ is $N(0, 1)$.

The values of Z corresponding to $X = 12.01$ and $X = 11.98$ are respectively,

$$Z = \frac{12.01 - 12}{0.02} = \frac{0.01}{0.02} = 0.5 \text{ and } Z = \frac{11.98 - 12}{0.02} = \frac{-0.02}{0.02} = -1$$

and so

$$P(\bar{X} > 12.01) = P(Z > 0.5) = 1 - P(Z < 0.5)$$

$$= 1 - 0.6915 = 0.3085$$

$$P(\bar{X} < 11.98) = P(Z < -1) = 0.1587$$

$$P(11.98 < \bar{X} < 12.01) = P(-1 < Z < 0.5)$$

$$= P(Z < 0.5) - P(Z < -1)$$

$$= 0.3085 - 0.1587 = 0.1498$$

Roughly about 14.98% of balls produced will have the mean diameter between 11.98 and 12.01 mm.

Example 3(f). It is found that the mark(score)in mathematics, of a large number of students over a number of years, has a distribution with mean 50 and s.d. 16 (but the exact distribution is not known). A random sample of 64 students of every batch for 20 batches is taken. For the sampling distribution of means, find the mean and s.d. Find the probability that (i) $\bar{X} > 54$, (ii) $\bar{X} > 56$ (iii) $\bar{X} \leq 50$ where $\bar{X}$ is the sample mean calculated for the sample of 64.

Here, mean $= 50$, s.d. $= 16$, $n = 64$.

$\bar{X}$ has mean 50 and s.d. $= \frac{16}{8} = 2$.

Though the distribution of the population is not known, we still can assume approximately normal distribution for the distribution of $\bar{X}$, as the sample size $n = 64$ is large (Theorem 8.3). Thus we can assume that

$$Z = \frac{\bar{X} - 50}{2} \text{ is approximately } N(0, 1).$$

The values of Z corresponding to $\bar{X} = 54$, 56, and 50 are respectively,

$$Z = \frac{54-50}{2} = 2,\ Z = \frac{56-50}{2} = 3,\ Z = \frac{50-50}{2} = 0$$

Thus $\quad P(\bar{X} > 54) = P(Z > 2) = 1 - P(Z < 2)$

$$= 1 - 0.9772$$

$$= 0.1228$$

$$P(\overline{X} > 56) = P(Z > 3) = 1 - P(Z \leq 3)$$
$$= 1 - 0.9987$$
$$= 0.0013$$
$$P(\overline{X} \leq 50) = P(Z \leq 0)$$
$$= 0.5$$

Note: If the original population distribution is normal, then $\overline{X}$ is also normal. It is approximately normal even if the population distribution is not normal. Now the question arises *how large should n* be in order that the distribution of $\overline{X}$ is approximately normal. Usually it is expected that $n \geq 30$ will give a good approximation.

8.4. SAMPLING DISTRIBUTION OF THE (SAMPLE) MEAN

Sampling (without replacement) from a finite population

When a sample (of size 1) is taken from a population with a very large number of units or elements, the effect of taking a subsequent sample is little in the sense that the population structure will not change appreciably for the second sample, even if the sample drawn is not replaced. Thus, if a sample of size n is drawn one by one from a population with N elements, where N is large compared to n, sampling without replacement can be treated as sampling with replacement. But when a sample is taken from a population with a finite number of elements, the removal of an element as a sample would effect the drawing of the subsequent sample and thus sampling without replacement cannot be treated as equivalent to sampling with replacement. The results obtained for sampling with replacement (or sampling without replacement from a large population) are not applicable without modification. We state the result below without proof.

Theorem 8.5. Suppose that a population consisting of N elements has mean μ and s.d. σ and that a sample of size n *without replacement* is taken, and the sample mean $\overline{X}$ is calculated based on this sample of size n. The distribution of the sample mean $\overline{X}$ (or the sampling distribution of mean) will have the same mean μ as the population. The variance of the sampling distribution of the mean will be

$$\frac{\sigma^2}{n}\left(\frac{N-n}{N-1}\right) \tag{4.1}$$

The factor $\frac{N-n}{N-1}$ is known as the finite population correction factor. When N is large compared to n the factor $\left(\frac{N-n}{N-1}\right)$ will be nearly equal to 1, and so the correction factor will be 1. Then the variance of the sampling distribution of the mean will be approximately equal to $\frac{\sigma^2}{n}$, as in the case of sampling with replacement.

Example 4 (a). Consider Example 3 (b); here $N = 3$, $\mu = 2$, $\sigma^2 = \frac{2}{3}$ and sampling *without replacement* is considered, the size of sample being $n = 2$.

Using the above formula (4.1) we get the variance of the sampling distribution of the mean to be

$$\frac{2}{3}\left(\frac{1}{2}\right)\left(\frac{3-2}{3-1}\right) = \frac{1}{6};$$

this is the result which was obtained by direct calculation in the Example 3 (b).

Example 4 (b). A population consists of 20 elements, has mean 8 and s.d. 2 and a sample of 5 elements is taken without replacement. Find the mean and s.d. of the sampling distribution of the mean. What will be the s.d. for samples of size 10 ?

Here $N = 20$, mean $= \mu = 8$, s.d. $= \sigma = 2$ and $n = 5$.

The sampling distribution of mean has mean 8; its variance is

$$\frac{\sigma^2}{n}\left(\frac{N-n}{N-1}\right) = \frac{4}{5}\left(\frac{20-5}{20-1}\right) = \frac{12}{19}$$

and so $$\text{s.d.} = \sqrt{\frac{12}{19}} = \frac{2\sqrt{3}}{\sqrt{19}} = \frac{2\sqrt{57}}{19}$$

With sample size 10, while the mean remains the same, the variance becomes

$$\frac{4}{10}\left(\frac{20-10}{20-1}\right) = \frac{4}{19} \text{ and s.d.} = \frac{2}{\sqrt{19}} = \frac{2\sqrt{19}}{19}.$$

Note:

(1) With the increase in the sample size, there is decrease in the variance (and s.d.) of the sample mean.

(2) The formula for the variance (4.1) given above is similar to the formula for the variance of hypergeometric distribution, the distribution of a random variable giving the number of successes in n trials where drawings are made without replacement.

8.5. SAMPLING DISTRIBUTION OF PROPORTION

Suppose that we are sampling with replacement from a Bernoulli random variable X with probability of success p at each trial. Denoting success by 1, and failure by 0, we can denote the distribution of X as:

Value of X	1	0
Probability	p	$q = 1 - p$

We have

$$E(X) = p \text{ and } \operatorname{var}(X) = pq$$

(where p is the proportion of successes per trial in the population or the population proportion of successes.)

A sample of size n drawn from this population with replacement corresponds to a set of n random variables $X_1, X_2, \ldots, X_n$ each having the same distribution as X. That is

$$X_i = 1 \text{ (if the trial is a success) with probability } p$$

$$X_i = 0 \text{ (if the trial is a failure) with probability } q = 1-p.$$

Consider the sample total $T = X_1 + \ldots + X_n$. Then T will give the total number of successes in n trials. The sample mean $\overline{X} = \dfrac{X_1 + \ldots + X_n}{n}$ will give the average number of successes in n trials or the proportion of successes in n trials. Thus the sample mean $\overline{X}$ is nothing but the proportion (or relative frequency) of successes in n trials. Thus whereas p is the *population* proportion of successes, (per trial) the sample mean $\overline{X}$ is the sample proportion of successes (per trial). And because of this, in this case, $\overline{X}$ is usually denoted by $\hat{p}$ (i.e., $\overline{X} = \hat{p}$). Distribution of sample proportion (p) is the same thing as the distribution of the sample mean $\overline{X}$ (for sampling from this population of Bernoulli distribution or from the random variable X).

Now since the population has mean $E(X) = \mu = p$ and variance, $\sigma^2 = pq$, the distribution of sample proportion (or sampling distribution of proportion) will have (by Theorem 8.1) mean equal to p and variance equal to $\dfrac{\sigma^2}{n} = \dfrac{pq}{n}$, and so standard deviation is equal to $\dfrac{\sigma}{\sqrt{n}} = \sqrt{\left(\dfrac{pq}{n}\right)}$.

The distribution of X is Bernoulli (and not normal), so the distribution of $\hat{p}$ will not be normal. But by Theorem 8.3, when n is large, the distribution will be approximately normal.

We put the result in the form of a Theorem.

Theorem 8.6. Suppose that a random sample of size n is taken with replacement from a population distribution of proportion of a certain character or attribute (such as success) with the probability of the attribute occurring being p. Then the sampling distribution of the (sample) proportion p will have mean p and s.d. $\sqrt{\left(\dfrac{pq}{n}\right)}$. Further, when n is large, the distribution of $\hat{p}$ will be approximately normal with mean p and s.d. $\sqrt{\left(\dfrac{pq}{n}\right)}$. That is, the distribution of

$$Z = \frac{\hat{p}-p}{\sqrt{\left(\dfrac{pq}{n}\right)}}$$

will be approximately normal with mean 0 and s.d. 1.

Example 5(a). Suppose that the probability of a male birth (in a certain city) is $\frac{1}{2}$. From the birth register a random sample of 100 births is considered and the sample proportion $\hat{p}$ of male births noted. What is the probability that (i) $\hat{p}$ will exceed 0.55 (ii) $\hat{p}$ will be less than 0.50, (iii) $\hat{p}$ will lie between 0.48 and 0.51 ?

Here p = population proportion (of male births) $= \frac{1}{2}$ and so $q = 1-p = \frac{1}{2}$.

As $n = 100$, the sample proportion $\hat{p}$ will have approximately normal distribution with mean $p = \frac{1}{2}$ and

$$\text{s. d.} = \sqrt{\frac{pq}{n}} = \sqrt{\left(\frac{\frac{1}{2}\cdot\frac{1}{2}}{100}\right)} = \frac{\frac{1}{2}}{10} = \frac{1}{20}$$

So $$Z = \frac{\hat{p}-p}{\left(\frac{1}{20}\right)} \text{ will be approximately } N(0, 1).$$

(i) $\hat{p} = 0.55$ corresponds to $Z = \frac{0.55 - 0.50}{\left(\frac{1}{20}\right)} = \frac{0.05}{\left(\frac{1}{20}\right)} = 1$

Thus

$$\begin{aligned} P(\hat{p} > 0.55) &= P(Z > 1) \\ &= 1-P(Z < 1) \\ &= 1-0.8413 \\ &= 0.1587 \end{aligned}$$

(ii) $\hat{p} = 0.50$ corresponds to $Z = \frac{0.50 - 0.50}{\left(\frac{1}{20}\right)} = 0$

Thus

$$P(\hat{p} < 0.50) = P(Z < 0) = 0.50$$

(iii) $\hat{p} = 0.48$ corresponds to $Z = \frac{0.48 - 0.50}{\left(\frac{1}{20}\right)} = \frac{-0.2}{\left(\frac{1}{20}\right)} = -4$

$\hat{p} = 0.51$ corresponds to $Z = \frac{0.48 - 0.50}{\left(\frac{1}{20}\right)} = \frac{0.1}{\left(\frac{1}{20}\right)} = 2$

Thus the required probability

$$\begin{aligned} &= P(0.48 < \hat{p} < 0.51) = P(-4 < Z < 2) \\ &= P(Z < 2)-P(Z < -4) \\ &= 0.9772-0 \\ &= 0.9772 \end{aligned}$$

Example 5(b). Find the probability that in tossing a fair (good) coin, the proportion of head lies between (i) 40% and 50%, when 36 tossings are made; (ii) 45% and 55%, when 121 tossings are made.

(i) Here $p = \frac{1}{2}$, as the coin is fair (good) and $n = 36$. We are to find

$$P\left(\frac{40}{100} < \hat{p} < \frac{50}{100}\right).$$

The distribution of $\hat{p}$ can be taken to be approximately normal (as the sample size $n > 30$): and the distribution of $\hat{p}$ will have mean $\frac{1}{2}$

and
$$\text{s.d.} = \sqrt{\frac{pq}{n}} = \sqrt{\left(\frac{\frac{1}{2}\cdot\frac{1}{2}}{36}\right)} = \frac{\frac{1}{2}}{6} = \frac{1}{12}$$

Thus
$$Z = \frac{\hat{p} - \frac{1}{2}}{\frac{1}{12}} \text{ is } N(0, 1)$$

Now $\hat{p} = 40\% = \frac{40}{100} = \frac{2}{5}$ corresponds to $Z = \frac{\frac{2}{5} - \frac{1}{2}}{\frac{1}{12}} = \frac{-1}{10} \times 12 = -1.2$

$$\hat{p} = 50\% = \frac{50}{100} = \frac{1}{2} \text{ corresponds to } Z = 0.$$

Thus
$$P\left(\frac{40}{100} < \hat{p} < \frac{50}{100}\right) = P(-1.2 < Z < 0)$$
$$= P(Z < 0) - P(Z < -1.2)$$
$$= 0.5 - 0.1151$$
$$= 0.3849$$

(ii) Here $p = \frac{1}{2}$ and $n = 121$. The distribution of $\hat{p}$ is approximately normal with mean $\frac{1}{2}$ and

$$\text{s.d.} = \sqrt{\frac{pq}{n}} = \sqrt{\left(\frac{\frac{1}{2}\cdot\frac{1}{2}}{121}\right)} = \frac{\frac{1}{2}}{11} = \frac{1}{22}$$

Thus $Z = \dfrac{\hat{p}-\frac{1}{2}}{\frac{1}{22}}$ is $N(0, 1)$.

$$\hat{p} = 45\% = \frac{45}{100} = \frac{9}{20} \text{ corresponds to } Z = \frac{\frac{9}{20}-\frac{1}{2}}{\frac{1}{22}} = \frac{-\frac{1}{20}}{\frac{1}{22}} = -1.1$$

$$\hat{p} = 55\% = \frac{55}{100} = \frac{11}{20} \text{ corresponds to } Z = \frac{\frac{11}{20}-\frac{1}{2}}{\frac{1}{22}} = 1.1$$

Thus the required probability is

$$P\left(\frac{9}{20} < \hat{p} < \frac{11}{20}\right)$$

$$= P(-1.1 < Z < 1.1) = P(Z < 1.1) - P(Z < -1.1)$$

$$= 0.8643 - 0.1357$$

$$= 0.7286$$

8.6. STANDARD ERRORS

Consider a random variable X, say, having Poisson distribution with mass function

$$P(X = k) = \frac{e^{-a}a^k}{k!}, \; k=0, 1, 2, \ldots ;$$

it contains a parameter a, which is the population mean as well as the population variance of the random variable X. For a random variable X having Binomial distribution the parameters are n and p; np is the population mean and npq is the population variance of X. In a broad sense, the mean, the variance, moments of higher order etc. of the distribution of the variable in the population as well as the constants occurring in the mass function or the p.d.f. of a variable are spoken of, generally, as the parameters of the population; commonly, however, by parameters, the constants occurring in the p.m.f. or p.d.f. are meant. The distribution of a r.v. is a *probability distribution*.

When we take a random sample of size, say n of a population we get a *frequency distribution* from which the mean, the variance etc. of the sample can be calculated. These may be considered as *estimators* of the values of the corresponding parameters of the population. Any such estimator calculated from the frequency distribution of the sample is called a *statistic; a statistic is a function of the sample values.* Any function of the n sample values used as an estimator of a parameter of the population is called a statistic.

If we take another random sample of the same size n, we get another set of sample values, and so another frequency distribution. This will give another value of the estimator of the population parameter, that is, another statistic. Thus, if a number of different random samples, each of the same size n, are taken we shall get a number of the same statistic. It is a matter of common experience that a statistic will vary from sample to sample. We shall then get a *frequency distribution* of the statistic. The distribution of the values of the statistic obtained from an infinite number (or a very large) number of random samples each of the given size n, is defined as the *sampling distribution* of that statistic. A sampling distribution is essentially a probability distribution, and is a continuous distribution. The standard deviation (s.d.) of the sampling distribution is termed as the *standard error* (s.e.) of that statistic for samples of given size. The p.d.f. of sampling distribution will be a function of n.

For example, suppose that μ and σ^2 are the population mean and population variance of the probability distribution of a r.v. X. If a random sample of size n is taken, we can calculate a sample mean $\bar{X}$; this is a statistic. It will have a sampling distribution. From Theorem 8.1, we get that the mean of the sampling distribution of the sample mean $\bar{X}$ is also μ and that the variance is $\dfrac{\sigma^2}{n}$, so that the s.d. of $\bar{X}$ is $\dfrac{\sigma}{\sqrt{n}}$. Thus the standard error of the sample mean is $\dfrac{\sigma}{\sqrt{n}}$. From Theorem 8.6, we see that the standard error of sample proportion is $\sqrt{(pq)/n}$.

The probability of a sample value lying outside $\pm$ 3 times s.e. of the true value is, in general, very small.

EXERCISES - 8

Sections 8.1-8.3

1. Explain what is meant by (i) a random sample (ii) sampling with replacement (ii) sampling without replacement.
2. Suppose that a population consists of 4 elements 1, 3, 5, 7. Samples of size 3 are taken from the population *with replacement.* Find the distribution of the sample mean. and also the mean and the variance of the distribution.
3. Consider Exercise 2 when samples are taken without replacement.
4. A population consists of 2 elements, 1 and 0 each of which occurs probability $\frac{1}{2}$. A random sample X_1, X_2, X_3, X_4 of size 4 is taken with replacement from the population. Find the distribution of $T = X_1 + X_2 + X_3 + X_4$, and its mean and variance. Explain how you get binomial distribution with parameters $n = 4$ and $p = \frac{1}{2}$ for T.
5. A random sample X_1, X_2 of size 2 with replacement is taken from the following population:

Value of X (k)	0	1	2	3
Probability $P(X = k)$	$\frac{1}{8}$	$\frac{3}{8}$	$\frac{3}{8}$	$\frac{1}{8}$

Find the distribution of $X_1 + X_2$, its mean and variance. Also find the distribution of the sample mean X, its mean and variance.

6. A population has the distribution as given below

Value of X (k)	0	1	2	3	Total
Probability $P(X = k)$	0.25	0.35	0.25	0.15	1.0

A random sample X_1, X_2, X_3 of size 3 is taken from the population with replacement. Find the distribution of $X_1+X_2+X_3$ and its mean and variance. Hence find the mean and variance of the sample mean $\bar{X}$, and verify that the results obtained by using the formula agree.

7. What do you understand by the Central Limit Theorem? Explain its importance in sampling theory.
8. Suppose that a fair coin is tossed 4 times and the average number of heads is noted. Find the distribution of the average number of heads, its mean and variance.
9. A fair die is thrown 4 times. Getting a 6 is considered as success. Find the distribution of the average number of times 6 occurs in the 4 throws; find also its mean and variance.
10. Suppose that the weight of bags of cement produced and packed in a cement factory is normally distributed with mean weight 35 kg and s.d. 1 kg. Samples of 10 bags are taken and their weights recorded and the average computed. What will be the distribution of the average weights of cement bags sampled? Find the probability that the average of 10 bags will (i) exceed 35 kg, (ii) be less than 35 kg, (iii) exceed 34 kg, (iv) be less than 34 kg and (v) be between 34 kg and 36 kg.
11. The length of pins produced by a machine has normal distribution with mean length 2.5 cm. and s.d. 0.2 cm. Samples of 16 pins are taken, their lengths noted and the mean length (of 16 pins) is calculated. What can you say about the distribution of the mean length of pins? State the theorems you apply. Find the probability that the sample mean of lengths of (16 pins) (i) exceeds 2.5 cm. (ii) is less than 2.5 cm. (iii) lies between 2.4 cm. and 2.6 cm.
12. If the heights of school children of a particular standard in a big city is normally distributed with mean 150 cm. and s.d. 4 cm., find the probability that the sample mean in a random sample of 64 students will differ from the mean height of the population of children by (i) more than 2 cm. (ii) by less than 2 cm.
13. Consider Exercise 12 when the form of distribution heights is not given but only the s.d. (4 cm.) is given. Do you arrive at the same results? Explain.
14. The weight of baby food of a certain brand contained in a tin is stated to be approximately 400 gms: in fact the weights of contents are normally distributed with a mean of 400 gms and s.d. of 10 gms. A sample of 50 tins of baby food is taken, the weights of contents are noted and the mean weights are calculated. Find the probability that the mean weight will (i) be less than 400 gms (ii) exceed 400 gms, (iii) lie between 380 and 420 gms.

Section 8.4

15. A population has 10 members with mean 50 and s.d. 3. A sample of size 4 is drawn. Find the mean and s.d. of the sampling distribution of the mean.
16. A population has 1000 members with mean 100 and s.d. 10. A sample of size 15 is drawn (i) with replacement, (ii) without replacement. Find the mean and variance of the sampling distribution of mean.

Section 8.5

17. Suppose that in a University with large number of students, 30% are female students. A random sample of 50 students is taken. Find the probability that the sample proportion of male students p (i) will exceed 70%, (ii) be less than 70% (iii) will lie between 65% to 75%.
18. From past experience it is found that 1% of the parts produced by a machine are defective. Find the probability that in random lot of 100 parts, the proportion of defectives will be (i) more than 1%, (ii) less than 1% (iii) between 1.5% to 2%.
19. From demographic records it is found that the ratio of male births to female births is as 100 : 96. A random sample of 64 persons is taken. Find the probability that the proportion of females in the sample will be (i) greater than $\frac{1}{2}$ (ii) less than $\frac{1}{2}$.
20. A machine produces a component for a transistor set. Of the total produce, 5 percent are defective. A random sample of 4 components is taken for examination from (i) a very large lot of produce, (ii) a box of 10 components.

Find the mean and the variance (and s.d.) of the average number of defectives found among the 4 components taken for examination.

Section 8.6

21. Explain clearly what is meant by a (i) probability distribution, (ii) frequency distribution, and (iii) sampling distribution. Bring out their differences.
22. Explain the terms parameter and statistic with suitable examples.
    What are the parameters of (i) Bernoulli distribution, (ii) geometric distribution, (iii) binomial distribution, and (iv) normal distribution?
23. Indicate the essential features of a sampling distribution.
24. What do you understand by standard error?
    Indicate its uses and importance.
25. (a) Calculate the standard error of the sample mean from a random sample of size 36 from $N(0, 10)$.
    (b) The average of daily absentees in a factory as compiled from a very long record is 5 per cent. The (sample) proportion of absentees from a random sample of 25 days is considered. Find the standard error of the sample proportion.

Chapter 9

Correlation and Regression

9.1. INTRODUCTION

Most of our discussions so far have been confined to a single variable. In statistical work we often have to deal with problems involving more than one variable. Let us consider, for the moment, two variables. Our interest lies in studying the relationship between the two variables.

For example, we may be interested in finding the relationship, if any, between

(1) the performance of students in two subjects, say, mathematics and statistics or accountancy and economics, or between two examinations, the entrance examination to a professional course and the final examination in the course;
(2) the amount spent on advertising and that received by sales;
(3) the heights of father and the eldest son;
(4) the measurement of chest and the height of individuals (in which a readymade shirt manufacturer may be interested);
(5) the index of wholesale prices and the index of agricultural production;
(6) the age of a mother at first birth and the interval between the first and the second births;

and so on.

We may be able to make *qualitative* statement of the relationship; like, say that tall fathers will usually have tall sons or that the sales rise with the increase in the amount spent in advertising. These statements are not precise enough to be of use to decision makers. A *quantitative* statement will be of much greater value and use. We are therefore to look for a *quantitative measure* of the relationship between two variables, and also for an appropriate mathematical or statistical form of the relationship. While the second question will be discussed later (when we discuss regression analysis), we shall discuss the first question now.

9.2. SCATTER DIAGRAM

As a first step we get a set of quantitative values for the two variables. The independent variable will be denoted by x and the dependent variable by y, and the corresponding pair of values will be denoted by (x_i, y_i) though sometimes it is difficult to determine which one is to be considered as the independent variable. If we have n such pairs of values, then $i = 1, 2, \ldots n$. For example, scores in mathematics and statistics of 50 students may be denoted by the pairs

(x_i, y_i), where $i = 1, 2, \ldots, 50$; (x_{10}, y_{10}) denote the scores of the 10th student, in whichever order you take the students. By considering the pairs of numbers as two dimensional coordinates and plotting the corresponding points on an ordinary graph paper, we get a set of points. The diagram is called a *scatter diagram*. The scatter diagram serves as a useful aid in the study of the relationship and also for assessing how marked the relationship is. Consider the following data of scores in mathematics and statistics of 20 students of a class. Scores in mathematics and statistics are denoted by x and y respectively.

x	56	73	65	80	35	62	36	40	92	45
y	75	80	56	82	45	65	30	25	90	50

x	45	75	82	90	35	78	65	45	55	60
y	30	80	80	90	36	36	85	56	50	45

The marks are out of a maximum of 100 for each subject. Plotting the points (x, y) on a graph paper, we get a scatter diagram of the data.

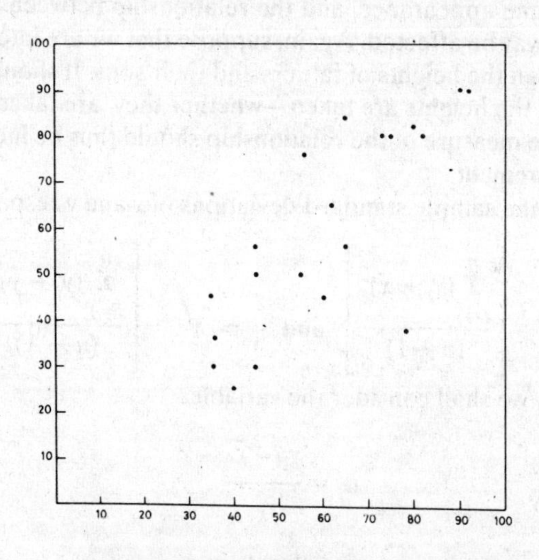

Fig. 9.1 Scatter Diagram of the above data

A glance at the scatter diagram reveals that there exists a tendency for small values of x to be associated with small values of y, and so also for larger values of x and y. Further, the scatter shows a marked tendency of the points (x_i, y_i) to lie near a *straight line*, or in other words, there appears to be some *linear relationship* between the two variables x and y. Again, this statement is qualitative; it does not indicate the degree or markedness of the relationship. To specify this, we are to devise some *quantitative measure*.

9.3. THE COEFFICIENT OF CORRELATION

Let us first consider the desirable properties that this quantitative measure should have. There are two such important properties:

(1) The desired measure should be independent of the choice of origin. For example, if instead of taking (0, 0) as the origin (i.e. the point through which both the x and y axes pass), we may draw the scatter diagram with axes chosen to pass through another point, say (20, 30); we would have then the same type of scatter diagram (so far as its scattering is concerned). This implies that the relationship does not depend on the choice of the origin, and the same picture would emerge if we consider $x - 20$ and $y - 30$ instead of x and y respectively. Therefore, the measure of the relationship should be such that it is independent of the origin.

We can therefore consider, instead of the absolute values of x and y, their deviations from the respective means $\bar{x}$ and $\bar{y}$. We consider deviations from $\bar{x}$ and $\bar{y}$ as was done in case of computing standard deviation. Thus $x - \bar{x}$ and $y - \bar{y}$ are taken for constructing the measure of the relationship, instead of the absolute values of x and y.

(2) The desired measure should be independent of the scale of measurement. Consider, for example, that the marks are doubled so that these may be considered as being awarded out of a maximum of 200. The scatter diagram should have the same appearance, and the relationship between the variables should not by any way be affected. Again, suppose that we are interested in the relationship between the heights of fathers and their sons. It should not matter then, in what units the heights are taken—whether they are taken in inches or in centimetres. The measure of the relationship should thus be independent of the scale of measurement.

Let s_x and s_y be the sample standard deviations of x and y respectively, then

$$s_x = \sqrt{\left[\frac{\sum_{i=1}^{n} (x_i - \bar{x})^2}{(n-1)}\right]} \quad \text{and} \quad s_y = \sqrt{\left[\frac{\sum_{i=1}^{n} (y_i - \bar{y})^2}{(n-1)}\right]}$$

instead of x_i and y_i we shall consider the variables

$$u_i = \frac{x_i - \bar{x}}{s_x}$$

$$v_i = \frac{y_i - \bar{y}}{s_y};$$

(u_i, v_i) will show the same pattern as shown by (x_i, y_i) so far as the scatter diagram is concerned. This can be observed by looking at the following figures of the scatter diagrams of the points (x_i, y_i) and of the points (u_i, v_i).

From the scatter diagram in Fig. 9.1 we find that most of the points lie in the first and third quadrants, and few points lie in the second and fourth quadrants. Further, the points lying in the first and third quadrants show a tendency to lie near a straight line path. If we consider the product $u_i v_i$, this product will be positive for points lying on the first and third quadrants and negative for points lying on the second and fourth quadrants. Consider the sum $\Sigma u_i v_i$ for products of pairs u_i and v_i. The points on the first and third quadrants contribute to positive values of the sum and the points on the second and fourth quadrants giving negative values of $u_i v_i$ tend to diminish the sum $\Sigma u_i v_i$. A large positive value of the sum $u_i v_i$ will indicate a *strong tendency of the points* (u_i, v_i) *to lie near a line*: or, in other words, a *strong linear trend* for the points (u_i, v_i). If the number of points were exactly double, with two points instead of one each, then the same pattern for the scatter diagram would emerge, but then the value of $u_i v_i$ would be exactly double; that is for the same strong linear trend, we shall get a value of $\Sigma\, u_i\, v_i$ which is two times as large. As we are interested in a measure of the linear relationship between x and y, it is necessary to divide the sum $\Sigma u_i v_i$ by the number n of points (x_i, y_i); but for theoretical reasons it is preferable to divide by $(n - 1)$ instead of n. Thus we arrive at the following definition of the measure of the linear relationship between the two variables. The measure is denoted by r and is called the *product moment correlation coefficient*. The concept was first formulated by Karl Pearson in 1896.

Definition : The product moment correlation coefficient r between values x_i and y_i, $i = 1, 2, \ldots, n$ of two variables is defined by

$$r = \frac{\sum_{i=1}^{n} \left(\frac{x_i - \bar{x}}{s_x}\right)\left(\frac{y_i - \bar{y}}{s_y}\right)}{(n-1)} \tag{3.1}$$

Putting the expressions for s_x and s_y, the above can also be written as

$$r = \frac{\sum_{i=1}^{n} (x_i - \bar{x})(y_i - \bar{y})}{\sqrt{\left[\sum_{i=1}^{n} (x_i - \bar{x})^2\right]\left[\sum_{i=1}^{n} (y_i - \bar{y})^2\right]}} \tag{3.2}$$

(3.1) involves $\bar{x}, \bar{y}, s_x$ and s_y and (3.2) involves $\bar{x}$ and $\bar{y}$ and may be convenient when these are given. For computational purpose, the following direct formula (which does not involve $\bar{x}, \bar{y}, s_x, s_y$ as in (3.1) and $\bar{x}, \bar{y}$ as in (3.2)) may be used :

$$r = \frac{n \sum_{i=1}^{n} x_i y_i - \left(\sum_{i=1}^{n} x_i\right)\left(\sum_{i=1}^{n} y_i\right)}{\sqrt{\left[\left\{n \sum_{i=1}^{n} x_i^2 - \left(\sum_{i=1}^{n} x_i\right)^2\right\}\left\{n \sum_{i=1}^{n} y_i^2 - \left(\sum_{i=1}^{n} x_i\right)^2\right\}\right]}} \tag{3.3}$$

Another formula for calculation of r based on an arbitrary origin, suitably chosen is given below. Suppose that

$$x_i' = x_i - a, \quad y_i' = y_i - b, \quad i = 1, 2, \ldots, n$$

then

$$r = \frac{n \Sigma x_i' y_i' - (\Sigma x_i')(\Sigma y_i')}{\sqrt{\left[\left\{n \Sigma (x_i')^2 - (\Sigma x_i')^2\right\}\left\{n \Sigma (y_i')^2 - (\Sigma y_i')^2\right\}\right]}} \tag{3.4}$$

Whereas (3.2) involves calculation at first of $\bar{x}, \bar{y}$ and (3.3) involves calculation with the raw values of x_i, y_i, formula (3.4) involves calculation based on deviations $x_i' = x_i - a$ and $y_i' = y_i - b$ from a suitable point (a, b). Some examples are considered below and a student will be able to find which formula is more convenient for calculation in the particular situation.

Before proceeding to show how to calculate r from a given set of pairs of values (x_i, y_i) we examine some interesting cases.

Let us consider the scatter diagrams shown in Fig. 9.2.

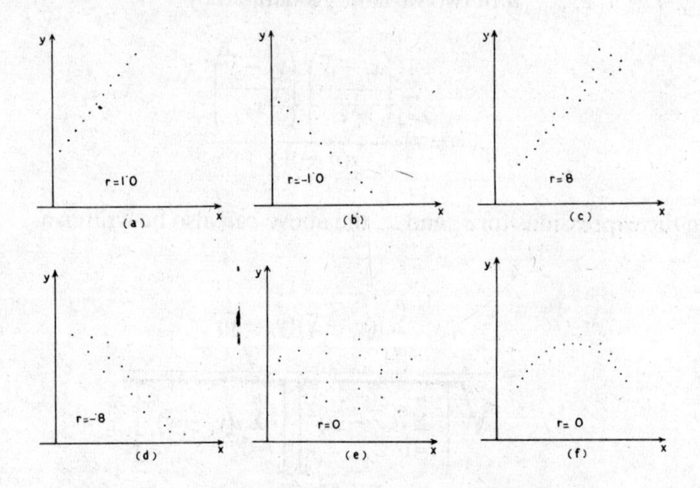

Fig. 9.2 Correspondence between the value of r and the amount of scatter

Fig. 9.2(a) shows that all the points of the scatter diagram lie on a straight line; further that y increases as x increases. The coefficient of correlation in this case is $r = 1$, the maximum value that the coefficient r can attain; we find that all the points lie on a line, and that both x and y increase or decrease together.

Fig. 9.2(b) shows that all the points lie on a straight line, and that y decreases as x increases. In this case $r = -1$, the minimum value that r can attain; we note that all the points lie on a line but y decreases as x increases and vice versa. Correlation is *positive* in (a) and *negative* in (b). Fig. 9.2(c) indicates that the points of the scatter diagram do not lie on a straight line but lie close to one; the correlation is positive and the coefficient r has a value equal to 0.8. Fig. 9.2(d) shows negative correlation of the same magnitude. Fig. 9.2(e) shows that the points do not show any tendency to lie on or near a line; r has the value 0 in this case. Fig. 9.2(f) shows that the points lie near a curve, which is not a straight line; the coefficient has the value 0 in the case also.

Example 3(a). The following table gives, to the nearest unit, the index numbers of agricultural production (x) with base 1970 = 100 and the index numbers of wholesale prices (y) with base 1970-71 = 100 for India. It is assumed that the quantities of interest are independent of time variation.

	1971	1972	1973	1974	1975
x	104	110	112	114	120
y	106	116	140	175	173

We have $\bar{x} = 112, \bar{y} = 142$

x_i	y_i	$x_i - \bar{x}$	$y_i - \bar{y}$	$(x_i - \bar{x})^2$	$(y_i - \bar{y})^2$	$(x_i - \bar{x})(y_i - \bar{y})$
104	106	−8	−36	64	1296	288
110	116	−2	−26	4	676	52
112	140	0	−2	0	4	0
114	175	2	33	4	1089	66
120	173	8	31	64	961	248
Total 560	710	0	0	136	4026	654

Using (3.2), we get

$$r = \frac{\Sigma(x_i - \bar{x})(y_i - \bar{y})}{\sqrt{\left[\left\{\Sigma(x_i - \bar{x})^2\right\}\left\{\Sigma(y_i - \bar{y})^2\right\}\right]}} = \frac{654}{\sqrt{136 \times 4026}} = \frac{654}{739.96} = 0.88$$

Consider an arbitrary origin (100, 140); then measured from this origin we get

$x_i' = x_i - 100$	$y_i' = y_i - 140$	$(x_i - 100)^2$	$(y_i - 140)^2$	$(x_i - 100)(y_i - 140)$
4	−34	16	1156	−136
10	−24	100	576	−240
12	0	144	0	0
14	35	196	1225	490
20	33	400	1089	660
$\Sigma x_i' = 60$	$\Sigma y_i' = 10$	$\Sigma x_i'^2 = 856$	$\Sigma y_i'^2 = 4046$	$\Sigma x_i' y_i' = 774$

$$r = \frac{n \Sigma x_i' y_i' - (\Sigma x_i')(\Sigma y_i')}{\sqrt{\left[\left\{n \Sigma (x_i')^2 - (\Sigma x_i')^2\right\}\left\{n \Sigma (y_i')^2 - (\Sigma y_i')^2\right\}\right]}}$$

$$= \frac{5 \times 774 - 60 \times 10}{\sqrt{\left[\left\{5 \times 856 - (60)^2\right\}\left\{5 \times 4046 - (10)^2\right\}\right]}}$$

$$= \frac{3870 - 600}{\sqrt{[4280 - 3600][20230 - 100]}} = \frac{3270}{\sqrt{(680 \times 20130)}}$$

$$= \frac{3270}{10 \times \sqrt{136884}} = \frac{327}{369.978} = 0.88.$$

Example 3(b). The following table shows the heights of a sample of 10 fathers and their eldest sons :

(to nearest cm)

Height of father :	170	167	162	163	167	166	169	171	164	165
Height of son :	168	167	166	166	168	165	168	170	165	168

Taking deviations from 165 cm in each case i.e., writing $x' = x - 165$, $y' = y - 165$, we can construct the following table.

x_i'	y_i'	$(x_i')^2$	$(y_i')^2$	$(x_i' y_i')$
5	3	25	9	15
2	2	4	4	4
−3	1	9	1	−3
−2	1	4	1	−2
2	3	4	9	6
1	0	1	0	0
4	3	16	9	12
6	5	36	25	30
−1	0	1	0	0
0	3	0	9	0
$\Sigma x_i' = 14$	$\Sigma y_i' = 21$	$\Sigma (x_i')^2 = 100$	$\Sigma (y_i')^2 = 67$	$\Sigma x_i' y_i' = 62$

Using formula (3.4), we get

$$r = \frac{10 \times 62 - 14 \times 21}{\sqrt{\left\{10 \times 100 - (14)^2\right\}\left\{10 \times 67 - (21)^2\right\}}}$$

$$= \frac{326}{429.0764} = 0.76$$

Properties of the coefficient of correlation

(1) r lies between -1 and $+1$

(2) r is independent of the choice of origin and the scale of measurement. The value of r calculated for the set (x_i, y_i) $i = 1, 2, \ldots, n$ will be the same as that of the set (x_i', y_i'), $i = 1, 2, \ldots, n$ where $x_i' = ax_i + b$, $y_i' = cx_i + d$ and a, c are non zero and both positive or both negative, and b, d any constants.

(3) r is a measure of the *linear* relationship; r is positive when both the variables increase or decrease together; r is negative when the values of one variable increase as the values of the other decrease and vice-versa.

We prove below (1) and (2).

(1) To prove that r lies between -1 and 1.

Proof : Suppose that the variables are measured from their means. Let $u_i = x_i - \bar{x}, v_i = y_i - \bar{y}$. Then from (3.2) we get

$$r = \frac{\Sigma u_i v_i}{\sqrt{\{\Sigma u_i^2\}\{\Sigma v_i^2\}}} \tag{3.5}$$

so that
$$r^2 = \frac{(\Sigma u_i v_i)^2}{\Sigma u_i^2 \, \Sigma v_i^2} \tag{3.6}$$

Cauchy-Schwarz inequality states that

$$\sum_{i=1}^{n} a_i^2 \sum_{i=1}^{n} b_i^2 \geq \left(\sum_{i=1}^{n} a_i b_i \right)^2 \tag{3.7}$$

For $n = 2$,
$$\Sigma a_i^2 \Sigma b_i^2 - (\Sigma a_i b_i)^2 \geq 0$$

$$\Rightarrow \quad (a_1^2 + a_2^2)(b_1^2 + b_2^2) - (a_1 b_1 + a_2 b_2)^2 \geq 0$$

or $(a_1^2 b_2^2 + a_2^2 b_1^2 - 2a_1 a_2 b_1 b_2) \geq 0$ or $(a_1 b_2 - a_2 b_1)^2 \geq 0$

For any n,

$$\sum_{i=1}^{n} a_i^2 \sum_{i=1}^{n} b_i^2 - \left(\sum_{i=1}^{n} a_i b_i \right)^2$$

$$= \sum_{i \neq j}^{n} (a_i b_j - a_j b_i)^2 \geq 0$$

which gives (3.7). Using Cauchy-Schwarz inequality (3.7) in (3.6) we get

$$r^2 = \frac{(\Sigma u_i v_i)^2}{\Sigma u_i^2 \Sigma v_i^2} \leq 1;$$

hence we get

$$-1 \leq r \leq 1.$$

(2) To prove that r is independent of the origin and the scale of measurement.

Proof: Let

$$u_i = \frac{x_i - x_0}{h}, \quad v_i = \frac{y_i - y_0}{k}$$

Then (x_i, y_i) differ from (u_i, v_i) in a change of origin and scale of measurement.

We have

$$x_i = hu_i + x_0,\ y_i = kv_i + y_0$$

and
$$s_x = hs_u,\ s_y = ks_v$$

Also
$$x_i - \bar{x} = h(u_i - \bar{u}), y_i - \bar{y} = k(v_i - \bar{v})\,.$$

We have the product moment correlation coefficient r between x and y given by

$$r = \frac{\Sigma(x_i - \bar{x})(y_i - \bar{y})}{(n-1)\, s_x s_y} \qquad \text{[from (3.1)]}$$

$$= \frac{\Sigma h(u_i - \bar{u})\, k\,(v_i - \bar{v})}{(n-1)\, hs_u ks_v}$$

$$= \frac{\Sigma(u_i - \bar{u})(v_i - \bar{v})}{(n-1)\, s_u s_v}$$

which is the product moment correlation coefficient between u and v. Thus u and v have the same correlation coefficient as x and y.

Notes :

1. The above two results can also be proved to hold in case of grouped data.
2. The second result is used for numerical calculation of r. See Example 8(a).

Interpretation of the value of r

The value of r is a measure of the strength of the *linear relationship between sets of values of* x_i and y_i. A large value of r indicates that there is a strong linear relationship between x_i and y_i, that is, the points (x_i, y_i) lie near a straight line. A large positive value of r indicates that the straight line has a positive slope and a large negative value of r indicates that the straight line has a negative slope. The value $r = +1$ or -1 indicates that all the points lie on a straight line. A small value of r indicates that there is no linear relationship between x_i and y_i. The points (x_i, y_i) do not lie near a line; it does not necessarily imply that there is no relationship between x_i and y_i, it simply means that even if a relationship exists, the relationship is not linear. Fig. 9.2(f) shows that x_i, y_i lie on a curve which is not a straight line and a relationship between x_i and y_i exists but it is not linear.

Another point to be noted is that a large value of r indicates a strong *mathematical* relationship which is *linear* in form but does not imply a *causal* relationship between them; in fact both the variables may be influenced by other variable or variables, and the influence may be such to give a strong mathematical (linear) relationship. For example, over a period of years, the coefficient of correlation between the number of commerce graduates and the number of crimes committed (in a number of cities) may turn out to be quite high, say around (0.9); this would simply imply that both the number of commerce graduates and the number of crimes committed vary in the same direction and that the relationship between them is nearly linear, but one cannot be said to be the cause of the other; the increase in the population (in the cities) may have caused both of them to vary in the same direction, and in fact the two may be completely unrelated or even negatively related. We thus get a high correlation due to a third variable; such a high correlation due to another variable is called *spurious correlation* and the variable which causes such high correlation is called *lurking variable.* Consider another example : there seems to be a high positive correlation between cigarette smoking and cancer. From this fact *alone* one cannot jump to the conclusion that cigarette smoking is *the or a cause* of cancer; in fact it may be that there is some genetic cause behind both; a person prone to cancer may get a pleasurable sensation by smoking due to his genetic constitution. The fact that there may be high positive correlation between cigarette smoking and cancer (of a certain type) would not indicate anything about existence of *causal* relationship between the two; causal relationship is to be based on etiological evidence.

9.4 LINEAR REGRESSION (SIMPLE LINEAR REGRESSION)

Suppose that the coefficient of correlation r is calculated from a set of pairs of observations (x_i, y_i), $i = 1, 2, \ldots, n$. The coefficient of correlation indicates whether the variables are linearly related and if so, how strong the relationship is. If $r = \pm 1$, then the points will lie on a straight line, a high value of r will imply that the points will lie near about a straight line. The question arises : what is this line, how can this line be found and what, if any, is the usefulness of this line? Methods employed to answer such questions are known as regression methods. The main objectives of regression analysis are the formulation and determination of the mathematical model or form of the relationship between the variables and the use of the model for prediction purposes. Regression methods will consist of determining the exact mathematical form of the relationship and then using the relation for prediction purposes. The word regression, which means reversion, was first used by Galton, to whom regression analysis is due. His study was on heights of fathers and sons and showed that the heights of sons tended or reverted towards the average rather than to the extreme values. Let us consider our data of Example 3(b) of heights of fathers and their eldest sons. Drawing the scatter diagrams we find that the points lie near a line and that x and y are approximately linearly related for the range of values of x. Suppose that we fit the line to these points by some method (which will be explained later here).

The scatter diagram and the fitted line can be shown in a graph; from it we can find the value of y to any given value of x in the range of values of x considered. For the value of x, say, $x = 167$ cm., the value of y is approximately 167.3 cm. This is the predicted value of y corresponding to the value of $x = 167$ cm. Now we find from the table that there are two values of y corresponding to

$x = 167$ cm.; $y = 167$ cm. and $y = 168$ cm.; and the value of y predicted from the graph of the fitted line is 167.3. Similarly we can find the predicted value of y corresponding to any value of x in the range considered. Note that the method may not be used for predicting the value of y for values of x outside the range considered. For example, if we consider the ages of 20 children (of ages between 6 to 16) and their heights and obtain the regression line of height (y) on age (x), shall we be justified in predicting the value of y corresponding to $x = 30$?

Further note that the predicted value of y corresponding to $x = 167$ cm. is different from the values of y corresponding to $x = 167$ cm. given in the table. What we have in the table is a *sample* of 10 pairs of measurements. If we had taken a large number of samples and taken the average value of x and the average value of y, the straight line fitted to these average values of x and y would have been of greater precision. It is assumed that there is, theoretically, a straight line that expresses the relationship between the theoretical mean (expected) values of x and y. The line that is fitted from the sample values of x, y as given in Table 9.1 is a (sample) straight line obtained as an estimate of the theoretical one; the value of y obtained from this (sample) line is expected to be more or less a reasonable value of y obtained from the theoretical line than the observed values of $y = 167$ cm. and $y = 168$ cm. given in the Table.

We write below the values of y, observed y and estimated (or predicted) $\hat{y}$ corresponding to values of x in the Table below.

x	170	167	162	163	167	166	169	171	164	165
y	168	167	166	166	168	165	168	170	165	168
$\hat{y}$	168.3	167.3	165.2	165.6	167.3	166.8	168.2	169.0	166.0	166.4

Once the line is fitted, the predicted values can be found. We shall again consider the values $\hat{y}$ later on. We shall now consider how the regression line is fitted.

9.5. FITTING OF REGRESSION LINE

Suppose that the variables x and y are linearly related and that with a view to predict the values of y corresponding to some given values of x we are to fit a straight line to the set of points (x_i, y_i), $i = 1, 2, \ldots, n$. Now the equation of a straight line can be written in the form

$$y = a + bx$$

in which a and b are parameters; the parameter a is the length of the intercept cut off from the y-axis by the line and b is the slope of the line.

The straight line is completely known when the values of the two parameters a and b are known. Fitting of a straight line to a set of points (x_i, y_i), $i = 1, 2, \ldots, n$ amounts to finding the values of a and b with the help of the values of x_i, y_i such that the line will fit the set of points well. We are concerned here with estimating the values of a and b.

9.5.1 The Method of Least Squares

There are several methods for estimating the values of the parameters a and b and thereby fitting a straight line; we shall adopt the best known among them, the *method of least squares*; This method ensures that the *errors of prediction*, for which the regression line is used, is the minimum possible.

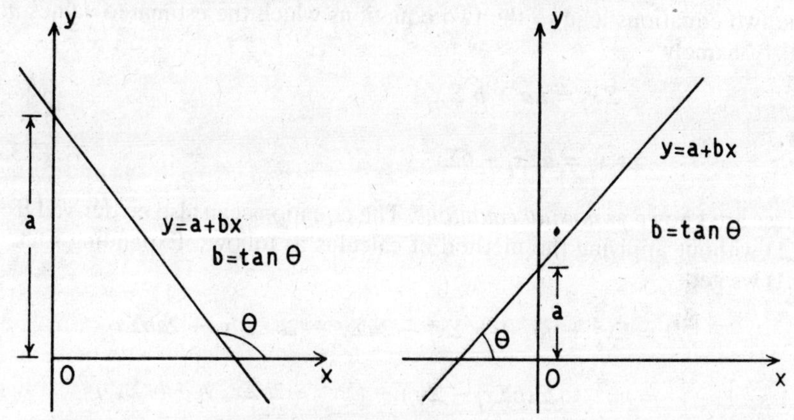

Fig. 9.3 Graph of the straight line $y = a + bx$

When we estimate or predict the value of y corresponding to a value of x by means of the fitted line, there is a difference between the observed value y and the estimated value $\hat{y}$; this difference $(y - \hat{y})$ is the error of prediction. The error $(y - \hat{y})$ is positive when the point lies above the line and is negative when it lies below the line. We are to find the line such that the total error of prediction for all the n points is the least or minimum. What should be the criterion of considering the total error? Simple addition of individual errors will not be helpful as the positive and negative values may tend to balance and even a poor fitting line may have very small total error; another way may be to consider the absolute value of the error, but the sums of absolute values are not very convenient to work with. This difficulty is obviated by considering the sum of the squares of individual errors. The values of the parameters a and b to be found should be such that the sum of the squares of errors $\sum_{i=1}^{n} (y_i - \hat{y}_i)^2$ is the least possible; in other words, values of a and b should be such as to minimise the sum of the squares of errors. The straight line $y = a + bx$ will be the straight line which will best fit the set of points. It will be the best fitting line under the criterion or sense of least squares.

The best fitting line will be such that

$$D = \sum_{i=1}^{n} (y_i - \hat{y}_i)^2 = \sum_{i=1}^{n} (y_i - a - bx_i)^2 \tag{5.1}$$

is the least or minimum. For the function D to have a minimum, it is necessary that its two partial derivatives vanish. Hence the estimated values of a, b for which D is minimum must satisfy the two equations

$$\frac{\partial D}{\partial a} = \Sigma\, 2(y_i - a - bx_i)(-1) = 0 \tag{5.2}$$

$$\frac{\partial D}{\partial b} = \Sigma\, 2(y_i - a - bx_i)(-x_i) = 0 \tag{5.3}$$

The two equations lead to the two equations which the estimated values $\hat{a}, \hat{b}$ satisfy, namely

$$\Sigma\, y_i = n\hat{a} + \hat{b}\,\Sigma\, x_i \tag{5.2a}$$

$$\Sigma x_i y_i = \hat{a}\Sigma x_i + \hat{b}\Sigma x_i^2 \tag{5.3a}$$

These are known as *normal equations*. The equations can also be derived from (5.1) without applying the method of calculus as follows. Expanding r.h.s. of (5.1) we get

$$D = \Sigma y_i^2 + na^2 + b^2\Sigma x_i^2 - 2a\Sigma y_i - 2b\Sigma x_i\, y_i + 2ab\Sigma x_i$$

$$= na^2 + 2a(b\Sigma x_i - \Sigma y_i) + (\Sigma y_i^2 - 2b\Sigma x_i\, y_i + b^2\Sigma x_i^2)$$

which is a quadratic in a of the form $pa^2 + qa + r$.

Now $$pa^2 + qa + r = p\left(a^2 + \frac{q}{p}a + \frac{r}{p}\right)$$

$$= p\left\{\left(a + \frac{q}{2p}\right)^2 + \frac{4pr - q^2}{4p^2}\right\}$$

is minimum when $\hat{a} + \dfrac{q}{2p} = 0$, i.e. for $\hat{a} = -\dfrac{q}{2p}$.

This leads to the first normal equation (5.2a).

Similarly, considering D as a quadratic in b, we can get the second normal equation (5.3a).

Though it does not indicate that D is the least, a straight forward method of arriving at the normal equations is as follows. We have

$$y_i = a + bx_i \tag{5.4}$$

Multiplying both sides by x_i, we get

$$x_i\, y_i = ax_i + bx_i^2 \tag{5.4a}$$

Summing both sides of (5.4) and (5.4a) we get the two normal equations

$$\Sigma y_i = na + b\,\Sigma x_i,$$

$$\Sigma\, x_i y_i = a\Sigma x_i + b\Sigma x_i^2$$

Solution of the normal equations

Solving (5.2a) and (5.3a) we get

$$\hat{a} = \frac{\Sigma y_i \Sigma x_i^2 - \Sigma(x_i)(\Sigma x_i y_i)}{n\Sigma x_i^2 - (\Sigma x_i)^2} \tag{5.5}$$

$$\hat{b} = \frac{n\Sigma x_i y_i - (\Sigma x_i)(\Sigma y_i)}{n\Sigma x_i^2 - (\Sigma x_i)^2} \tag{5.6}$$

These are known as the least squares estimates of a and b respectively. Using the mean values,

$$\bar{x} = \sum_{i=1}^{n} \frac{x_i}{n}, \quad \bar{y} = \sum_{i=1}^{n} \frac{y_i}{n},$$

$\hat{a}$ can also be expressed as

$$\hat{a} = \bar{y} - \hat{b}\,\bar{x} \tag{5.5a}$$

Thus (5.5) and (5.6) (or alternatively (5.5a) and (5.6)) give the estimated values of a and b. The fitted line has the equation

$$\hat{y} = \hat{a} + \hat{b}x \tag{5.7}$$

or,

$$\hat{y} - \bar{y} = \hat{b}(x - \bar{x}) \quad \text{[using (5.5a)]} \tag{5.7a}$$

As $x = \bar{x}$ and $y = \bar{y}$ satisfy the equation of the line, the line passes through the point $(\bar{x}, \bar{y})$. The formula (5.6) is independent of the choice of origin and scale. It can be applied if x_i and y_i are replaced by

$$u_i' = \frac{x_i - c}{d}, \quad v_i' = \frac{y_i - c}{d'}$$

where c, d, c', d' are constants (and d, d' have the same sign). The formula (5.5) is not independent of the choice of origin and scale; however, when x, y are known (5.5a) can be used instead of (5.5). The value of y obtained by putting the value of x in $y = a + bx$ is the predicted or estimated value of y and is denoted by $\hat{y}$.

If we use deviations from the mean $x_i' = x_i - \bar{x}$ and $y_i' = y_i - \bar{y}$, then since $\Sigma x_i' = \Sigma y_i' = 0$, we get from (5.6).

$$\hat{b} = \frac{n\,\Sigma(x_i - \bar{x})(y_i - \bar{y})}{n\,\Sigma(x_i - \bar{x})^2} = \frac{\Sigma x_i y_i - \bar{x}\,\bar{y}}{\Sigma x_i^2 - n\overline{x^2}} \tag{5.6a}$$

and thence $\hat{a}$ from (5.5a). Finally we get

$$\hat{y} - \bar{y} = \hat{b}(x - \bar{x}) \tag{5.7a}$$

which is the fitted regression line of y on x.

The constant $\hat{b}$ is known as the regression coefficient of y on x. When $\hat{b}$ is positive, it gives the amount of increase (decrease) in y corresponding to the unit increase (decrease) in x; when $\hat{b}$ is negative it gives the amount of decrease (increase) in x corresponding to a unit increase (decrease) in x.

Let us consider some examples as to how to find the best fitting line.

Example 5(a). Consider the data of Example 3(a).

Working with $x_i - \bar{x}$ and $y_i - \bar{y}$ for x_i, y_i respectively we get from (5.6) and (5.5a)

$$\hat{b} = \frac{5 \times 654 - 0 \times 0}{5 \times 136} = 4.8088 \cong 4.81$$

$$\hat{a} = \bar{y} - \hat{b}\bar{x} = 142 - (4.81)(112) = -396.72$$

so that $$\hat{y} = -396.72 + (4.81)x$$

is the regression line.

The predicted values $\hat{y}$ of y corresponding to the values of x can be easily found. For example, for $x = 104$, $y = -396.72 + (4.81) \times 104 = 103.52$ and so on. We write below the values of x, the observed values y and the predicted values $\hat{y}$.

x	104	110	112	114	120
y	106	116	140	175	173
$\hat{y}$	103.52	132.38	142.00	151.62	180.48

Example 5(b). Consider the data of Example 3(b). Using $x_i - 165$ and $y_i - 165$ for x_i and y_i respectively, we get from (5.6)

$$\hat{b} = \frac{10 \times 62 - 14 \times 21}{10 \times 100 - (14)^2} = \frac{326}{804} = 0.40547 \cong 0.41.$$

Since $\bar{x} = 165 + \dfrac{14}{10} = 166.4$ and $\bar{y} = 165 + \dfrac{21}{10} = 167.1$, we get from (5.5a)

$$\hat{a} = \bar{y} - \hat{b}\,\bar{x} = 167.1 - (0.41)(166.4) = 98.88$$

Thus $$\hat{y} = 98.88 + (0.41)x$$

is the regression line of y on x.

The values of x, y (observed) and y (predicted) are given below:

Also noted are the squares of errors $(y - \hat{y})^2$.

x	170	167	162	163	167	166	169	171	164	165
y	168	167	166	166	168	165	168	170	165	168
$\hat{y}$	168.6	167.4	165.3	165.7	167.4	166.9	168.2	169.0	166.1	166.5
$y - \hat{y}$	−0.6	−0.4	0.7	−0.3	0.6	−1.9	−0.2	1.0	−1.1	1.5
$(y - \hat{y})^2$	0.36	0.16	0.49	0.09	0.36	3.61	0.04	1.00	1.21	2.25

We have

$$\sum_{i=1}^{10} (y_i - \hat{y}_i)^2 = 0.36 + 0.16 + \ldots + 2.25 = 9.57$$

If we fit another line, say, given by

$$_1\hat{y} = 100 + (0.5)\,x$$

predicted values can be obtained by putting the value of x.

We have

y	168	167	166	166	168	165	168	170	165	168
$_1\hat{y}$	185.0	183.5	181	181.5	183.5	183	184.5	185.5	182	182.5
$y - {_1\hat{y}}$	−17.0	−16.5	−15.0	−15.5	−15.5	−18	−16.5	−15.5	−17.0	−14.5

We can easily see that $\Sigma(y - {_1\hat{y}})^2$ will be greater than $\Sigma(y - \hat{y})^2$ obtained from the best fitted line. This verifies that the error of prediction for the best fitted line obtained by the method of least squares is smaller than the corresponding error for any other line.

9.5.2 Explained and Unexplained Variation : Coefficient of Determination

After a regression line is drawn, it may be of interest to assess how useful is the line for predicting the value of y from that of x. Consider that a regression of line y on x is to be drawn from the pairs of observation (x_i, y_i) $i = 1, 2, \ldots, n$. Suppose that the equation is

$$\hat{y} = \hat{a} + \hat{b}\,x$$

where $\hat{a}, \hat{b}$ are obtained as solutions of the normal equations (5.6) and (5.5a) or (5.5).

Let the variables be measured from their means. Denote

$$\bar{x} = \frac{1}{n}\Sigma x_i,\ \bar{y} = \frac{1}{n}\Sigma y_i$$

$$S_x^2 = \Sigma(x_i - \bar{x})^2,\ S_y^2 = \Sigma(y_i - \bar{y})^2 \qquad (5.8)$$

$$S_{xy} = \Sigma(x_i - \bar{x})(y_i - \bar{y})$$

Then from (5.6a) and (5.7a)

$$\hat{b} = \frac{S_{xy}}{S_x^2} \qquad (5.9)$$

$$\hat{y}_i - \bar{y} = \hat{b}\,(x_i - \bar{x}) \qquad (5.10)$$

and from (3.2)

$$r = \frac{S_{xy}}{S_x S_y} \qquad (5.11)$$

$$= \hat{b}\frac{S_x}{S_y} \qquad \text{[using (5.9)]}. \tag{5.11a}$$

The sum of the squares of errors in fitting a line is given by

$$D = \Sigma\,(y_i - \hat{y}_i)^2$$

$$= \Sigma\left\{(y_i - \bar{y} - \hat{b}\,(x_i - \bar{x})\right\}^2, \qquad \text{[using (5.10)]}$$

$$= \Sigma(y_i - \bar{y})^2 + (\hat{b})^2\Sigma(x_i - \bar{x})^2 - 2\,\hat{b}\,\Sigma\,(y_i - \bar{y})(x_i - \bar{x})$$

$$= S_y^2 + (\hat{b})^2 S_x^2 - 2\hat{b}\,S_{xy}$$

$$= S_y^2 + (\hat{b})^2 S_x^2 - 2\hat{b}\,(\hat{b}S_x^2), \qquad \text{[using (5.9)]}$$

Thus
$$D = S_y^2 - (\hat{b})^2 S_x^2 \tag{5.12}$$

Again using (5.11a), we have

$$D = S_y^2 - \left(r \cdot \frac{S_y}{S_x}\right)^2 S_x^2$$

$$= S_y^2 - r^2 S_y^2 \tag{5.13}$$

Thus
$$\frac{D}{S_y^2} = 1 - r^2. \tag{5.14}$$

Since $\frac{D}{S_y^2} \geq 0$, it follows from (5.14) that

$$1 - r^2 \geq 0, \text{ i.e. } r^2 \leq 1 \text{ or } -1 \leq r \leq 1.$$

Thus (5.14) also shows that r lies between -1 and $+1$
Again it can be seen that

$$S_y^2 = \Sigma(y_i - \bar{y})^2 = \Sigma(y_i - \hat{y}_i)^2 + \Sigma(\hat{y}_i - \bar{y})^2 \tag{5.15}$$

$$= D + \Sigma(\hat{y}_i - \bar{y})^2 \tag{5.15a}$$

To prove (5.15) we write

$$\Sigma(y_i - \bar{y})^2 = \Sigma\left\{(y_i - \hat{y}_i) + (\hat{y}_i - \bar{y})\right\}^2$$

$$= \Sigma(y_i - \hat{y}_i)^2 + \Sigma(\hat{y}_i - \bar{y})^2 + 2\Sigma(y_i - \hat{y}_i)(\hat{y}_i - \bar{y}) \tag{5.16}$$

Now the term

$$2\Sigma(y_i - \hat{y}_i)(\hat{y}_i - \bar{y}) = 2[\Sigma\hat{y}_i(y_i - \hat{y}_i) - \bar{y}\Sigma(y_i - \hat{y}_i)] = 0$$

since each of the terms vanishes as can be seen by applying the normal equations (5.4a and b). Thus (5.15) holds. The second term

$$\Sigma(\hat{y}_i - \bar{y})^2$$

which gives the sum of the squares of the errors of the predicted values $\hat{y}_i$ (based on the line of regression) from the mean value $\bar{y}$ is called *explained variation.* The first term $D = \Sigma(y_i - \hat{y}_i)^2$ is called the *unexplained variation.*
The ratio

$$\frac{\Sigma(\hat{y}_i - \bar{y})^2}{S_y^2} \tag{5.17}$$

is the proportion of the explained variation to the total variation S_y^2 and is called the *coefficient of determination.* This gives a measure of the usefulness of the line of regression in predicting y from x.

It follows that the coefficient of determination lies between 0 and 1. If all the points lie on the regression line obtained, then the coefficient of determination becomes unity.

We shall now proceed to show that the coefficient of determination equals r^2. Dividing both sides of (5.15a) by S_y^2 and transposing, we get the coefficient of determination

$$\frac{\Sigma(\hat{y}_i - \bar{y})^2}{S_y^2} = 1 - \frac{D}{S_y^2}$$

$$= r^2 \qquad \text{[from (5.14)]} \tag{5.18}$$

When r is known it can at once be said that the proportion of the explained variation to the total variation is equal to r^2 in magnitude, i.e. the sum of the squares of the error of prediction has been reduced by about r^2 per cent by using the regression line. This also gives a relation between correlation and regression analyses. The relation (5.18) can be used to find the coefficient of determination, when r is known, even without obtaining the regression equation.

Consider the data of Example 5(b).
We have

$$\bar{y} = 167.1, y_1 - \bar{y} = 0.9, \ldots, y_{10} - \bar{y} = 0.9$$

so that

$$S_y^2 = \sum_{i=1}^{10} (y_i - \bar{y})^2 = (0.9)^2 + (0.1)^2 + \ldots + (0.9)^2$$

$$= 22.90$$

Again we have $\Sigma(y_i - \hat{y})^2 = 9.57$

so that $\Sigma(\hat{y}_i - \bar{y})^2 = 22.90 - 9.57 = 13.33$

The coefficient of determination equals

$$\frac{\Sigma(\hat{y}_i - \bar{y})^2}{\Sigma(y_i - \bar{y})^2} = \frac{13.33}{22.90} = 0.5820$$

That is, the regression line accounts for 58.2% of the total variation. Further, $r = \sqrt{0.5820} = 0.76$

This tallies with the value of r found in Example 3(b).

Note : Direct calculation of $\Sigma(\hat{y}_i - \bar{y})^2$ gives the value 13.85. Then $r = 0.78$. The relation (5.18) holds only in case of simple linear regression.

9.6 TWO REGRESSION LINES : RELATION WITH r

So far we have considered the regression of y on x. This is based on the assumption that x is the independent variable and y is the dependent variable. The regression line is used to estimate or predict the value of y corresponding to a given value of the independent variable x. When y is considered as the independent variable and x as the dependent variable then we have to find the regression line of x on y for predicting the value x for a given value of y; the predicted value will be denoted by $\hat{x}$. The equation of the regression line of x on y is of the form.

$$x = c + dy \tag{6.1}$$

The estimated parameters $\hat{c}, \hat{d}$ are obtained as solutions of the normal equations:

$$\Sigma x_i = \hat{c}\,.\,n + \hat{d}\,\Sigma y_i \tag{6.2}$$

$$\Sigma x_i y_i = \hat{c}\,\Sigma y_i + \hat{d}\,\Sigma y_i^2 \tag{6.3}$$

Solving the above equations, we get

$$\hat{d} = \frac{n\,\Sigma x_i y_i - \Sigma x_i\,\Sigma y_i}{n\,\Sigma y_i^2 - (\Sigma y_i)^2} \tag{6.4}$$

and

$$\hat{c} = \frac{(\Sigma x_i)(\Sigma y_i^2) - (\Sigma y_i)(\Sigma x_i y_i)}{n\,\Sigma y_i^2 - (\Sigma y_i)^2}; \tag{6.5}$$

$\hat{c}$ can also be found from

$$\hat{c} = \bar{x} - \hat{d}\,\bar{y} \tag{6.5a}$$

and the estimated or predicted value $\hat{x}_i$ corresponding to given y_i can be found

from the equation.

$$\hat{x} = \hat{c} + \hat{d}\,y \qquad (6.6)$$

by putting the value of y on the r.h.s.

This is the regression line of x on y : to be used for prediction or estimation of x_i for given y_i when y is considered as the independent variable and x is considered as the dependent variable. Here d is the regression coefficient of x on y. It is positive when y and x increase or decrease together and is negative when one increases the other decreases. It can be easily seen that the two regression coefficients (calculated from the same set of observations (x_i, y_i)) will *be of the same sign*. The amount of increase (or decrease) in x when d is positive corresponding to a unit increase (decrease) in y is equal to d, and the reverse is the case when d is negative. We shall have similar properties for the reression line of x on y. We can define the coefficient of determination in the same way. The value of this coefficient will be equal to r^2.

Further, it can be shown that the product of the two regression coefficients is equal to r^2 i.e. $bd = r^2$. Thus it is possible to find r when the two regression lines are known or simply when the two regression coefficients are known; r will have the same sign as the regression coefficients.

The two regression equations are identical *if and only if* all the points of the scatter diagram lie on a straight line. When this happens, the coefficient of correlation r has the value $+1$ or -1; the linear correlation is perfect.

Example 6(a). **Find the equations of the two regression lines from the set of observations given in Example 3(a).**
Here we have $x_i - \bar{x}$ and $y_i - \bar{y}$ so that

$$\hat{b} = \frac{5 \times 654}{5 \times 136} = 4.8088$$

$$\hat{a} = \bar{y} - \hat{b}\,\bar{x} = 142 - 4.8088 \times 112 = 142 - 538.5856 = -396.5856$$

$$\hat{d} = \frac{5 \times 654}{5 \times 4026} = 0.1624$$

$$\hat{c} = \bar{x} - \hat{d}\,\bar{y} = 112 - (0.1624) \times 142 = 112 - 23.0608 = 88.9392.$$

The regression line of y on x is given by

$$y = -396.59 + (4.81)\,x$$

or $$y - 142 = 4.81(x - 112)$$

since $$\bar{x} = 112 \text{ and } \bar{y} = 142.$$

The regression line of x on y is given by

This is $$x = 88.94 + (0.16)\,y$$

or $$x - 112 = (0.16)(y - 142)$$

It may be noted that $\hat{b}, \hat{d}$ are of the same sign and that $\hat{b}\hat{d} = 0.780949$ and $r = \sqrt{0.780949} = 0.88$. This verifies the value of r found in example 3(a). Verify that the two regression lines pass through the point $(\bar{x}, \bar{y}) \equiv (112, 142)$, i.e. they intersect at (112, 142).

It can be seen that when the two regression equations are given, the coefficient of correlation r and the means $\bar{x}$ and $\bar{y}$ can be found by using the relations $\hat{b}\hat{d} = r^2$ (for finding r) and $\hat{a} = \bar{y} - \hat{b}\,\bar{x}$ and $\hat{c} = \bar{x} - \hat{d}\,\bar{y}$ (for finding $\bar{x}$ and $\bar{y}$).

Example 6(b). Suppose that the regression line of y on x is given by $y = 26.98 - 1.35x$ and that of x on y is given by $x = 16.38 - 0.45y$. Find the coefficient of correlation between x any y and also $\bar{x}$ and $\bar{y}$.

We have
$$r^2 = (-1.35)(-0.45) = 0.6075$$

so that
$$r = \pm\sqrt{0.6075} = \pm 0.78.$$

Since r has the same sign as that of the regression coefficients, we take the negative square root, so that $r = -0.78$

Again,
$$26.98 = \hat{a} = \bar{y} - \hat{b}\,\bar{x} = \bar{y} - (-1.35)\bar{x}$$

and
$$16.38 = \hat{c} = \bar{x} - \hat{d}\,\bar{y} = \bar{x} - (-0.45)\,y.$$

Solving the two simultaneous equations, we get $\bar{x} = 10.8$ and $\bar{y} = 12.4$.

9.7 STATISTICAL MODEL

We have been discussing about the coefficient of correlation and the regression lines with respect to a set of observed values (x_i, y_i), $i = 1, 2, \ldots, n$. We get a value of r and also two regression lines. These are obtained from the sample observed and are sample estimates. If we take another sample, we get another value of r and also two other regression lines. In fact the value of r and the regression lines are sample estimates of the corresponding theoretical value of r and the regression lines respectively. The sample (x_i, y_i) is supposed to be drawn from a population of values of two random variables X and Y. We recall that the theoretical or population correlation coefficient between the random variables X and Y is defined as follows:

Let us assume that the random variables assume discrete values. We assume that the r.v. X assumes values $x_1, \ldots, x_k$ with corresponding probability (distribution) $a_i = P(X = x_i), i = 1, 2, \ldots, k$ and the r.v. Y assumes values $y_1, \ldots, y_r$ with corresponding probability (distribution) $b_j = \hat{P}(Y = y_i), j = 1, 2, \ldots, r$. This idea is extended to the joint probability distribution of the two variables X and Y; this requires that (X, Y) assumes the values (x_i, y_j) with probability

$$f_{ij} = Pr(X = x_i, Y = y_j),$$
$$i = 1, 2, \ldots, k, \;\; j = 1, 2, \ldots, r$$

We get $a_i = \sum_{j=1}^{r} f_{ij}$ and $b_j = \sum_{i=1}^{k} f_{ij}$.

The population correlation coefficient ρ between the random variables X and Y is defined by

$$\rho = \frac{E(XY) - E(X)E(Y)}{\sigma_X \sigma_Y}$$

where E denotes expectation and σ denotes the standard deviation.

The above is the probability model for the coefficient of correlation. The sample correlation coefficient r computed from a sample of observations (x_i, y_i) $i = 1, 2, \ldots, n$ is an estimate of the population correlation coefficient between the two random variables X, Y.

Suppose that we consider X as the independent variable and Y as the dependent variable, and that the regression line of y on x is computed from a set of sample observations. This line is an estimate of the theoretical or population regression line. If the relation between the two variables is exactly linear, they are connected by a linear relationship

$$y = \alpha + \beta x;$$

this corresponds to the situation that, when x is given, y can be exactly determined. It is however reasonable to assume that this will not be so and that random errors or disturbances will be associated with the variables. Assuming that the random variable X is independent of error, we can write the model as

$$y_i = \alpha + \beta x_i + e_i$$

where

(i) $e_1, e_2, \ldots$ are random and unknown errors, and these errors are themselves random variables; it is assumed that the variables e_i are independently and identically distributed with mean 0 and variance σ^2.

(ii) the observation y_i which corresponds to a given x_i is a random sample of one observation from a normal population with mean $\alpha + \beta x_i$ and variance σ^2.

Our assumption is that the independent variable is free from error whereas the dependent variable is associated with uncontrollable error. Now cases arise where both the variables are subject to error. The model will then involve bivariate normal distribution.

There are many regression problems where the x values can be chosen in advance. For example, if an experimenter wishes to ascertain the effect of one factor on a certain product, then he would wish to run the experiment on a set of chosen values or the strength of the factor, with a view to find out the effect on the product. Here x values are chosen in advance and the y values are recorded corresponding to the chosen x values. On the other hand when we consider a sample of students and take their heights (x) and weights (y), x values are obtained at random.

The regression methods outlined above are very versatile; they can be applied both in the case when the values of x are chosen in advance and when the values of x are obtained by random selection.

But this is not the case with correlation analysis. Correlation methods require that both X and Y are random variables and that observations are obtained by random selection of x values, i.e. x's are not chosen in advance. Thus, though it would be appropriate to use r^2 as the measure of the explained variation (as also of the usefulness of the regression line for prediction purpose), it would not be appropriate to treat the square root of r^2, (obtained from regression equation as explained variation), as the coefficient of correlation between the variables; this should be done only when it is known that x values are also obtained by random selection.

An important point to be noted is that the regression methods discussed for prediction of y values are valid even when the x's are chosen in advance. But the values y_i of y are supposed to be independent. This assumption does not hold when dealing with time series data. For example, when we wish to determine the relationship between wholesale prices (x) and cost of living index numbers (y) from a set of values of (x_i, y_i) over a number of months (or years), the consecutive values y_i, y_{i+1} may not be independent. The investigator has to take caution in such cases of dealing with correlation and regression analyses of time series data.

We have so far discussed the problem of predicting y values on the basis of x values when x and y are connected by a linear relationship of the form $y = a + bx$. There may not always be a satisfactory linear relationship; in fact the relationship between x and y may be of a higher degree, say, of the form $y = a + bx + cx^2$. We are then concerned with nonlinear correlation and regression. The coefficient of correlation, which is a measure of the linear relationship between the variables, will not be useful then; however, second degree equation can be fitted by using the method of least squares. Regression methods for predicting the values of y can be used. This is briefly discussed in section **9.9.1.**

Again, instead of two variables, more than two variables may be involved; a linear relationship between y and two or more variables may be more useful in some situations. We shall then be concerned with multiple regression. We propose to discuss this briefly in section **9.9.2.**

9.8. CORRELATION AND REGRESSION FROM GROUPED DATA

Suppose that the values of x, y are grouped in a frequency distribution table, with x values grouped in k classes and y values grouped in m classes, as given below; and with class frequency f_{ij} for the ith y – class interval and jth x– class interval.

x / y	Class interval								Total
	mid value		x_1					x_k	
class interval	mid value	y_1	f_{11}					f_{1k}	g_1
					f_{ij}				g_i
		y_m	f_{m1}					f_{mk}	g_m
Total			h_1		h_j			h_k	N

Suppose that $$\sum_{j=1}^{k} f_{ij} = g_i,\ i = 1, 2, \ldots, m$$

and

$$\sum_{i=1}^{m} f_{ij} = h_j,\ j = 1, 2, \ldots, k$$

are the marginal frequencies. We have

$$N = \sum_{i,j} f_{ij} = \sum_{i=1}^{m} g_i = \sum_{j=1}^{k} h_j \text{ as the total frequency.}$$

Suppose that $x_1, \ldots, x_k$ are the mid-values of the k x-classes and that $y_1, \ldots, y_m$ are the mid values of the m y-classes. The formula for product moment correlation r can then be written as

$$r = \frac{N \Sigma f_{ij} x_i y_j - \left(\sum_{i=1}^{k} h_j x_j\right)\left(\sum_{i=1}^{m} g_i y_i\right)}{\sqrt{\left[N \sum_{j=1}^{k} h_j x_j^2 - \left(\sum_{j=1}^{k} h_j x_j\right)^2\right]} \sqrt{\left[N \sum_{i=1}^{m} g_i y_i^2 - \left(\sum_{i=1}^{m} g_i y_i\right)^2\right]}} \tag{8.1}$$

Suppose that

$$u_j = \frac{x_j - u}{c},\quad j = 1, 2, \ldots, k$$

$$v_i = \frac{y_i - v}{d}, i = 1, 2, \ldots, m;$$

then replacing x_j by u_j and y_i by v_i in the above formula we get r.

Suppose that the regression line of y on x, is given by

$$y = a + bx$$

The parameters *a, b* are estimated by

$$\hat{b} = \frac{N\Sigma f_{ij}x_i y_j - \left(\sum_{i=1}^{k} h_j x_j\right)\left(\sum_{i=1}^{m} g_i y_i\right)}{N\sum_{j=1}^{k} h_j x_j^2 - \left(\sum_{j=1}^{k} h_j x_j\right)^2} \tag{8.2}$$

and $\hat{a} = \bar{y} - \hat{b}\,\bar{x}$; where $\bar{x}, \bar{y}$ are the means of x and y respectively.

If we use $\quad u_j = \dfrac{x_j - u}{c} \quad$ and $\quad v_i = \dfrac{y_i - v}{d}$

then $\hat{b}$ can be obtained from (8.2) by replacing x_j by u_j and y_i by v_i. We then have

$$\bar{x} = u + \frac{c\left(\sum_{j=1}^{k} u_j h_j\right)}{N}$$

$$\bar{y} = v + \frac{d\sum_{i=1}^{m} v_i g_i}{N} \tag{8.3}$$

so that $\qquad \hat{a} = \bar{y} - \hat{b}\,\bar{x}.$

The regression line is $(y - \bar{y}) = \hat{b}\,(x - \bar{x})$. $\qquad$ (8.4)

Similarly the regression line of x on y can be found out.

We shall illustrate the procedure by an example.

Example 8(a). The frequency distribution of heights (x) and weights (x) of 100 students are as given below.

x : Height (in cm)

y : weight (in kg)	150–154	155–159	160–164	165–169	170–174	175–179	Total
51–53	1	1					2
54–56	1	2	1				4
57–59		2	2	5			9
60–62		1	15	23	1		40
63–65			6	18	1		25
66–68			1	3	7	1	12
69–71				1	3	4	8
Total	2	6	25	50	12	5	100

Calculate the coefficient of correlation between x and y and also find the regression line of y on x.

Let x_i, y_j be the mid values of the corresponding x and y classes.

Put $$u_i = \frac{x_i - 162}{5} \text{ and } v_j = \frac{y_j - 64}{3}$$

Here $$i = 1, 2, \ldots, 6; j = 1, 2, \ldots, 7;$$

and f_{ij} is the frequency of the cell (i, j).

$$\sum_{i=1}^{6} f_{ij} = h_j \text{ are the 7 marginal } x\text{-class frequencies}$$

$$h_1 = 2, h_2 = 4, \ldots, h_7 = 8$$

$$\sum_{j=1}^{7} f_{ij} = g_i \text{ are the 6 marginal } y\text{-class frequencies}$$

$$g_1 = 2, g_2 = 6, \ldots, g_6 = 5$$

Write $$U_j = \sum_{i=1}^{6} u_i f_{ij}, \; V_i = \sum_{j=1}^{7} v_j f_{ij}$$

Note that $$\sum_{j=1}^{7} U_j = \sum_{j=1}^{7} \left\{ \sum_{i=1}^{6} u_i f_{ij} \right\} = \sum_{i=1}^{6} u_i \left\{ \sum_{j=1}^{7} f_{ij} \right\} = \sum_{i=1}^{6} g_i u_i$$

and $$\sum_{i=1}^{6} V_i = \sum_{i=1}^{6} \left\{ \sum_{j=1}^{7} v_j f_{ij} \right\} = \sum_{j=1}^{7} v_j \left\{ \sum_{i=1}^{6} f_{ij} \right\} = \sum_{j=1}^{7} h_j v_j.$$

For computation purpose we construct the following table (with x, y denoting the mid value of the corresponding class).

y	x	152	157	162	167	172	177					
	u_i / v_j	−2	−1	0	1	2	3	h_j	$h_j v_j$	$h_j v_j^2$	U_j	$v_j U_j$
52	−3	1	1					2	−6	18	−3	9
55	−2	1	2	1				4	−8	16	−4	8
58	−1		2	2	5			9	−9	9	3	−3
61	0		1	15	23	1		40	0	0	24	0
64	1			6	18	1		25	25	25	20	20
67	2			1	3	7	1	12	24	48	20	40
70	3				1	3	4	8	24	72	19	57
	g_i	2	6	25	50	12	5	100	50	188	79	131
									$\Sigma h_j u_j$		ΣU_i	
	$g_i u_i$	−4	−6	0	50	24	15	79	$\Sigma g_i u_i$			
	$g_i u_i^2$	8	6	0	50	48	45	157				
	V_i	−5	−9	4	22	24	14	50	ΣV_i			
	$u_i V_i$	10	9	0	22	48	42	131				

Verify that

$$\Sigma V_i = \Sigma_j h_j v_j = 50; \ \Sigma U_j = \Sigma_j g_i u_i = 79$$

$$\Sigma v_j U_j = \Sigma u_i V_i = \Sigma_i \Sigma_j u_i v_j f_{ij} = 131;$$

these work as checks on computation.

Writing u_i for x_i and v_j for y_j in (8.1) we get

$$r = \frac{100 \times 131 - 50 \times 79}{\sqrt{\left\{100 \times 188 - (50)^2\right\}\left\{100 \times 157 - (79)^2\right\}}} = \frac{9150}{\sqrt{(16300 \times 9459)}}$$

$$= \frac{915}{1241.6992} = 0.7367 \cong 0.74$$

Writing u_i for x_i and v_j for y_j in (8.2) we get

$$\hat{b} = \frac{100 \times 131 - 50 \times 79}{100 \times 188 - (50)^2} = \frac{9150}{16300} = 0.5613$$

This is the regression coefficient of y on x. Using (8.3) we get

$$\bar{x} = 162 + 5 \times \frac{79}{100} = 165.95 \text{ cm}, \ \ \bar{y} = 61 + 3 \times \frac{50}{100} = 62.50 \text{ kg}.$$

and finally substituting in (8.4) we get the line of regression of y on x given by

$$y - 62.50 = 0.5613\ (x - 165.95).$$

Again, the regression coefficient of x on y is given by

$$\hat{d} = \frac{9150}{9459} = 0.9673$$

so that the regression line of x on y is given by

$$x - 165.95 = 0.9673\ (y - 62.50).$$

We close our discussion on (simple) linear regression here.

9.9 FURTHER DISCUSSION ON REGRESSION

9.9.1 Non Linear Regression

We have discussed at some length about the linear relationship between two variables on the basis of a set n pairs of values (x_i, y_i). The line of regression of y on x enables us to obtain estimates of the value of the dependent variable y from the values of the independent variable x. Often it is found that the estimates used for prediction of the values of dependent variable produce poor results.

One reason behind this may be that the relationship between the two variables may be far removed from the linear, and that curvilinear relationship may be more appropriate. This means that instead of considering a linear relationship of the form $y = a + bx$, a relationship of the second degree of the form $y = a + bx + cx^2$ (or even of higher degree) may be appropriate. We can apply the method of least squares to find the estimates of a, b and c for the relationship $y = a + bx + cx^2$. The best fitting parabola will be such that

$$D = \Sigma(y_i - \hat{y}_i)^2 = \Sigma(y_i - a - bx_i - cx_i^2)^2 \tag{9.1}$$

is the least or minimum.

For D to be minimum, it is necessary that its *three* partial derivaties vanish. The estimated values of a, b and c for which D is minimum must satisfy the three equations :

$$\frac{\partial D}{\partial a} = \Sigma\, 2(y_i - a - bx_i - cx_i^2)(-1) = 0 \tag{9.2}$$

$$\frac{\partial D}{\partial b} = \Sigma\, 2(y_i - a - bx_i - cx_i^2)(-x_i) = 0 \tag{9.3}$$

$$\frac{\partial D}{\partial c} = \Sigma\, 2(y_i - a - bx_i - cx_i^2)(-x_i^2) = 0 \tag{9.4}$$

These lead to the three normal equations

$$\Sigma y_i = n\,\hat{a} + \hat{b}\,\Sigma x_i + \hat{c}\,\Sigma x_i^2 \tag{9.2a}$$

$$\Sigma x_i y_i = \hat{a}\,\Sigma x_i + \hat{b}\,\Sigma x_i^2 + \hat{c}\,\Sigma x_i^3 \tag{9.3a}$$

$$\Sigma x_i^2 y_i = \hat{a}\,\Sigma x_i^2 + \hat{b}\,\Sigma x_i^3 + \hat{c}\,\Sigma x_i^4 \tag{9.4a}$$

whose solutions will yield $\hat{a}$, $\hat{b}$ and $\hat{c}$.

The regression equation (which is non linear) can then be written down.

9.9.2 Multiple Linear Regression

There may be yet another reason for which prediction of values of the dependent variable y from the values of the independent variable x may yield poor results. The single variable x may not be closely related to y. It may happen that there are more than one variable, which, when taken together may serve as a more satisfactory basis for predicting the values of the dependent variable. For example, weights of children may not only be related to ages, but may be related to both ages and heights. The variable y may be related to k number of variables $x_1, x_2, \ldots, x_k$ instead of just one. The question then arises : what kind of relationship is to be tried? It is found from experience that many sets of variables are approximately linearly related. Further, linear relationships are easy to work with. We shall therefore consider linear relationship between y and $x_1, x_2, \ldots, x_k$. Without loss of generality, we take $k = 2$. We consider a linear relationship between y and x_1, x_2.

Suppose that n sets of values of y, x_1, x_2 are given; denote these n sets of values by (y_i, x_{1i}, x_{2i}), $i = 1, 2, \ldots, n$.

Let us denote the linear relationship by

$$y_i = a + bx_{1i} + cx_{2i} \tag{9.5}$$

This will be the linear regression line of y on x_1 and x_2. The constants a, b, c may be determined by the method of least squares. The best fitting line is to be such that the sum of squares

$$D = \Sigma(y_i - \hat{y}_i)^2 = \Sigma(y_i - a - bx_{1i} - cx_{2i})^2 \tag{9.6}$$

will be the least or minimum.

The estimated values $\hat{a}, \hat{b}, \hat{c}$ satisfy the three normal equations

$$\frac{\partial D}{\partial a} = \Sigma\, 2(y_i - a - bx_{1i} - cx_{2i})(-1) = 0$$

$$\frac{\partial D}{\partial b} = \Sigma\, 2(y_i - a - bx_{1i} - cx_{2i})(-x_{1i}) = 0$$

$$\frac{\partial D}{\partial c} = \Sigma\, 2(y_i - a - bx_{1i} - cx_{2i})(-x_{2i}) = 0$$

These lead to the three normal equations

$$\Sigma y_i = n\hat{a} + \hat{b}\,\Sigma x_{1i} + \hat{c}\,\Sigma x_{2i} \tag{9.7}$$

$$\Sigma x_{1i} y_i = \hat{a}\,\Sigma x_{1i} + \hat{b}\,\Sigma x_{1i}^2 + \hat{c}\,\Sigma x_{1i} x_{2i} \tag{9.8}$$

$$\Sigma x_{2i} y_i = \hat{a}\,\Sigma x_{2i} + \hat{b}\,\Sigma x_{1i} x_{2i} + \hat{c}\,\Sigma x_{2i}^2 \tag{9.9}$$

All summations Σ are to be taken over from $i = 1, 2, \ldots, n$. The solutions of (9.7) to (9.9) will yield $\hat{a}, \hat{b}, \hat{c}$. It may be noted that calculations will be easier when the variables are measured from their sample means, namely, $\bar{y}, \bar{x}_1$ and $\bar{x}_2$, i.e., take $y_i - \bar{y}$ rather that y_i, $x_{1i} - \bar{x}_1$ rather than x_{1i} and $x_{2i} - \bar{x}_2$ rather than x_{2i}.

Thus writing $\quad y_i' = y_i - \bar{y}, \; x_{1i}' = x_{1i} - \bar{x}_1, \; x_{2i}' = x_{2i} - \bar{x}_2$

where $\quad \bar{x}_1 = \frac{1}{n}\Sigma x_{1i}, \; \bar{x}_2 = \frac{1}{n}\Sigma x_{2i}, \; \bar{y} = \frac{1}{n}\Sigma y_i$

we get that (9.7) reduces to $n\,\hat{a} = 0$, i.e. $\hat{a} = 0$

The other two equations become

$$\Sigma x_{1i}'\, y_i' = \hat{b}\,\Sigma (x_{1i}')^2 + \hat{c}\,\Sigma (x_{1i}')(x_{2i}') \tag{9.8a}$$

$$\Sigma x_{2i}'\, y_i' = \hat{b}\,\Sigma (x_{1i}')(x_{2i}') + \hat{c}\,\Sigma (x_{2i}')^2 \tag{9.9a}$$

from which $\hat{b}$ and $\hat{c}$ can be obtained.

The results can be generalised to cover more variables ($k \geq 3$).

Example 9(a). Suppose that an experiment was conducted to determine the effect of temperature (x_1) and manure (x_2) on the yield (y) of a variety. Suppose that the variables are measured from their respective means and that the following values are obtained:

$$\Sigma (x_{1i}')^2 = 38.4, \quad \Sigma (x_{2i}')^2 = 3.4$$

$$\Sigma x_{1i}' y_i' = 29.76, \quad \Sigma x_{2i}' y_i' = 8.94$$

$$\Sigma x_{1i}' x_{2i}' = 9.6$$

To find the regression line of y on x_1 and x_2. We have from (9.8a) and (9.9a)

$$29.76 = \hat{b} \times (38.4) + \hat{c} \ (9.6)$$

$$8.94 = \hat{b} \times (9.6) + \hat{c} \ (3.4)$$

Solving the above equations, we get

$$\hat{b} = 0.4, \ \hat{c} = 1.5$$

so that the regression line is given by

$$y_i' = (0.4)\, x_{1i}' + (1.5)\, x_{2i}';$$

suppose that $\bar{y} = 55.2, \ \bar{x}_1 = 20.1, \ \bar{x}_2 = 6.4$

then we get

$$(y_i - 55.2) = (0.4)(x_{1i} - 20.1) + (1.5)(x_{2i} - 6.4)$$

or $$y_i = 0.4\, x_{1i} + 1.5\, x_{2i} + 37.56$$

which is the required regression line in terms of the original variables.

EXERCISES-9

Sections 9.1-9.3

1. What is a scatter diagram? Mention its uses.
2. Explain the concept of correlation. Can it be used to measure *any* relationship between two variables?
3. What are the desirable properties of a measure of linear relationship between two variables? Discuss
4. Does the existence of high correlation between two variables indicate a causal relationship between them? Explain.
5. Define product moment correlation coefficient between two sets of observations $x_i, y_i, i = 1, 2, \ldots, n$.
6. Show that Karl Pearson's product moment correlation coefficient is independent of the change of origin and scale.
7. How would you interpret the observed values of the coefficient of correlation? What do you infer about the relationship if (i) $r = 0$, (ii) $r = +1$ and $r = -1$.
8. Find the coefficient of correlation between the sales (x) and expenses (y) of a firm for 6 months.

Sales (x) :	50	55	60	65	60	70
Expenses (y) :	12	12	15	14	13	14

9. The following table gives the wholesale price index (x) and the index of agricultural production (y) for a period of 5 years.

x	100	104	102	103	106
y	111	109	113	107	110

Draw the scatter diagram and find the coefficient of correlation between x and y.

10. The following table gives the marks obtained in Statistics and Accountancy by a sample of 7 students:

Statistics :	40	62	55	35	48	88	76
Accountancy :	50	70	65	30	45	92	84

Draw the scatter diagram and find the coefficient of correlation between the marks obtained in Statistics and Accountancy.

11. The following data give the proportionate changes in output (x) and employment (y) for a number of industries.

Industries :	I	II	III	IV	V	VI	VII
Output (x) :	230	160	145	140	140	135	100
Employment (y) :	15	60	20	30	25	25	35

Find the coefficient of correlation between x and y. Interpret the result.

12. The following table gives the year-to-year changes in index numbers of output and productivity for a certain industry in a country during 1956-57 to 1961-62.

	56-57	57-58	58-59	59-60	60-61	61-62
Output change (x)	6.5	3.0	9.3	8.3	2.7	6.2
Productivity change (y)	8.7	2.1	9.7	2.3	4.0	3.2

Find the coefficient of correlation between x and y.

13. The following table gives results of chemical analyses of 6 samples of Fe-Ti Oxides: x represents the Fe 0 content and y represents measure of degree of oxidation.

Sample	1	2	3	4	5	6
x	49.4	41.0	29.3	26.4	20.5	26.7
y	23.6	22.7	22.1	21.6	21.1	20.9

14. The following table gives the percentage of National Income from agriculture (x) and the percentage of illiterates (y) of groups of countries with different per capita incomes.

Income group	I	II	III	IV	V	VI
x	11	11	15	30	33	41
y	2	6	19	30	49	71

Draw the scatter diagram and calculate the coefficient of correlation between x and y. Would you infer that there is some causal relationship between x and y?

15. Calculate the coefficient of correlation between the index of weekly wages (x) and the cost of living index (y) of a certain type of workers for a period of 5 weeks.

x	110	120	100	130	115
y	109	110	107	113	111

Sections 9.4-9.8

16. Explain the concept of regression. Indicate the significance of regression analysis.
17. Define regression coefficients. What do they imply?
18. What are regression lines? Why should there be two regression lines? Explain.
19. When are the two regression lines (i) identical, (ii) perpendicular to each other? When the two regression lines are identical, what will be the value of the coefficient of correlation?
20. A set of n pairs of observations $x_i, y_i, i = 1, 2, \ldots, n$ is given. Explain how you would find the regression line of y on x.
    Find the equations of the lines of regression of y on x and of x on y for the data given in Exercise numbers noted below. Draw the lines on graph paper.
21. Exercise no. 9
22. Exercise no. 10
23. Exercise no. 11
24. Exercise no. 12
25. Exercise no. 14
26. Consider the regression lines obtained in Exercise 23. (a) Estimate the change in employment when the change in output is (i) 150, (ii) 200; (b) and also the change in output when the change in employment is (i) 40, (ii) 50.
27. The two regression coefficients are found to be 0.6 and 1.4. What is the coefficient of correlation?
28. Suppose that the two regression lines are given by
    $$y = 0.45x - 21.50$$
    and
    $$x = 0.87y + 60.25$$
    Find the coefficient of correlation between x and y.
29. The regression lines between two variables x and y are given by
    $2x + 3y - 6 = 0$ and $5x + 7y - 12 = 0$
    Is it possible to identify the two regression lines? Examine.
30. Consider Exercise 29.
    (a) Find (i) r and the (ii) mean values of x and y (b) Is it possible to find s.d. of x and y? Can you say anything about the ratio of the two standard deviations?
31. How would you interpret the value of r^2? If the coefficient of correlation between two variables x and y is 0.8, what percentage of the total variation remains unexplained by the regression equation.
32. Explain the statistical model behind linear regression between two variables. What are the basic assumptions for the validity of the model?

Section 9.9

Given that

$$\Sigma f_{ij} x_i y_j = 90, \quad \Sigma h_j x_j = 20, \quad (h_j = \Sigma f_{ij}) \quad \Sigma h_j x_j^2 = 80.$$
$$\Sigma g_i y_i = 12 \ (g_i = \Sigma f_{ij}), \quad \Sigma g_i y_i^2 = 169 \text{ and } N = 10$$

33. Calculate the coefficient of correlation between x and y.
34. Find the two lines of regression.
35. Find the coefficient of correlation and two lines of regression for the data given below.

		x : marks in statistics				
		0-20	21-40	41-60	61-80	81-100
	0-20	3	–			
y:	21-40		4	2		
marks in	41-60		1	16	5	2
accountancy	61-80			10	18	5
	81-100			8	10	16

● 36. Find the coefficient of correlation between the heights and weights of 90 persons as given below.

		x : heights (in cm)		
	141-150	151-160	161-170	171-180
y : weights (in kg) 41-50	1	1	—	—
51-60	3	4	2	1
61-70	1	8	12	4
71-80	—	3	15	5
81-90	—	1	20	3
91-100	—	—	5	1

37. Find the equations of the two regression lines for the data given in Exercise 36. Estimate the weight of an individual whose height is 177 cm and the height of an individual whose weight is 65 kg.

38. Find the coefficient of correlation between the mother's age at first birth (x) and the interval between the first and second births (y) from the following data involving a sample of size 143.

			y		
x	less than 1 yr	1-2 yrs	2-3 yrs	3-4 yrs	4-5 yrs
15-16	—	6	3		
16-17	4	3	3	2	
17-18	3	9	1		
18-19	2	12	7	1	
19-20	5	12	3		
20-21	8	7	3	2	2
21-22	4	8	2		
22-23	1	14	1		
23-24	1	7	3		
24-25			2		
25-26		1		1	

Make suitable assumptions for computation.

39. Find the equations of the two regression lines for the data given in Exercise 38.

40. Find the coefficient of correlation and the line of regression of y on x from the following data:

			x		
y	51-60	61-70	71-80	81-90	91-100
101-150	15	12	3	—	—
151-200	6	18	8	2	—
201-250	—	12	12	3	9
251-300	—	5	18	6	3
301-350	—	—	8	16	2

Find also the equation of the line of regression of y on x.

Chapter 10

Statistical Inference

10.1. INTRODUCTION

One of the principal objectives of statistical analysis is to draw inference about the population on the basis of data collected by sampling from the population. In other words, one is required to draw inference (or to generalise) about the population from sample data; the inference to be drawn relates to some parameters of the population, such as the mean, standard deviation or some other feature like the proportion of an attribute occurring in the population. We have spoken earlier about the parameters of a population distribution. For example, n and p are the parameters of binomial distribution, the mean λ that of Poisson distribution and the mean μ and s.d. σ are the parameters of normal distribution. Suppose it can be assumed that the daily number of absentees in a factory has Poisson distribution : we do not bother at the moment about the basis of this assumption. It may be possible to formulate a probabilistic or statistical model of the phenomenon of the number of absentees from theoretical considerations. On the other hand experience could be the guide for such an assumption. Suppose that Poisson distribution is a reasonable assumption for this phenomenon. To specify completely the Poisson distribution, one would be required to specify the parameter λ. Now all that we have are records of daily number of absentees for a large number of days; we take a sample of such data and our problem is to infer about the parameter λ of the population Poisson distribution, that is, we are faced with the problem of drawing inference about the population parameter from the sample.

The two most important types of problems of inference in statistics are (i) estimation of parameter or parameters and (ii) testing of statistical hypothesis or hypotheses.

In the absence of the complete data or information about the population, it would not be possible to determine the exact or true value of a parameter. It would be worthwhile to obtain from the sample data an *estimate* of the unknown true or exact value of the parameter or an interval of values in which the parameter lies, and also to determine a procedure for determining the *accuracy* of the estimate. This type of inference is known as *estimation of parameters.* There are two types of estimation :

(a) *point estimation* : here the objective is to find a *single* value for the unknown parameter

(b) *interval estimation* : here the objective is to find an *interval* of plausible values in which the unknown parameter lies.

10.2. POINT ESTIMATION

Suppose that we are concerned with the estimation of a parameter of a population from a given sample of the population.

The procedure of point estimation consists of determining a single quantity from the sample values given such that the single number is fairly close to the unknown value of the parameter of the population. The parameter of the population could be the population mean, and/or standard deviation or some other measure related to the population, for example, the proportion of occurrence of a certain trait in the population. Suppose that the sample (of size n) drawn from the population is denoted by $X_1 \ldots, X_n$, and that the unknown parameter is denoted by θ. The point estimation of θ will be based on the sample observations $X_1, X_2, \ldots, X_n$. It will be a function of the sample observations $X_1, \ldots, X_n$, that is, a *statistic.* The statistic to be used for point estimation of θ is called a point estimator and is denoted by $\hat{\theta}$. When an actual set of sample values is given, we can compute a numerical value, which is called *point estimate* of θ. The point estimate has a definite numerical value. The estimator $\hat{\theta}$ of the parameter θ is a function of the sample observations $X_1, \ldots, X_n$, and will assume different numerical values corresponding to different sets of sample observations $X_1, \ldots, X_n$. For a given set of sample observations, we get a point estimate of θ; this is one of the possible values of $\hat{\theta}$.

We have noted earlier that X_i are identically and independently distributed r.v. having the same distribution as the population. Any function of these samples will also be a r.v. As such the point estimator $\hat{\theta}$ will also be a random variable and it will have a probability distribution. For example, if the parent population is normal with mean μ and s.d. σ, then the sample mean $\overline{X} = \dfrac{X_1 + ... + X_n}{n}$ is also a random variable having normal distribution with mean μ and s.d. $\sigma/\sqrt{n}$ (Theorem 8.3). The sample mean $\overline{X}$ is a point estimator $\hat{\mu}$ of the population mean μ. The point estimator $\hat{\mu}$ can assume an infinite number of values corresponding to the infinite set of (the numerical) sample values that $X_1, \ldots, X_n$ take. From one given set of sample values, that is, a *particular* set of numerical values one can compute one *particular* value of the estimator $\hat{\mu}$ and this value is a *point estimate* of μ. Besides the sample mean $\overline{X} = \dfrac{X_1 + ... + X_n}{n}$ there *may* be other types of estimator of μ, based on some other function of the same set of sample observations $X_1, \ldots, X_n$; in fact the sample median is also an estimator of μ. The question then arises: which of the sample estimators is to be preferred and why. This raises another question: what should be the basis of selecting an estimator, or in other words what should be the *criteria of a good estimator.* These questions form the subject matter of the theory of estimation.

Without going into details, we would like to state below two desirable properties that an estimator should possess.

An estimator being a random variable, will have a mean and variance.

(1) *Unbiasedness:* an estimator $\hat{\theta}$ of θ is said to be *unbiased* if $E(\hat{\theta}) = \theta$, i.e. its mean is equal to the parameter value, otherwise it is said to be biased. The property of unbiasedness ensures that there will not be *over estimation* or *under estimation*.

(2) *Minimum variance*: an estimator $\hat{\theta}$ is a function of the sample observations $X_1, X_2, \ldots, X_n$. There can be many functions of $X_1, X_2, \ldots, X_n$ which could be used to estimate θ. The variance of an estimator $\hat{\theta}$ is a measure of the variability or spread. An estimator with a smaller variance will have greater concentration near the parameter to be estimated. It will therefore be appropriate to select the estimator with the smallest variance. This would ensure greater accuracy.

10.3. ESTIMATION OF THE POPULATION MEAN : POINT ESTIMATION

Suppose that we have a sample $X_1, X_2, \ldots, X_n$ of size n from a population with mean μ and s.d. σ and that μ is unknown. We examine the question of estimation of μ from the sample. We assume that the sample size n is *large*. Two cases arise: the s.d. may or may not be known.

Assume first that σ is known. We have from Theorem 8.4, $E(\bar{X}) = \mu$ and $\sigma_{\bar{x}} = \text{s.d.}(\bar{X}) = \frac{\sigma}{\sqrt{n}}$. If the parent population is normal, then $\bar{X}$ is exactly normal. Even when nothing is known about the distribution of the parent population, we get from Theoerem 8. 4 that when the sample size n is large, $\bar{X}$ is approximately normal. A sketch of the distribution of $\bar{X}$ is given in Fig. 10.1.

We have

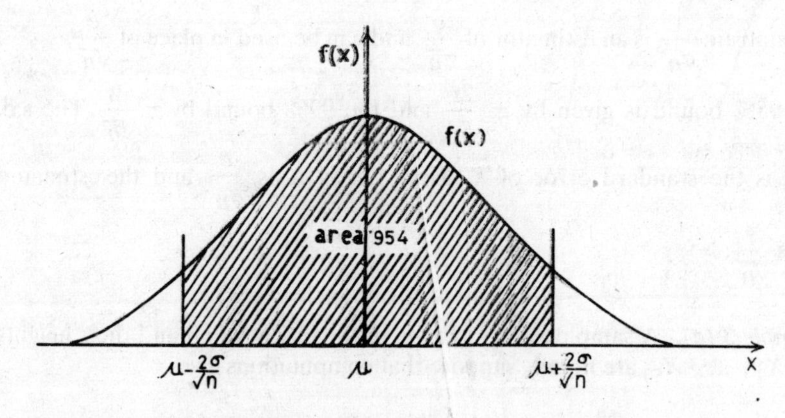

Fig. 10.1. Distribution of the sample mean (of size n) from normal population with mean μ and s.d. σ

$$P\left\{\mu - \frac{2\sigma}{\sqrt{n}} < \bar{X} < \mu + \frac{2\sigma}{\sqrt{n}}\right\} = 0.954$$

Now $\mu - \frac{2\sigma}{\sqrt{n}} < \overline{X} < \mu + \frac{2\sigma}{\sqrt{n}}$ can also be written as

$$-\frac{2\sigma}{\sqrt{n}} < \overline{X} - \mu < \frac{2\sigma}{\sqrt{n}}$$

so that

$$P\left\{-\frac{2\sigma}{\sqrt{n}} < \overline{X} - \mu < \frac{2\sigma}{\sqrt{n}}\right\} = 0.954 \tag{3.1}$$

Further $$P\left\{-\frac{3\sigma}{\sqrt{n}} < \overline{X} - \mu < \frac{3\sigma}{\sqrt{n}}\right\} = 0.997 \tag{3.2}$$

In fact, we can get, from Table of normal distribution, the probability of $\overline{X} - \mu$ lying in any interval with end points expressed as multiples of the s.d. $\frac{\sigma}{\sqrt{n}}$. Thus when σ is *known* we can find the probability of the difference of $\overline{X} - \mu$ lying within k times $\frac{\sigma}{\sqrt{n}}$. The 95% bound is given by $\pm \frac{2\sigma}{\sqrt{n}}$ and the 99% bound by $\pm \frac{3\sigma}{\sqrt{n}}$.

Secondly, assume that σ is *not known*, then we have to use an estimate of σ: a reasonable estimate of unknown population s.d. σ is the sample standard deviation s given by

$$7739 = \sqrt{\frac{\sum_{i=1}^{n} (X_i - \overline{X})^2}{(n-1)}}$$

The statistic $\frac{s}{\sqrt{n}}$ is an estimator of $\frac{\sigma}{\sqrt{n}}$ and can be used in place of $\frac{\sigma}{\sqrt{n}}$. The 95% bound is given by $\pm \frac{2s}{\sqrt{n}}$ and the 99% bound by $\pm \frac{3s}{\sqrt{n}}$. The s.d. of $\overline{X}$ is the standard error of $\overline{X}$. Thus s.e. of $\overline{X}$ is $\frac{\sigma}{\sqrt{n}}$ and the estimated s.e. is $\frac{s}{\sqrt{n}}$.

Example 3 (a). A sample of 50 students of the city is taken and their heights $X_1, X_2, \ldots, X_{50}$ are noted. Suppose that computatrons give

$$\sum_{i=1}^{50} X_i = 8000, \quad \sum_{i=1}^{50} (X_i - \overline{X})^2 = 1225.$$

Now $\overline{X} = \frac{\Sigma X_i}{n} = \frac{8000}{50} = 160$ cm is an estimate of the population mean μ.

The sample s.d., $s = \sqrt{\dfrac{\Sigma (X_i - \overline{X})^2}{n-1}} = \sqrt{\dfrac{1225}{49}} = \sqrt{25} = 5$ is an estimate of the population s.d. σ.

Again, $\dfrac{s}{\sqrt{n}} = \dfrac{5}{\sqrt{50}} = 0.7$ (cm) is an estimate of $\dfrac{\sigma}{\sqrt{n}}$. It is the estimated s.e. of $\overline{X}$. It follows that the probability that the difference between the sample and population means is within (i) $2 \times \dfrac{s}{\sqrt{n}} = 1.4$ cm is 0.954 and (ii) $3 \times \dfrac{s}{\sqrt{n}} = 2.1$ cm is 0.997.

Example 3 (b). Suppose that a big dairy produces a certain milk product which is marketed in tin containers. It has been noted that the means of the contents of tins vary from batch to batch but s.d. of contents remain fairly constant at a value $\sigma = 0.10$ gm. Suppose that a sample of 25 tins from a batch is taken to find an accurate estimate of the mean of the batch.

(i) How accurate is the sample mean as a point estimate of the batch (population) mean μ ?

(ii) Suppose that the manager is not satisfied with the accuracy of his estimate based on 25 samples. If he desires that his estimate will not be in error by more than 0.025 gm. with a probability of 0.95, how large should his sample be?

(i) As the population is normal, the sample mean $\overline{X}$ will also be normal and the s.d. of $\overline{X}$ will be equal to

$$\text{s.d. } (\overline{X}) = \sigma_{\overline{X}} = \frac{\sigma}{\sqrt{n}} = \frac{0.10}{\sqrt{25}} = 0.02 \text{ gm.}$$

Again, from (3.1)

$$P\left\{-2 \times 0.02 < \overline{X} - \mu < 2 \times 0.02\right\} = 0.954$$

or, $$P\left\{-0.04 < \overline{X} - \mu < 0.04\right\} = 0.954$$

It follows that the probability is 0.954, that the difference $\overline{X} - \mu$ between the sample mean $\overline{X}$ and the population mean μ will not exceed 0.04 gm; in other words, the manager can be confident that in 95.4 percent of the cases, the sample mean will be within 0.04 gm of the population (batch) mean μ. This gives a measure of the accuracy of the estimate.

(ii) Here it is required that

$$P\left\{-0.025 < \overline{X} - \mu < 0.025\right\} = 0.95$$

or $$P\left\{-\frac{0.025}{\sigma_{\overline{X}}} < \frac{\overline{X} - \mu}{\sigma_{\overline{X}}} < \frac{0.025}{\sigma_{\overline{X}}}\right\} = 0.95$$

From Table of normal distribution we find that, for the above to hold, we must have

$$\frac{0.025}{\sigma_{\overline{X}}} = 1.96$$

This gives
$$\frac{0.025}{1.96} = \sigma_{\overline{X}} = \frac{\sigma}{\sqrt{n}} = \frac{0.10}{\sqrt{n}}$$

or
$$\sqrt{n} = \frac{0.10 \times 1.96}{0.025} = 7.84$$

Squaring, we get $n = 61.5$. Thus a sample size of 62 is required. In other words, if he obtains his batch mean from a sample of 62 tins, he should expect that in 95 percent of the cases, his estimate will be within 0.025 gm of the true batch (population) mean.

10.4. INTERVAL ESTIMATION : ESTIMATION OF MEAN

We have considered point estimation of a parameter; there is another procedure of estimation known as *interval* estimation. Supppose that a sample of values of a population is given. Interval estimation consists in finding an interval, determined by the two end numbers; these two numbers are obtained from computation from the sample values and the interval is such that it is expected to contain the true value of the parameter in its interior. The numerical values of the two numbers computed from sample values determine the interval which is known as interval estimate. The interval estimate constructed, by finding the two numbers (the end points of the interval), is such that the probability of the parameter lying in the interval can be obtained and thus the accuracy with which the parameter is being estimated can be determined .The length of the interval would indicate the accuracy of the estimate, a small interval would indicate a low confidence level and a big interval would indicate a high confidence level.

Consider now the interval estimation of the mean μ of a population from sample observations. *First* suppose that the s.d. of the population is *known*. Suppose that the sample size is large. Now the distribution of the sample mean $\overline{X}$ is approximately normal with mean μ and s.d. $\frac{\sigma}{\sqrt{n}}$. We can therefore find the probability of $\overline{X} - \mu$ lying within multiples of $\pm \frac{\sigma}{\sqrt{n}}$. In particular we get that the probability of $\overline{X} - \mu$ lying within $\pm 2\frac{\sigma}{\sqrt{n}}$ is 0.954.

In other words,

$$P\left\{-\frac{2\sigma}{\sqrt{n}} < \overline{X} - \mu < \frac{2\sigma}{\sqrt{n}}\right\} = 0.954$$

The inequality

$$\overline{X} - \mu < \frac{2\sigma}{\sqrt{n}}$$

is equivalent to
$$\mu > \overline{X} - \frac{2\sigma}{\sqrt{n}}$$

and the inequality

$$-\frac{2\sigma}{\sqrt{n}} < \overline{X} - \mu$$

is equivalent to
$$\mu < \overline{X} + \frac{2\sigma}{\sqrt{n}}$$

so that

$$\overline{X} - \frac{2\sigma}{\sqrt{n}} < \mu < \overline{X} + \frac{2\sigma}{\sqrt{n}} \Big\} = 0.954$$

That is, the true value μ of the parameter lies in the interval $\left(\overline{X} - \frac{2\sigma}{\sqrt{n}}, \overline{X} + \frac{2\sigma}{\sqrt{n}}\right)$ with probability 0.954. Now the sample values $X_1, ..., X_n$ being given and being known, the end points $\overline{X} - \frac{2\sigma}{\sqrt{n}}$ and $\overline{X} + \frac{2\sigma}{\sqrt{n}}$ of the interval can be computed and the interval determined. The interval gives an interval estimate of μ and the probability that μ lies in this interval is known. We are 95.4% confident that this interval contains the true value of the parameter μ. Similarly, we get

$$P\left\{\overline{X} - \frac{3\sigma}{\sqrt{n}} < \mu < \overline{X} + \frac{3\sigma}{\sqrt{n}}\right\} = 0.997.$$

We are 99.7% confident that the interval

$$\left(\overline{X} - \frac{3\sigma}{\sqrt{n}}, \overline{X} + \frac{3\sigma}{\sqrt{n}}\right)$$

contains the true value of μ.

Let us now interpret the statement $P\left(\overline{X} - \frac{2\sigma}{\sqrt{n}} < \mu < \overline{X} + \frac{2\sigma}{\sqrt{n}}\right) = 0.954$. In practice we get only one sample of observations $X_1, \ldots, X_n$; on the basis of this we construct the confidence limits $\overline{X} - \frac{2\sigma}{\sqrt{n}}$ and $\overline{X} + \frac{2\sigma}{\sqrt{n}}$ and the confidence interval $\left(\overline{X} - \frac{2\sigma}{\sqrt{n}}, \overline{X} + \frac{2\sigma}{\sqrt{n}}\right)$ is obtained. Suppose that we have another sample and we then get another interval and so on for any other samples. Suppose that we make a large number of experiments and from each of these we get a sample of observations and then compute the corresponding interval $\left(\overline{X} - \frac{2\sigma}{\sqrt{n}}, \overline{X} + \frac{2\sigma}{\sqrt{n}}\right)$ then we shall find that in 95.4 percent of the cases the true value of μ will lie in the intervals constructed. In view of this, the interval

$\left(\overline{X} - \dfrac{2\sigma}{\sqrt{n}}, \overline{X} + \dfrac{2\sigma}{\sqrt{n}}\right)$ is also known as the 95.4 percent *confidence interval* for μ, and the end points of the interval $\overline{X} - \dfrac{2\sigma}{\sqrt{n}}$ and $\overline{X} + \dfrac{2\sigma}{\sqrt{n}}$ are known as *confidence limits* for μ with *confidence coefficient* 0.954.

An advantage of interval estimation is that given σ, it is always possible to construct the interval from the sample observations. Another advantage is that a probability statement of the true unknown value μ lying in the interval can be made; this indicates how closely the true value μ is being estimated: whereas a point estimate says nothing about the closeness of the estimate $\overline{X}$ from μ or how good the estimate $\overline{X}$ is.

Secondly, suppose that σ is *not known*; then we have to find a sample estimate of σ, that is, an estimated value of σ based on the sample of observations given and then replace the value of σ by this sample estimate of σ and proceed as before.

Example 4(a). The sample mean computed from a random sample of size 50 drawn from a population has the value 52.5. Suppose that the population standard deviation is known to be equal to 16. Find

(i) an estimate of the population mean

(ii) a 95% confidence interval of the population mean.

The sample mean 52.5 is an estimate of the population mean. The sample mean $\overline{X}$ computed from a sample of 50 has s.d. equal to $\dfrac{16}{\sqrt{50}} = \dfrac{16}{7.071} = 2.26$. The sample mean is approximately normally distributed with mean μ and s.d. 2.26. From table of normal distribution we get that 95% of the values of $\overline{X} - \mu$ will lie within $\pm$ 1.96 times the s.d. Thus

$$0.95 = P\left\{-1.96 \times 2.26 < \overline{X} - \mu < 1.96 \times 2.26\right\}$$
$$= P\left\{-4.43 < \overline{X} - \mu < 4.43\right\}$$
$$= P\left\{\overline{X} - 4.43 < \mu < \overline{X} + 4.43\right\}$$

so that 95% confidence interval of μ is given by $(\overline{X} - 4.43, \overline{X} + 4.43)$. Putting the value of $\overline{X} = 52.5$, we get that the 95% confidence limits are

$$\overline{X} - 4.43 = 52.5 - 4.43 = 48.07$$

and

$$\overline{X} + 4.43 = 52.5 + 4.43 = 56.93$$

so that the 95% confidence interval for the population mean is (48.07, 56.93). We are 95% certain that the true parameter μ lies in (48.07, 56.93).

Example 4 (b). Consider a random sample of size 50 from a population with mean μ and s.d. σ (σ *not* known). The computations from the 50 sample observations yield $\overline{X} = 52.5$ and $\sum_{i=1}^{50} (X_i - \overline{X})^2 = 4900$. Find a 95% confidence interval for μ.

We have to find and use an approximate value or estimate of $\frac{\sigma}{\sqrt{n}}$. Now

$$s = \sqrt{\left[\sum_{i=1}^{50} (X_i - \overline{X})^2/(n-1)\right]} = \sqrt{\frac{4900}{49}} = 10$$

so that an estimate of $\frac{\sigma}{\sqrt{n}}$ is given by $\frac{s}{\sqrt{n}} = \frac{10}{7.071} = 1.41.$

We have from table of normal distribution

$$P\left\{-1.96 < \frac{\overline{X} - \mu}{\sigma/\sqrt{n}} < 1.96\right\} = 0.95$$

$$\text{or, } P\left\{-1.96 \times 1.41 < \overline{X} - \mu < 1.96 \times 1.41\right\} = 0.95$$

Thus the 95% confidence limits of μ are

$$\overline{X} - 1.96 \times 1.41 = 52.5 - 2.76 = 49.74$$

and $$\overline{X} + 1.96 \times 1.41 = 52.5 + 2.76 = 55.26$$

so that the 95% confidence interval for μ is (49.74, 55.26).

10.5. ESTIMATION OF THE PARAMETER p

Let us now turn our attention to the problem of estimation of the parameter p, the proportion of a certain characteristic in the population. Suppose we are interested in estimating the proportion p of the population of school children who wear spectacles. Suppose that a random sample of size n of school children is found to contain m spectacle users, then $\frac{m}{n}$ gives an estimate of the unknown proportion p. Denote the estimator of p by $\hat{p}$, i.e. $\hat{p} = \frac{m}{n}$.

For large samples, the estimator $\hat{p}$ is approximately normal with mean p and s.d. $\sqrt{\frac{pq}{n}}$ where $q = 1 - p$.

In practice p is *unknown* and we are concerned with estimation of p. The s.d. $\sqrt{\frac{pq}{n}}$ is unknown then and has to be estimated.

The statistic

$$\sqrt{\frac{\hat{p}\hat{q}}{n}} \quad \text{where } \hat{p} = \frac{m}{n}, \ \hat{q} = 1 - \hat{p}$$

is an estimator of $\sqrt{\frac{pq}{n}}$

Thus 95 percent confidence limits for p are

$$\hat{p} - 1.96 \times \sqrt{\frac{\hat{p}\hat{q}}{n}}$$

and
$$\hat{p} + 1.96 \times \sqrt{\frac{\hat{p}\hat{q}}{n}}$$

and the 95% confidence interval is given by

$$\left(\hat{p} - 1.96 \times \sqrt{\frac{\hat{p}\hat{q}}{n}},\ \hat{p} + 1.96 \times \sqrt{\frac{\hat{p}\hat{q}}{n}}\right)$$

Similarly, we can find confidence interval corresponding to a given probability level.

Example 5 (a). Suppose that a random sample of 60 school children gives 40 spectacle users. Find 95 confidence limits of the spectacle users of the population of school children and the 95% confidence interval.
Here we have $n = 60$, $m = 40$, so that an estimate of p is given by

$$\hat{p} = \frac{40}{60} = 0.67$$

so that $\hat{q} = 1 - \hat{p} = 0.33$. Thus an estimate of the s. d. is given by

$$\sqrt{\left(\frac{\hat{p}\hat{q}}{n}\right)} = \sqrt{\left(\frac{0.67 \times 0.33}{60}\right)} = \sqrt{0.003685} = 0.06$$

The 95% confidence limits are given by

$$\hat{p} - 1.96 \times 0.06 = 0.67 - 0.12 = 0.55$$
$$\hat{p} + 1.96 \times 0.06 = 0.67 + 0.12 = 0.79$$

and the 95% confidence interval for the population p is (0.55, 0.79).

10.6. SMALL SAMPLE RESULT

The results derived so far are valid for large samples. A sample of size 30 or more is *generally* considered to be a large sample. When samples are small then the results of these sections do not hold good. For a large sample, the distribution of the sample mean $\overline{X}$ is approximately normal with mean μ and s.d. $\frac{\sigma}{\sqrt{n}}$, where μ, and σ are the mean and s.d. respectively of the population. This fact has been used in deriving the results. When samples are small, this assumption of approximate normality of the distribution of $\overline{X}$ is not true; in fact another distribution is to be used and the result modified accordingly.

We shall discuss the small sample theory in Chapter 11.

10.7. TESTING OF STATISTICAL HYPOTHESIS

We shall now turn our attention to the second most important aspect of statistical inference – the testing of statistical hypothesis and the associated test of significance. Suppose that we wish to infer about some parameter of the population, say of the population mean μ, on the basis of a sample of the population, then we may consider the sample mean computed from the sample. There may be some difference between the population and the sample means. The question arises whether this difference is real or whether it arises due to a variation arising from the sampling process. If a second sample is taken, the sample mean computed may be different from the sample mean computed from the first sample. The question arises whether the difference between the sample means computed from the two samples is real or is due to a variation arising from the sampling process. By real difference we mean the one which is inherent in the very nature of the data as opposed to a *chance* difference caused by the process of sampling. Investigations of such questions of statistical inference are within the domain of testing of statistical hypotheses. Tests are used to answer whether the differences indicated are real or due to chance. Such tests are called tests of *statistical significance.*

Two types of problems arise in testing statistical significance. These are indicated below :

1. Problem relating to the comparison of a characteristic of the population and that of the sample. For example, one may wish to ascertain whether the difference between the two values, the population mean and the sample mean, is real or is due to chance;
2. Problem relating to the comparison of a characteristic obtained from two samples. For example, one may wish to know whether the difference between the two values (means) computed from two samples is real or due to chance.

Let us consider the procedure of establishing whether the difference is statistically significant or not. To do so we first formulate a hypothesis about the difference. A hypothesis may be *defined* as a statement about one or more populations. It is usually concerned with the parameters of the population or populations about which a statement is made.

Ordinarily, this hypothesis states that the true difference between the two values is *zero*, that is, there is *no difference.* This formulation is known as *null hypothesis.* An *alternative hypothesis* is also formulated. Based on the data available, we adopt a procedure to reject or not to reject (accept) the null hypothesis. The rejection of the null hypothesis implies that the difference found is statistically significant. Non rejection (or 'acceptance') of the null hypothesis implies that the observed difference is due to chance. The value obtained on the basis of the sample may differ from sample to sample and may not be exactly equal. Such a value obtained from a sample may be different also from the value of the corresponding parameter of the population. The difference is not statistically significant and may arise due to chance. This is what is implied by the acceptance of the null hypothesis. It is to be noted that for a null hypothesis to be accepted, the observed difference need not be exactly zero but that such a difference is likely to arise due to chance. The null hypothesis is rejected when the observed difference appears to be large enough to have been caused by chance, and it is considered that a real difference exists.

The question now arises: how to decide whether the difference is real or due to chance and how big a difference is considered to be statistically significant. The procedure to be adopted is being described below. We indicate below the principal steps in testing a hypothesis:

(1) *Assumption*: Identify the probability model which is appropriate to the situation and the parameter. Make the appropriate assumption on the basis of the model.

(2) *Hypothesis*: Formulate the null hypothesis H_0 and the alternative hypothesis H_1. The null hypothesis H_0 is to be tested against the alternative hypothesis H_1.

(3) *Test Statistic*: Choose the test statistic, and specify its sampling distribution.

(4) *Decision rule* : We have to find the acceptance and rejection regions, on the basis of the level of significance specified. The null hypothesis H_0 is rejected if the computed value of the test statistic falls in the *rejection region*, and is not rejected ("accepted") if it falls in the *acceptance region*. The decision as to which values fall in the rejection region and which values fall in the *acceptance region* is made on the basis of the level of significance, denoted by α. The value α is the probability of rejecting a *true null hypothesis*. Rejection of a true null hypothesis constitutes an error, and α is a measure of the error. It is reasonable to make the error small; in practice α is taken to be 0.01, 0.05, 0.10 and more generally α is taken to be .05. When the rejection region consists of two regions, we get what is called a *two-sided test*; then each of the two regions will be associated with probability $\alpha/2$. When the region consists of only one region either to the right or left, we get a one-sided test, with associated probability α.

(5) *Computed test statistic:* From the data given, we compute a value of the test statistic. Locate the position of this statistic.

(6) *Statistical decision:* If the computed value of the test statistic lies on the rejection region, the null hypothesis is rejected. We then conclude that the data are not compatiable with the null- hypothesis. This also implies statistical significance at the level of significance α.

If it lies in the acceptance region, the null hypothesis is *not rejected* (or "accepted"); in this case we do not say that the null hypothesis is true but that it *may be* true. When speaking of nonrejection or acceptance of the null hypothesis, we have to bear this limitation in mind. We then conclude that the data (on which the test is based) do not provide sufficient evidence to cause rejection of the null hypothesis. This in turn implies that the result is not statistically significant at the level of significance α.

For a clear understanding, the essential steps are elaborately discussed above. In solving problems, we need not, however, deal at length with each of the above steps.

One and two-sided tests

Suppose that we are testing a hypothesis about the mean μ of a population. Further, suppose that the null-hypothesis H_0 and the alternative hypothesis H_1 are of the form:

$$H_0 : \mu = \mu_0$$
$$H_1 : \mu \neq \mu_0$$

where μ_0 is a *given value*. Let α be the level of significance. Our rejection region will then consist of two parts, comprising extremely large values and extremely small values of the test statistic on each side of μ_0, each part being associated with probability $\alpha/2$. Such a test will be called a *two-tailed* or *two-sided* test (See Fig 10.2(a)).

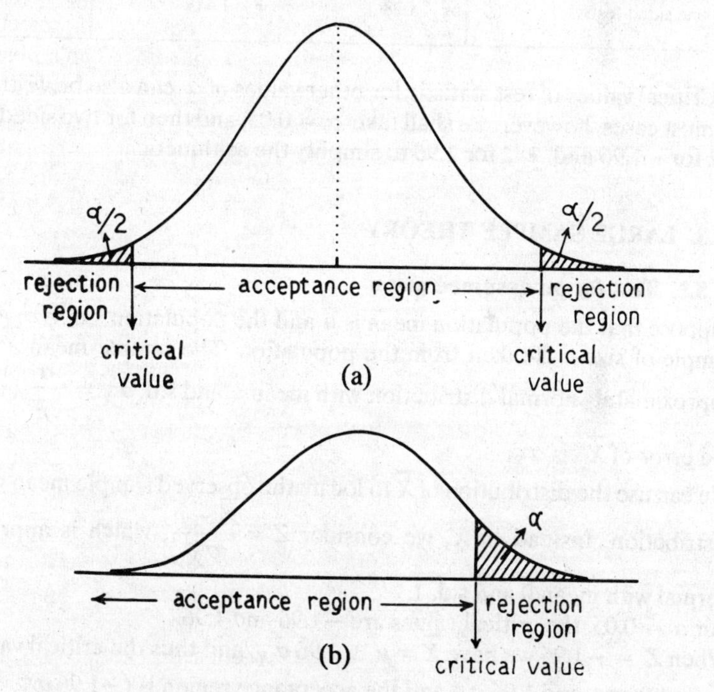

Fig. 10.2 Acceptance and rejections of (a) two sided test (upper figure) (b) one sided test (lower figure)

On the other hand suppose that the null hypothesis H_0 and the alternative hypothesis H_1 are of the form:

$$H_0 : \mu = \mu_0$$
$$H_1 : \mu > \mu_0$$

where μ_0 is a given value. Then the rejection region will consist of only one part, comprising extremely large values of the test statistic on the right hand side of μ_0, with associated probability α (See Fig. 10.2 (b)). Such a test is called a *one-tailed* or *one-sided* test. Similar is the case with alternative hypothesis $H_1 : \mu < \mu_0$. When the test statistic Z has a standard normal distribution, the critical values of Z for one-tailed and two-tailed tests for given values of α (that is, for levels of significance), can be obtained from the table of normal distribution.

The critical values for some levels of significance are given below:

Level of significance	0.10	0.05	0.01
Critical values of test statistics for two sided tests	−1.645 and 1.645	−1.96 and 1.96	−2.58 and 2.58
Critical values of test statistic for one sided tests	−1.28 or 1.28	−1.645 or 1.645	−2.33 or 2.33

Critical values of test statistic for other values of α can also be written down. In most cases, however, we shall take $\alpha = 0.05$, and then for two sided test take −2 for −1.96 and + 2 for 1.96 to simplify the arithmetic.

10.8. LARGE SAMPLE THEORY

10.8.1. Test for an assumed mean

Suppose that the population mean is μ and the population s.d. is σ. A random sample of size n is taken from the population. The sample mean $\overline{X}$ will have approximately normal distribution with mean μ and s.d. $\sigma_{\overline{X}} = \dfrac{\sigma}{\sqrt{n}}$; the standard error of $\overline{X}$ is $\sigma_{\overline{X}}$.

We can use the distribution of $\overline{X}$ to locate the observed sample mean within this distribution. Instead of $\overline{X}$, we consider $Z = \dfrac{\overline{X} - \mu}{\sigma_{\overline{X}}}$, which is approximately normal with mean 0 and s.d. 1.

For $\alpha = 0.05$, the critical values are −1.96 and 1.96.

When $Z = -1.96$ we have $\overline{X} = \mu \pm 1.96\,\sigma_{\overline{X}}$; and thus the critical values for $\overline{X}$ are $-1.96\,\sigma_{\overline{X}}$ and $1.96\,\sigma_{\overline{X}}$ and the acceptance region is $(-1.96\,\sigma_{\overline{X}},\ 1.96\,\sigma_{\overline{X}})$.

We assume that the population mean and s. d. are known.

Let us consider the following example :

Example 8 (a). A company producing light bulbs finds that the mean life span of the population of bulbs is 1200 hours with a standard deviation of 125 hours. A sample of 100 bulbs produced in a lot is found to have a mean life span of 1150 hours. Test whether the difference between the population and sample means is statistically significant.

The *assumption* is that the life span of the population of bulbs has normal distribution with mean 1200 hrs and s.d. 125 hrs.

The *hypotheses* are

$$H_0 : \mu = 1200$$

and

$$H_1 : \mu \neq 1200$$

Under H_0, the *test statistic* $\overline{X}$ is normal with mean 1200 and s.d

$\sigma_{\overline{X}} = \dfrac{\sigma}{\sqrt{n}} = \dfrac{125}{100} = 12.5$ hrs. In other words the statistic

$Z = \dfrac{\overline{X} - 1200}{\sigma_{\overline{X}}} = \dfrac{\overline{X} - 1200}{12.5}$ is standard normal with mean 0 and s.d. 1.

The *decision rule* involves finding of *acceptance* and *rejection regions* based on the level of significance. We assume $\alpha = 0.05$, and the test is two-sided. The critical values corresponding to $\alpha/2 = 0.025$ are -1.96 and 1.96 and the acceptance region is the interval $(-1.96, 1.96)$ and the two rejection regions consist of the intervals to the right and left of the acceptance region; the rejection regions are $(-\infty, -1.96)$ and $(1.96, \infty)$.
The computed test statistic gives $\overline{X} = 1150$ so that

$$Z = \frac{1150 - 1200}{12 - 5} = \frac{-50}{12 - 5} = -4$$

The *statistical decision* is based on the location of this computed value. Now the computed value -4 is less than -1.96 and lies in the rejection region $(-\infty, -1.96)$. The decision is that the null hypothesis is *rejected*. We conclude that the difference between the population and sample means is statistically significant at 5% level of significance. We conclude that the sample mean may not have been the mean of a random sample from the population under consideration.

Example 8 (b). The marks secured by a sample of 36 students of a college in P.U. class give a sample mean of 55 and standard deviation of 10. Test the hypothesis that the mean μ of the population (consisting of the population of students of P.U. class of all colleges under the University) is 50 against the alternative hypothesis that the mean is different from 50. Here the population s.d. is not known and has to be estimated.

We are to test $\qquad H_0 : \mu = 50$

against $\qquad H_1 : \mu \neq 50$

Under the null hypothesis $\overline{X}$ has mean μ and s. d. $\sigma_{\overline{X}} = \dfrac{10}{\sqrt{36}} = \dfrac{10}{6} = \dfrac{5}{3}$.

Here sample s.d. is taken as an estimate of population s.d. σ.
We have

$$Z = \frac{\overline{X} - \mu}{\sigma_{\overline{X}}} = \frac{\overline{X} - 50}{(5/3)}$$

which is approximately normal with mean 0 and s.d. $\frac{5}{3}$. The computed value of $\overline{X}$ is 55 so that the computed test statistic is

$$Z = \frac{55 - 50}{(5/3)} = 5 \times \frac{3}{5} = 3$$

As it is greater than 1.96 it lies on the right hand rejection region. Thus the null hypothesis is rejected.

Note: Since the population s.d. is not known but is estimated, the test is only *approximate*. An exact test for such a situation is given in Chapter 11.

Example 8 (c). A manufacturer of a certain electronic tube claims that the average life span of tubes will exceed 1000 hrs. From past experience the s.d. is known to be 120 hrs. A retailer is willing to order for a large consignment if the manufacturer's claim is true. The retailer gets a sample of 36 tubes tested and finds that the sample mean life span is 1040 hrs. (a) Should the retailer order for the consignment (a) at $\alpha = 0.05$ level of significance (b) at $\alpha = 0.01$ level of significance?

Here

$$H_0 : \mu = 1000$$

$$H_1 : \mu > 1000$$

and the test is one-sided. The test statistic is

$$Z = \frac{\overline{X} - 1000}{\sigma_{\overline{X}}} = \frac{\overline{X} - 1000}{\frac{120}{\sqrt{36}}} = \frac{\overline{X} - 1000}{20}$$

(a) The critical value for this one-sided test at $\alpha = 0.05$ level is 1.645 and the critical region is $(1.645, \infty)$. The computed test statistic is $\frac{1040 - 1000}{20} = +\frac{40}{20} = 2$. Since this computed value lies in the critical region, the null hypothesis is rejected at 5% level of significance and the alternative hypothesis is accepted. The manufacturer's claim that $\mu > 1000$ is accepted and the retailer should order the consignment. He would be sure in 95 percent of cases that the mean life will exceed 1000 hrs.

(b) The critical value for this one-sided test at $\alpha = 0.01$ level is 2.33 and the critical region is $(2.33, \infty)$ The computed test statistic 2 does not lie in the critical region and so the null-hypothesis is *not* rejected. 'Acceptance' of null hypothesis amounts to rejection of alternative hypothesis. The manufacturer's claim that $\mu > 1000$ is rejected. The retailer should not order the consignment at 1% level, that is, if he is to be sure that only in one percent of the cases, the mean life will not exceed 1000 hr.

10.9 TEST FOR AN ASSUMED PROPORTION

For a large sample, the sample proportion $\bar{p}$ is approximately normal with the mean equal to the population proportion p and s.d. $\sigma_{\bar{p}} = \sqrt{\left(\frac{pq}{n}\right)}$.
Thus

$$Z = \frac{\bar{p} - p}{\sigma_{\bar{p}}}$$

is approximately normal with mean 0 and s.d. 1. The probability that the observed sample proportion will be within $p \pm 2\sigma_{\bar{p}}$ is 0.95.

Example 9 (a). The proportion of defective items in a sample of 36 is found to be 0.07. Test the nypothesis that the population proportion is $p = 0.05$ against the alternative that $p > 0.05$.
Here

$$H_0 : p = 0.05$$

$$H_1 : p > 0.05$$

Under null-hypothesis, that is, with $p = 0.05$, the s.e. of $\bar{p}$ equals

$$\sigma_{\bar{p}} = \sqrt{\frac{p(1-p)}{n}} = \sqrt{\frac{(0.05)(0.95)}{36}} = \frac{0.218}{6} = 0.036$$

Now $$Z = \frac{\bar{p} - p}{\sigma_{\bar{p}}} = \frac{0.07 - 0.05}{0.036} = \frac{0.02}{0.036} = 0.55$$

We have $P\{Z > Z_1\} = 0.05$ for $Z_1 = 1.645$ for the one-sided test. The value 0.55 being less than 1.645, the null hypothesis $H_0 : p = 0.05$ is accepted at 5% level. On the basis of the evidence there is no reason to reject the null hypothesis at 5% significance level.

10.10. COMPARISON OF MEANS OF TWO SAMPLES

Let us consider that two samples of sizes n_1 and n_2 respectively are taken one from each of two populations with means μ_1 and μ_2 and s.d. σ_1 and σ_2. Then $D = \mu_1 - \mu_2$ is the *true* difference between the population means and if $\bar{X}_1$ and $\bar{X}_2$ are the two sample means then $\bar{D} = \bar{X}_1 - \bar{X}_2$ is the difference between the two sample means. To find whether this difference is statistically significant we have to consider the sampling distribution of $\bar{D}$. The sampling distribution of $\bar{D}$ has mean $D = \mu_1 - \mu_2$. Since the variance of $\bar{X}_1$ is $\frac{\sigma_1^2}{n_1}$ and that of $\bar{X}_2$ is $\frac{\sigma_2^2}{n_2}$ we have var $(\bar{D})$ = var $(\bar{X}_1)$ + var $(\bar{X}_2) = \frac{\sigma_1^2}{n_1} + \frac{\sigma_2^2}{n_2}$; so that $\bar{D}$ has mean D and s.d. $\sigma_{\bar{D}} = \sqrt{\left(\frac{\sigma_1^2}{n_1} + \frac{\sigma_2^2}{n_2}\right)}$. The sampling distribution of $\bar{D}$ for large samples is approximately normal. Thus the sampling distribution of $\bar{D}$ and its mean and standard error are known. We have then to locate where the *observed* difference between the sample means computed on the basis of the two samples lies. If this lies within $\pm 2\sigma_{\bar{D}}$ of the mean D, that is within the interval $(D - 2\sigma_{\bar{D}}, D + 2\sigma_{\bar{D}})$, the observed difference is not considered significant; where as if the observed difference lies outside the interval $(D - 2\sigma_{\bar{D}}, D + 2\sigma_{\bar{D}})$, the observed difference is considered to be statistically significant at the 5% level of significance.

When σ_1, σ_2 are not known, these are approximated by the corresponding sample standard deviations. The test will then be approximate.

Example 10 (a). It is required to test whether two similar machines A_1 and A_2 in a factory are producing articles of equal average weights. From past experience it is found that the standard deviations of weights of products of machines of A_1 and A_2 are 0.13 kg. and 0.15 kg. respectively. Two samples of sizes 40 and 50 respectively of the weights of products of two machines are taken. It is found that the sample mean weights are 1.05 kg and 1.10 kg. Test whether the differences of weights of products of two machines are statistically significant.

The hypotheses are $\quad H_0 : \mu_1 = \mu_2$

and $\quad H_1 : \mu_1 \neq \mu_2$

Here $n_1 = 40$, $n_2 = 50$, $\bar{X}_1 = 1.05$, $\bar{X}_2 = 1.10$

$$\sigma_1 = 0.13, \ \sigma_2 = 0.15$$

The standard error $\sigma_{\bar{D}}$ of the difference of means is given by

$$\sigma_{\bar{D}} = \sqrt{\left\{\frac{(0.13)^2}{40} + \frac{(0.15)^2}{50}\right\}} = \sqrt{(0.004225 + 0.00450)}$$

$$= \sqrt{0.008725} = 0.0295$$

Under hypothesis H_0, (when $\mu_1 = \mu_2$), $D = 0$ so that the sampling distribution of $\bar{D}$ has mean 0 and s.d. 0.0295; we have $D \pm 2$ (s.d.) $= \pm .059$ Now the observed difference between the sample means equals $\bar{X}_1 - \bar{X}_2 = 1.05 - 1.10 = -.05$. This lies within the interval $(-0.059, 0.059)$. Hence the null-hypothesis $\mu_1 = \mu_2$ is accepted. That is, the observed difference is not statistically significant at 5% level of significance. On the basis of the evidence there is no reason to believe that the products of the two machines differ in mean weights.

10.11. COMPARISON OF PROPORTIONS FROM TWO SAMPLES

Let us now consider statistical significance of the difference between two proportions computed from samples of sizes n_1 and n_2 taken from the two populations with population proportions p_1 and p_2 respectively. Then $P = p_1 - p_2$ is the true difference between the population proportions and $\bar{p}_1$ and $\bar{p}_2$ are the two sample proportions, and $\bar{P} = \bar{p}_1 - \bar{p}_2$ is the difference between the sample proportions. To test whether this difference is significant we need to consider the sampling distribution of $\bar{P}$. The sampling distribution of $\bar{P}$ has mean P, and

$$\text{var }(\bar{P}) = \text{var }(\bar{p}_1) + \text{var }(\bar{p}_2)$$

$$= \frac{p_1 q_1}{n_1} + \frac{p_2 q_2}{n_2}$$

so that $$\sigma_{\bar{P}} = \text{s.e. of } \bar{P} = \sqrt{\left(\frac{p_1 q_1}{n_1} + \frac{p_2 q_2}{n_2}\right)}$$

The sampling distribution of $\bar{P}$ for large samples is approximately normal. Thus the sampling distribution of $\bar{P}$ and its mean and standard error are known. We have to locate where the observed difference between the sample proportions of the two samples lies. If it lies within $\pm\, 2\sigma_{\bar{P}}$ of the mean P, that is within the interval $(P - 2\sigma_{\bar{P}},\ P + 2\sigma_{\bar{P}})$, then the observed difference is not considered significant, whereas, if it lies outside this interval, the observed difference is said to be statistically significant at 5% level of significance.

When p_1 and p_2 are not known, these are approximated by their corresponding sample values.

Example 11 (a). The proportions of literates between groups of people of two districts A and B are to be tested. Of the 100 persons from each of the districts selected at random, 32 of district A and 40 of district B are literates. Test whether the observed proportion of literates is statistically significant.

We have to decide between the two hypotheses

$$H_0 : p_1 = p_2$$

and $$H_1 : p_1 \neq p_2$$

Here $n_1 = n_2 = 100$.

Under the null hypothesis $P = p_1 - p_2 = 0$ and

$$\sigma_{\bar{P}} = \sqrt{\left(\frac{pq}{n_1} + \frac{pq}{n_2}\right)} = \sqrt{\left[pq\left(\frac{1}{n_1} + \frac{1}{n_2}\right)\right]}$$

As an approximate value of p, we use the average proportion between two samples.

The average proportion $\frac{32 + 40}{200} = 0.36 (= p)$, then $q = 1 - p = 0.64$

so that

$$\sigma_{\bar{P}} = \sqrt{\left[\left(\frac{(0.36)(0.64)}{100} + \frac{(0.36)(0.64)}{100}\right)\right]} = \frac{(0.48)\sqrt{2}}{10} = 0.068$$

The sampling distribution of $\bar{P}$ has mean 0 and standard error 0.068. We have to locate where the observed difference of sample proportions lies. If it lies within the interval $(-2 \times 0.68,\ 2 \times 0.68)$ that is within $(-0.136, 0.136)$ then the observed difference is not significant at 5% level and hypothesis H_0 is accepted. We have $\bar{p}_1 = \frac{32}{100} = 0.32$, $\bar{p}_2 = \frac{40}{100} = 0.40$ so that $\bar{P} = \bar{p}_1 - \bar{p}_2 = 0.32 - 0.40 = -0.08$. Since it lies within $(-0.136, 0.136)$, the observed sample difference is not statistically significant at 5% level, there is no adequate basis for thinking that the proportions of the illiterates in the two districts are not equal.

EXERCISES - 10

Sections 10.1-10.2

1. Explain what is meant by statistical inference. What are the important types of problems of inference ?
2. What is point estimation ? Distinguish between an estimator and an estimate.
3. What are the criteria of a good estimator ? Explain.

Section 10.3

4.* Consider the problem of finding an estimate of the population μ based on a random sample of size n from the population. Compute a point estimate of μ in the following cases :
   (i) $n = 60$, $\Sigma x_i = 350$, $\Sigma (x_i - \bar{x})^2 = 256$
   (ii) $n = 100$, $\Sigma x_i = 985$, $\Sigma (x_i - \bar{x})^2 = 435$
   (iii) $n = 150$, $\Sigma x_i = 1235$, $\Sigma (x_i - \bar{x})^2 = 480$
5. Find the estimated standard error in each of the cases considered in Exercise no. 4.
6. Compute a 95% and 99% error bound for the estimation of μ in each of the cases of Exercise no. 4.
7. It is found that all packets of 1 kg of washing powder do not contain exactly 1 kg, some contain slightly more and some slightly less. Further, it is seen that the s.d. of the weight contained in the packets is 5 gm. If a random sample of 25 packets is taken, how accurate would the sample mean as an estimate of the mean of the population of all packets? Give a measure of the accuracy.
8. Consider Exercise no. 7.
   Suppose that the production manager is not satisfied with his accuracy based on a sample of 25 samples. He desires that his estimate should not be in error by more than 1 gm with a probability of 0.95. How large should his sample be ?
9. Samples of size n are taken for estimation of mean of a certain population. It is desired to reduce the error involved in estimation of the population mean by (i) 50% (ii) 25%. What would be the required sample size and by what factor should the sample size be increased or decreased ?
10. Consider Exercise No. 9 with (i) $n = 25$, (ii) $n = 36$, (iii) $n = 100$. What should be the sample size, taking which, the error involved in estimation of the population mean could be reduced by (i) 50%, (ii) 25% ?

Section 10.4

11. Explain what is meant by interval estimation.
12. What do you mean by (i) 95% (ii) 99% confidence limits and confidence intervals for the estimation of the population mean? What is confidence coefficient?
13. The sample mean is computed from a random sample of size 25 drawn from a population having s.d. 10. Find (i) 95%, (ii) 99% confidence intervals of the population mean.
14. If $n = 50$ and if
    $0.95 = P\left\{-5.88 < \overline{X} - \mu < 5.88\right\}$
    find σ.
15. Given that $\overline{X} = 10.52$ and
    $0.95 = P\left\{-6.60 < \mu < 14.44\right\}$
    find σ (i) if $n = 30$ (ii) if $n = 8$.

Section 10.5

16. How would you proceed to estimate the proportion p of a certain characteristic in a population ?
    Find 95% confidence interval for p.
17. A small proportion p of parts produced by a machine is defective. A random sample of 100 parts shows 3 defective parts.
    Find a point estimate of p and (i) 95% (ii) 99% confidence interval for p.
18. Given that 95% confidence limits of p are 0.65 and 0.85 find the point estimate of p.
19. Consider Exercise 18. Find the sample size n.
20. In a dice-playing experiment, 6 appeared 18 times in 100 throws. Find the 95% confidence limits of the probability of throwing a six.
21. In 50 tosses of a coin, head appeared in 30 tosses. Find (i) 90%, (ii) 95%, (iii) 99% confidence intervals for the probability of throwing a head.

Section 10.7

22. What is meant by test of (statistical) significance ? Explain.
23. Discuss the problems that arise in testing statistical significance.
24. Explain the expressions 'null hypothesis', 'alternative hypothesis', 'acceptance region 'rejection region'.
25. Indicate the principal steps in testing a statistical hypothesis.
26. Discuss what is meant by one and two sided tests. What are the critical values for a one-sided test and a two-sided test at (i) 0.025% (ii) 0.15% level of significance ?

Section 10.8

27. Explain how you would proceed to test the difference between the sample and population means when the sample is large.
28. Given that $\overline{X} = 80$, $\sigma = 15$ and $n = 60$ test the hypothesis that $\mu = 85$ against $\mu \neq 85$.
29. Given that $\overline{X} = 50$, $\sigma = 10$ and $n = 100$ test the hypothesis that $\mu = 65$ against $\mu \neq 65$.
30. Records of the last several years show that the average marks secured by students of a certain class in Mathematics has mean 58 with a standard deviation of 10. In a certain year extra coaching was given to 50 students (of the class for that year) and the sample mean turned out to be 70. Test the hypothesis that the coaching has helped in raising the average marks of the class.
31. The mean life of a certain tube produced by a company is known to be 900 hours. Certain improvements are unertaken and a sample of 100 such tubes tested : they give a mean life of 965 hours with a s.d. of 125 hours. Test the hypothesis that the mean life span is $\mu = 900$ hrs. against the alternative $\mu > 900$ hrs.

Section 10.9

32. Given that $n = 50$, $p = 0.06$. Test the hypothesis $p = 0.05$ against the hypothesis that $p = 0.05$.
33. Given that $n = 100$, $p = 0.12$. Test the hypothesis that $p = 0.10$ against the hypothesis that $p \neq 0.10$.

34. A pair of coins is tossed 100 times. It is found that heads in both the coins appear 30 times. Test the hypothesis that the pair of coins is unbiased using (i) a two-sided test, (ii) a one-sided test at the 0.05 level of significance.
35. A machine produces spare parts for television. The manufacturer's claim is that only 2% of the parts are defective. A sample of 120 parts shows 5 defectives. Test the truth of the manufacturer's claim at 0.05 and at 0.01 levels of significance.

Section 10.10

36. Given that $n_1 = 50$, $n_2 = 60$, $\bar{x}_1 = 5.30$, $\bar{x}_2 = 5.82$, $\sigma_1 = 0.5$, $\sigma_2 = 0.6$, test the hypothesis that $\mu_1 = \mu_2$ against the alternative that $\mu_1 \neq \mu_2$.
37. Two machines manufacture similar parts. Two samples of sizes 100 and 150 lengths of parts produced by the two machines are tested. It is found that the mean of the samples of the machines are respectively 0.85 cm and 0.90 cm. It is also known from past experience that the s.d. of the parts produced by the machines are 0.18 cm and 0.20 cm. Test whether the differences of the lengths of the products of the two machines are significant.
38. Two batches of trainees are trained by two different methods. After training, an appropriate test is given to test their ability. It is found that the mean score of the first batch of 40 trainees is 65 with a s.d. of 15, while for the second batch of 60, the mean is 75 with a s.d. of 20. Test the hypothesis of the equality of means (Assume that the two batches are random samples from a large population).

Section 10.11

39. Given that $n_1 = 50$, $p_1 = 0.05$, $n_2 = 60$, $p_2 = 0.08$ test the hypothesis that $p_1 = p_2$ against $p_1 \neq p_2$.
40. Two machines *A*, *B* produce the same kind of parts. *A* sample of 100 of parts produced by each of the machines is tested. It is found that the number of defectives for the sample of products of machine *A* is 6 and that of machine *B* is 10. Test the hypothesis of equality of proportion of defective parts produced by the two machines against the alternative of inequality.

Chapter 11

Further Tests of Significance

11.1. SMALL SAMPLE THEORY

Introduction : Some of the statistical techniques discussed earlier in Sections 10.7-10.10 can be used only when the samples are large. The large sample results do not hold good for small samples. We shall discuss here methods which do not require the assumption that the sample or samples are large. The methods which will hold good for small samples will also be applicable for large samples. But small sample methods involve more information or assumptions than large sample methods. Thus small sample methods cannot altogether replace the large sample methods.

Consider the large sample technique of testing the significance of the mean (Section 10.8). For large sample, the distribution of the sample mean $\overline{X}$ is approximately normal with mean μ and s.d. $\sigma/\sqrt{n}$, where μ is the mean and σ is the s.d. of the population. This fact has been used in deriving the results.

When samples are small, this assumption of approximate normality of the distribution of $\overline{X}$ is *not* true; in fact another distribution is to be used and the test modified accordingly.

For large samples, the statistic

$$Z = \frac{\overline{X} - \mu}{\sigma/\sqrt{n}}$$

is approximately normal with mean 0 and s.d. 1. Here we can use s as an estimate of σ, when σ is not known. Use of s as an estimate of σ does not appreciably change the distribution of Z in large samples; but it does make a difference in the case of small samples (for samples of size, say, less than 30). For small samples, we shall use the statistic

$$t = \frac{\overline{X} - \mu}{s/\sqrt{n}} = \frac{(\overline{X} - \mu)\sqrt{n}}{s} \tag{1.1}$$

The distribution of the statistic is known as 'Student's t-distribution', named after W.S. Gosset who published the work under the pseudonym 'Student' in 1908.

11.2. STUDENT'S t-DISTRIBUTION

Theorem 11.1. If $X_1, X_2, \ldots\ldots, X_n$ is a random sample from a normal pupulation $N(\mu, \sigma)$ and $\overline{X} = \frac{1}{n}\Sigma X_i$ and $s = \sqrt{\frac{\Sigma (X_i - \overline{X})^2}{n-1}}$, then $$t = \frac{(\overline{X} - \mu)\sqrt{n}}{s}$$

is said to have student's t-distribution with $(n - 1)$ degrees of freedom.

The distribution (which is a sampling distribution) depends on the size of the sample, i.e. for each value of $(n - 1)$ there is a different t distribution, $(n - 1)$ being termed as the 'degrees of freedom' of the distribution. The shapes of the density functions of t-distribution are symmetric around 0 but their tails are more spread out than those of the density function of the standard normal distribution $N(0, 1)$. As n increases, the t-distribution approaches the $N(0, 1)$ distribution. For $n \geq 30$, t-distribution can be approximated by a $N(0, 1)$ distribution. Values of associated probabilities of t-distribution are given in the Appendix.

Applications of t-distribution

11.2.1. Test for an assumed mean

The following examples will illustrate how to use the t test in case of small samples.

Example 2 (a). A random sample of 16 newcomers gave mean height of 1.67 m and a s.d. of 0.16 m. The mean height of the students of the previous years is known to be 1.600 m.

Is the mean height of the newcomers significantly different from the mean height of the student population of the previous years ?

The problem may be considered as a problem of testing of hypothesis :

$H_0 = 1.60$ against $H_1 \neq 1.60$.

Here $n = 16, \overline{X} = 1.67, \mu = 1.60, s = 0.16$

$$\text{Thus } t = \frac{(\overline{X} - \mu)\sqrt{n}}{s} = \frac{(0.07) \times 4}{0.16} = 1.75.$$

From t-table with $\nu = n - 1 = 15$ d.f., $P(|t| > 1.75) = 0.10$, which is greater than 0.05. The calculated value of t is therefore not significant. On the basis of the data we have no evidence to suspect that the mean height of the newcomers is significantly different from the mean height of the student population of the previous years. In other words, on the basis of evidence, there is no reason to reject the null hypothesis H_0.

Example 2 (b). A census of retail stores in a particular year gave the mean annual turnover of Rs. 1,50,000. A random sample of 10 stores in the following year gave a mean turnover of Rs. 1,57,000 and a s.d. of Rs. 11,180. Has the mean turnover changed in the following year ?

The problem may be considered as a problem of testing the hypothesis

$$H_0 : \mu = 1{,}50{,}000 \text{ against } H_1 : \mu \neq 1{,}50{,}000$$

$$\text{Here } n = 10, \mu = 1{,}50{,}000, \overline{X} = 1{,}57{,}000.$$

$$s = 1{,}118$$

$$t = \frac{(\overline{X} - \mu)\sqrt{n}}{s} = 1.97.$$

From t-table with $\nu = n - 1 = 9$ d.f. the 5 per cent point of t is 2.262. The calculated value of t being less than the 5 per cent of the table value is not significant. There is no evidence to believe that the mean turnover has changed in the following year.

On the basis of the evidence before us, there is no reason to reject the null hypothesis.

It may be noted that if the large sample test is used, we would have got significant at 5% level, the calculated value of the statistic being greater than 1.96. This illustrates the fact that large sample method will lead to significant results more often than justified (when small sample method is used). This is explained by the fact that the t-distribution has a thiçker tail (or slightly larger dispersion) than the standard normal distribution.

11.2.2. Comparison of means of two samples

We shall now examine how t-distribution can be used to test whether the two random samples may be regarded to have been drawn from (a) the same normal population (b) two different normal populations having the same variance.

Theorem 11.2

(a) Let X_i $(i = 1, 2, \ldots\ldots, n_1)$ and X'_j $(j = 1, 2, \ldots\ldots, n_2)$ be the values of two random samples of sizes n_1 and n_2 respectively from the *same normal population* $N(\mu, \sigma)$.

If $\quad \overline{X}_1 = \frac{1}{n_1}\Sigma X_i, \ \overline{X}_2 = \frac{1}{n_2}\Sigma X'_j$

and

$$s^2 = \frac{\sum_{i=1}^{n_1} (X_i - \overline{X}_1)^2 + \sum_{j=1}^{n_2} (X'_j - \overline{X}_2)^2}{n_1 + n_2 - 2}$$

then the statistic

$$t = \frac{\overline{X}_1 - \overline{X}_2}{s\sqrt{\left(\frac{1}{n_1} + \frac{1}{n_2}\right)}}$$

has a t-distribution with $\nu = n_1 + n_2 - 2$ degrees of freedom. This may be used to test whether two sample means differ significantly.

(b) Let X_i $(i = 1, 2, \ldots\ldots, n_1)$, X'_j $(j = 1, 2, \ldots\ldots, n_2)$ be the values of two random samples from two different normal populations $N(\mu_1, \sigma)$ and $N(\mu_2, \sigma)$ having means μ_1 and μ_2 but the same s.d. σ. Then the statistic

$$t = \frac{(\overline{X}_1 - \overline{X}_2) - (\mu_1 - \mu_2)}{s\sqrt{\left(\frac{1}{n_1} + \frac{1}{n_2}\right)}}$$

has a t-distribution with $\nu = n_1 + n_2 - 2$ degrees of freedom.

Example 2(c). Fertilizers A and B were tried respectively on 10 and 8 randomly chosen experimental plots. The yields in the plots were as given below. Test whether there is a difference in the effects of the fertilizers as reflected in the mean yields.

Fertilizer	Yield									
A	8.0	7.6	8.2	7.8	8.3	8.4	8.2	7.8	7.7	8.0
B	7.4	8.1	7.6	8.1	7.5	7.6	7.3	7.2		

This may be treated as a problem of testing the hypothesis

$$H_0 : \mu_1 = \mu_2$$

against

$$H_1 : \mu_1 \neq \mu_2.$$

Here

$$n_1 = 10,\ \overline{X}_1 = 8.0,\ \Sigma\ (X_i - \overline{X}_1)^2 = 0.66$$

$$n_2 = 8,\ \overline{X}_2 = 7.6,\ \Sigma\ (X'_j - \overline{X}_2)^2 = 0.80$$

$$s^2 = \frac{\Sigma\ (X_i - \overline{X}_1)^2 + \Sigma\ (X'_j - \overline{X}_2)^2}{n_1 + n_2 - 2} = \frac{1.46}{16} = 0.09125$$

so that

$$s\sqrt{\left(\frac{1}{n_1} + \frac{1}{n_2}\right)} = 0.14328$$

and

$$t = \frac{0.4}{0.14328} = 2.79 \text{ for } \nu = n_1 + n_2 - 2 = 16\ d.f.$$

For 16 d.f., the probability of getting a value of t as high as 2.79 is less than 0.05. Hence the calculated value of t is significant at 5% level. We reject the null hypothesis H_0 and accept the alternative hypothesis. There is a significant difference in the mean yields.

Note : The t-distribution eliminates the inaccuracy of large sample methods when applied in case of small samples. Another important factor is that though the basic assumption is that the population from which the samples are drawn is normal, the t-distribution is not very much affected even if there is considerable departure from normality of the basic variable. Such a property is known as *robustness* : A test or an estimate is usually called *robust* if it performs well under modifications of the underlying assumption(s).

The distribution has other applications also.

11.3. THE F DISTRIBUTION

Introduction : For applying t-distribution to test the hypothesis $H_0 : \mu_1 = \mu_2$ against $H_1 : \mu_1 \neq \mu_2$ it was necessary to assume the equality of the variances, that is, that the samples were drawn from two normal populations N_1 (μ, σ) and N_2 (μ_2, σ), of equal variance.

The equality of the variances, which was not tested there, can be tested by means of a statistic called F statistic.

Theorem 11.3

If X_i, $i = 1, 2, \ldots\ldots, n_1$ and X'_j, $j = 1, 2, \ldots\ldots n_2$

are two independent random samples from two normal populations $N(\mu_1, \sigma_1)$ and $N(\mu_2, \sigma_2)$ respectively, then the distribution of the statistic.

$$F = \frac{s_1^2/\sigma_1^2}{s_2^2/\sigma_2^2} = \frac{\dfrac{\Sigma (X_i - \overline{X}_1)^2}{(n_1 - 1)\,\sigma_1^2}}{\dfrac{\Sigma (X'_j - \overline{X}_2)^2}{(n_2 - 1)\,\sigma_2^2}} \tag{3.1}$$

is knwon as F distribution with $n_1 - 1, n_2 - 1$ degrees of freedom.
Under the null hypothesis $H_0 : \sigma_1 = \sigma_2$ the distribution of

$$F = \frac{\Sigma\ (X_i - \overline{X}_1)^2/(n_1 - 1)}{\Sigma\ (X'_j - \overline{X}_2)^2/(n_2 - 1)} \tag{3.2}$$

is F with $n_1 - 1, n_2 - 1$ degrees of freedom. The proof is omitted.

Here the numerator and the denominator are unbiased estimates of the variances σ_1^2, σ_2^2 of the two populations respectively.

The density curve of the F distribution lies entirely in the positive quadrant and is symmetric with a long tail to the right; its shape depends on the quantities $n_1 - 1$ and $n_2 - 1$. The percentage points of the distribution are given in a table in the Appendix.

Note : The larger of the two quantities is to be taken in the numerator and the smaller in the denominator of the expression of F.

The F distribution can be applied to test the equality of two variances. We consider an example below.

Example 3(a). Consider the data of Example 2(c), where the yields under fertilizers A and B constitute two random samples drawn from two normal populations $N(\mu_1, \sigma_1)$, $N(\mu_2, \sigma_2)$ respectively.

Here we are to test the hypothesis

$$H_0 : \sigma_1 = \sigma_2$$

against

$$H_1 : \sigma_1 \neq \sigma_2$$

Estimated variances are

$$\hat{\sigma}_1^2 = s_1^2 = \frac{\Sigma (X_i - \bar{X}_1)^2}{n_1 - 1} = \frac{0.66}{9} = 0.0733$$

$$\hat{\sigma}_2^2 = s_2^2 = \frac{\Sigma (X'_j - \bar{X}_2)^2}{n_2 - 1} = \frac{0.80}{7} = 0.1143.$$

Taking the larger quantity in the numerator, we get

$$F = \frac{\hat{\sigma}_2^2}{\hat{\sigma}_1^2} = \frac{0.1143}{0.0733} = 1.56 \text{ with d.f. } (7, 9).$$

For this two tailed test, the critical value of F at 5% is 4.75. This can be obtained by interpolation for 2.5% level from the entries against $\nu_1 = 7$, $\nu_2 = 9$ at 5% and 1% in the table for F distribution given in the Appendix. Since the calculated value of F is less than the critical table value at 5%, there is no basis to reject the null hypothesis $H_0 : \sigma_1 = \sigma_2$, and the assumption of equal variance is a reasonable one.

It may be noted that the upper tail of the F distribution is to be used for testing $H_0 : \sigma_1 = \sigma_2$ against $H_1 : \sigma_1 > \sigma_2$ and lower tail for testing $H_0 : \sigma_1 = \sigma_2$ against $H_1 = \sigma_1 < \sigma_2$.

It is also to be noted that the test based on F distribution is *not robust.* Departure from normality of the two distributions will considerably affect the F distribution – and the test will not be usable.

11.4. THE CHI SQUARE DISTRIBUTION

Introduction : We have considered the sampling distribution of the sample mean in Chapter 8 (see Theorems 8.2 and 8.4). The sampling distribution of the mean of a random sample of size n from a normal population $N(\mu, \sigma)$ is normal $N\left(\mu, \frac{\sigma}{\sqrt{n}}\right)$. We shall now consider the sampling distribution of the sample variance.

The χ^2 distribution was first obtained by Helmert in 1875 and rediscovered by Karl Pearson in 1900.

Theorem 11.4. Let $X_i, i = 1, 2, \ldots, n$ be a random sample from a normal population $N(\mu, \sigma)$ and let $\bar{X} = \frac{1}{n} \Sigma X_i$ be the sample mean and $s^2 = \frac{1}{n-1} \Sigma (X_i - \bar{X})^2$ be the sample variance.
Then the distribution of the statistic

$$\chi^2 = \frac{(n-1)s^2}{\sigma^2} = \frac{\Sigma(X_i - \bar{X})^2}{\sigma^2} \tag{4.1}$$

is called a chi-square (in notation χ^2) distribution with $\nu = (n - 1)$ degrees of freedom.

The quantity chi-square is always positive and so the density curve of the χ^2 distribution lies entirely in the positive quadrant. The shape of the curve is asymmetric having a long tail on the right. Its shape depends on the quantity $\nu = (n - 1)$ which is its d.f. As ν increases, the shape of the curve approaches that of a normal curve. For $\nu > 30$, it is sufficient to assume that $\sqrt{2\chi^2} \to N(\sqrt{2n-1}, 1)$. Table of percentage points of χ^2 distribution, given in the Appendix, covers for values of ν upto 30; beyond this point, the above approximation holds and tables of normal distribution can be used.

It may be noted that the assumption of normality is a requirement for the result to hold. The χ^2 test *does not* possess the *robustness* property.

Application of χ^2 distribution

11.4.1. Testing a hypothetical value of σ

We illustrate this test by means of an example.

Example 4(a). Packages made by a standard method for a long time indicate that $\sigma_0 = 11.75$ for the variation of weight. A less expensive and less time consuming new method is tried; it shows that for a sample of size 20, the quantity

$$\sum_{i=1}^{20} (X_i - \bar{X})^2 = 2856.08$$

Before switching on this new and less time consuming method, it is desired to test whether this method results in an increase of the variability of the quality of the packages in terms of their weights.

Here we wish to test the hypothesis

$$H_0 : \sigma = \sigma_0 = 11.75$$

$$\text{against } H_1 : \sigma > \sigma_0 = 11.75.$$

We have $n = 20, \nu = n - 1 = 19$ and the statistic

$$\chi^2 = \frac{\Sigma (X_i - \overline{X})^2}{\sigma_0^2} = \frac{2856.08}{(11.75)^2} = 20.69.$$

The table value of χ^2 for 19 d.f. at 5% level is 30.14 and the calculated value is not significant. On the basis of the data there is no justification of rejecting the null hypothesis H_0 i.e., that the variability in terms of weight of the packages has not increased.

For testing against the two sided alternative $H_1 : \sigma \neq \sigma_0$ at the 5% level, generally two equal tails of 2.5% area of the χ^2 distribution is taken as the critical region.

Some other applications of the χ^2 distribution are considered below.

11.5. LARGE SAMPLE TEST

11.5.1. Goodness of fit test

Chi-square test as a goodness of fit test was first devised by Pearson. Consider a population which may be partitioned (or classified) into a number of k classes or cells, and let p_i be the probability that a member belongs to class i, $i = 1, 2, \ldots\ldots k$ with $\Sigma p_i = 1$. Suppose that a random sample of size n is drawn from the population : some members will fall in class 1, some in class 2 and so on and a frequency distribution can be constructed. Let n_i be the frequency of class i, i.e. n_i is the number of members falling into class i, $\Sigma n_i = n$, where some n_i may be 0. The number n_i is a random variable; it is the number of members falling in class i, out of a total of n members, the probability of a member falling in class i being p_i. Thus n_i is a binomial variate with parameters n and p_i so that

$$E(n_i) = np_i = e_i \text{ (say)}, i = 1, 2, \ldots\ldots k$$

is the expected frequency of the i th class. Now the problem is to find out whether a set of observed frequencies n_i, $i = 1, 2, \ldots\ldots$ is compatiable with the expected values $E(n_i) = e_i$. We shall use the following result; the proof is omitted (as it is beyond the scope of this book).

Theorem 11.5. If a random sample of size n is taken from a population whose members can be classified into k classes, and if $n_1, \ldots\ldots n_k$ $\left(\sum_{i=1}^{k} n_i = n \right)$ are the observed frequencies while $e_i = E(n_i)$ are the expected frequencies, then as $n \to \infty$, the distribution of the random variable

$$\sum_{i=1}^{k} \frac{(n_i - e_i)^2}{e_i}$$

approaches that of χ^2 with $(k-1)$ d.f.

We consider an example below.

Example 5 (a). A coin is thrown 50 times where Head occurs 29 times and Tail 21 times. To test whether the coin is defective so as to show Head more often than Tail.
Let p be the probability of throwing Head. Then the problem is of testing

$$H_0 : p = ½$$

$$\text{against } H_1 : p \neq ½$$

Here $n = 50, n_1 = 29, n_2 = 21, k = 2, e_1 = e_2 = 50 \times \frac{1}{2} = 25$

$$\sum_{i=1}^{2} \frac{(n_i - e_i)^2}{e_i} = \frac{(29-25)^2}{25} + \frac{(21-25)^2}{25} = 0.64 + 0.64 = 1.28.$$

The table value of χ^2 with 1 d.f. at 5% is 3.84 and the calculated value is less than the table value. On the basis of the evidence there is no reason to reject the null hypothesis that $p = ½$, i.e. that the coin is fair.
Example 5(b). The percentage of male students in a big college is 65. A class consisting of 30 students in a particular year has 16 male and 14 female students. Can this be considered as a random sample from the student population in the college ?

$$\text{We have } p_1 = P(\text{a student is male}) = 0.6$$

$$p_2 = P(\text{a student is female}) = 0.4$$

$$\text{We test the hypothesis } H_0 : p_1 = 0.6, p_2 = 0.4$$

$$n = 30, n_1 = 16, n_2 = 14,$$

$$e_1 = 30 \times 0.6 = 18, e_2 = 30 \times 0.4 = 12$$

$$\sum_{i=1}^{2} \frac{(n_i - e_i)^2}{e_i} = \frac{(16-18)^2}{18} + \frac{(14-12)^2}{12} = 0.22 + 0.33 = 0.55.$$

The value is not significant at 5 per cent level. There is no evidence to reject the hypothesis.

Example 5(c). A die is thrown 90 times and the number of faces shown are as indicated below.

Face (i)	1	2	3	4	5	6
Frequency (n_i)	18	14	13	15	14	16

Test whether the die is 'fair' (or honest).
The probability p_i of showing the face i is 1/6, $i = 1, 2, \ldots 6$. Here we are to test the hypothesis

$$H_0 : p_1 = p_2 = \ldots = p_6 = \frac{1}{6}$$

We have $e_i = E(n_i) = np_i = 90 \times \frac{1}{6} = 15$ so that

$$\sum_{i=1}^{6} \frac{(n_i - e_i)^2}{e_i} = \frac{(15-18)^2}{15} + \frac{(14-15)^2}{15} + \frac{(13-15)^2}{15}$$

$$+ \frac{(15-15)^2}{15} + \frac{(14-15)^2}{15} + \frac{(16-15)^2}{15} = 1.07.$$

For 5 d.f. the calculation is not significant, the probability of getting a value as high as 1.07 is more than 0.95. Hence there is no reason to reject the null hypothesis $\left(p_i = \frac{1}{6}\right)$ or that the die is fair on the basis of the evidence.

Example 5(d). A census of all the handloom establishments in the area (having a very large number of such establishments) taken in a year reveals that the percentage distribution of the establishments according to the number of employees is as given in column (3) of the Table below.
A random sample of 200 handloom establishments taken in a subsequent year indicates a distribution as given in column (2) of the table below :
Is there any evidence to show that there is a change in the structure of the establishments as regards the number of workers employed ?

TABLE 11.1

No. of Employees	Observed frequencies (n_i)	Percentage distribution	Expected frequency (e_i)
Less than 5	48	36.5	73
5–10	107	47.5	95
11–15	28	9.5	19
More than 15	17	6.5	13
	200	100.0	200

Here we test the hypothesis that

$$H_0 : p_1 = 0.365, p_2 = 0.475, p_3 = 0.095, p_4 = 0.065$$

We have

$$e_1 = 200 \times 0.365 = 73 \text{ and so on and}$$

$$\sum_{i=1}^{4} \frac{(n_i - e_i)^2}{e_i} = \frac{(48-73)^2}{73} + \frac{(107-95)^2}{95} + \frac{(28-19)^2}{19} + \frac{(17-13)^2}{13}$$

$$= 8.56 + 1.52 + 4.26 + 1.23 = 15.57.$$

The calculated value is highly significant being greater than the critical value of χ^2 for 3 d.f. at 1 per cent which is 11.34. On the basis of the evidence we reject the hypothesis that the structure as indicated in column (3) has remained unchanged for the following year.

11.5.2. Some observation on the use of χ^2 test

Limitations of the test

Chi-square test used as goodness of fit test has certain limitations.

The statistic $\dfrac{\Sigma (n_i - e_i)^2}{e_i}$ is approximately distributed like χ^2. The approximation is good for large n and for $e_i \geq 5$ and $k \geq 5$. When k (the number of classes) is less than 5, it is best to have e_i larger than 5. If the expected frequency of any class or cell is less than 5, it is better to combine such a cell with some other cell, so that the total number in the cells combined exceeds 5.

The reason behind is that the approximation of the binominal distribution to the normal for large n does not hold for small values of p or q.

Generality of the test

On the other hand, the test can be applied in even under some general conditions. In the statement of Theorem 11.5, it is assumed that $e_i = E(n_i) = np_i$ are known. This implies that the cell probabilities are known or given. In many cases, these cell probabilities are not exactly known but are known to be functions of some unknown parameters. These unknown parameters are then to be estimated from which the cell probabilities and the expected frequencies e_i can be estimated. Then the degrees of the freedom are reduced by the number of unknown parameters estimated. If there are k classes or cells, and 2 unknown parameters are estimated to find e_i, then the degrees of freedom of the χ^2 distribution would be $k-1-2$. One fact that may be noted that only two special methods, namely, the method of maximum likelihood and the method of minimum chi-square are recommended for estimation of the parameters.

11.5.3. Goodness of fit of distributions

We consider a slightly more general problem than that described in the preceding section. Suppose that we are given an observed frequency distribution (or empirical distribution) and we are interested in finding a standard probability distribution that fits the data. The question then arises whether the fit is satisfactory. This, in turn, involves finding a goodness of fit test. The χ^2 distribution may be used as a goodness of fit test.

The problem is similar to that described in section 11.5.1. Two cases may arise: First, the parameters of the distribution to be fitted may be given. Secondly, the parameters of the distribution are not given but may have to be estimated; in this case the d.f. of the χ^2 distribution is to be reduced by the number of parameters to be estimated.

We consider below an example of fitting a standard distribution to an empirical distribution.

Example 5 (e). The number of days with k absentees in a factory in a working year of 300 days are as given below :

No. of Absentees (k)	No. of days with absentees $n_{(k)}$
0	42
1	80
2	78
3	56
4	25
5	13
6	5
7	1
	300

Fit a Poisson distribution to the above empirical distribution. In order to find the expected frequencies we need to know the corresponding cell probabilities; these depend on the unknown parameter a of the Poisson distribution with pmf

$$p_k = P(X = k) = \frac{e^{-a} a^k}{k!}, k = 0, 1, 2, \ldots$$

It is known that the mean of the distribution is a. Thus we can estimate the mean of the distribution by the mean $\overline{X}$ of the empirical distribution. Thus a is estimated by $\overline{X}$. We have

$$\overline{X} = \frac{606}{300} = 2.02 \cong 2.$$

For ease of calculation let us take $a = 2$, we can then find Poisson pmf from the Table of Poisson distribution given in the Appendix. Thus $e_k = np_k$, $(n = 300)$ can be calculated and the value of χ^2 can be calculated by using the formula

$$\sum_k \frac{(n_k - e_k)^2}{e_k}$$

The calculations are shown in Table below.

No. of absentees	Observed frequency (n_k)	Poisson prob.	Expected frequency (e_k)
0	42	0.135	40
1	80	0.271	81
2	78	0.271	81
3	56	0.180	54
4	25	0.090	27
5	13	0.036	11
6	5 } 6	0.012	4 } 6
7	1	0.003	1
More than 7	0	0.002	1
	$n = 300$	1.000	300

Note that the expected frequencies in the last 3 cells are less than 5; so we pool the figures of the last three cells as one cell; then the total number of cells becomes 7, i.e. $k = 7$. One parameter a is estimated, thus the d.f. of the χ^2 distribution will be $k-1-1=5$. We get

$$\chi^2 = \sum_{k=0}^{6} \frac{(n_k - e_k)^2}{e_k} = \frac{(42-40)^2}{40} + \frac{(80-81)^2}{81} + \ldots + \frac{(6-6)^2}{6} = 0.81 \text{ for 5 d.f.}$$

The calculated value of χ^2 is not significant. Thus Poisson distribution with mean 2 gives a good fit.

Suppose that, the problem was stated as : Fit a Poisson distribution with mean 2 to the given data, then we proceed in the same way with the only difference that the degrees of freedom would be $k-1=6$, and the calculated value is to be compared with the critical value of χ^2 for 6 d.f.

Other distributions, like binomial, normal and so on can be fitted and their goodness of fit tested in the same way. We now consider another application of χ^2.

11.6. CONTINGENCY TABLES

Suppose that we have a sample of n observations classified in accordance with two criteria. For example, we may have a classification of a random sample of 300 buildings as indicated in the table below :

Table 11.2

	Location		
Nature of occupancy	**Big City**	**Small Town**	**Total**
Owner-occupied	52	62	114
Rented to offices etc.	97	22	119
Rented to private parties	36	31	67
Total	185	115	300

Here we have a classification by two criteria or characteristics — one location (two categories) and the other, the nature of occupancy (three categories). Such a two-way table is called a *contingency table*. The above is a 3 × 2 contingency table; the horizontal classification has 3 categories and the vertical has 2 categories. The table has 3 rows and 2 columns and 3 × 2 = 6 distinct cells. The purpose behind the construction of such a table is for studying the relation between the two variables of classification. We wish to find out whether there is no relationship between the two variables, i.e. the two characteristics appear to occur independently of each other; or whether there is some relationship between the two variables, i.e. whether certain levels of one characteristics tend to be contingent or to be associated with some levels of the other.

In the above case our interest lies in ascertaining whether there is no relationship between the two characteristics – location and nature of occupancy.

The χ^2 test can be used to test the hypothesis that there is no relationship between the two characteristics or variables of classification.

11.6.1. Application of χ^2 test

Consider a general $r \times c$ contingency table having r rows and c columns, i.e. one characteristics is classified into r categories and the other into c categories. Assume that the contingency table has been constructed from a random sample of n observations from a population.

Let p_{ij} be the probability that an individual selected from the population under consideration will be an element of the cell corresponding to the ith row and jth column of the contingency table : here

$$i = 1, 2, \ldots r, \quad j = 1, 2, \ldots, c.$$

Let
$$p_{i.} = \sum_{j=1}^{c} p_{ij}\ , \quad p_{.j} = \sum_{i=1}^{r} p_{ij} \tag{6.1}$$

be the probability that the individual will be a member of the i th row and of the j th column respectively. When the two variables are independent then

$$p_{ij} = (p_{i.}) \times (p_{.j}) \quad \left[\text{Note that } \sum_{i,j} p_{ij} = 1\right]$$

Suppose that of the n observations selected from the population on the basis of a random sample, n_{ij} are found in the cell corresponding to the i th row and j th cell. Then n_{ij} is a binomial random variable, and

$$e_{ij} = E(n_{ij}) = np_{ij} \tag{6.2}$$

Here the null hypothesis is that the two variables are independent : the null hypothesis can then be expressed as

$$H_0 : p_{ij} = (p_{i.}) \times (p_{.j}) \tag{6.3}$$

$$i = 1, 2, \ldots, r, \quad j = 1, 2, \ldots, c$$

and so under null hypothesis, we have

$$e_{ij} = np_{ij} = n(p_{i.})(p_{.j}) \tag{6.4}$$

Then applying Theorem 11.5, we find that, as $n \to \infty$, the distribution of

$$\chi^2 = \sum_{i=1}^{r} \sum_{j=1}^{c} \frac{(n_{ij} - e_{ij})^2}{e_{ij}} \tag{6.5}$$

approaches χ^2 distribution with $(rc - 1)$ d.f.

Under null hypothesis H_0, using (6.4) we get that the distribution of

$$\chi^2 = \sum_{i=1}^{r} \sum_{j=1}^{c} \frac{n_{ij} - np_{i\cdot}\, p_{\cdot j}}{n(p_{i\cdot})\,(p_{\cdot j})} \tag{6.6}$$

is, for large n, approximately χ^2 with $(rc - 1)$ d.f.
In practice the probabilities p_{ij} are not known and need have to be estimated. For using (6.6) we have to estimate

$$p_{i\cdot}\,,\ i = 1, 2, \ldots, \text{ and } p_{\cdot j} = 1, 2, \ldots, c.$$

Since $\sum_{c=1}^{r} p_{i\cdot} = 1$, if $(r - 1)$ of the $p_{i\cdot}$'s are estimated, then the remaining one will be known; similarly if $(c - 1)$ of the $p_{i\cdot}$'s are estimated the remaining one will be known. Thus the total number of unknowns to be estimated is $(r - 1) + (c - 1) = r + c - 2$. Thus when these are estimated the d.f. of the χ^2, given by (6.6), will be $\nu = rc - 1 - (r + c - 2) = rc - r - c + 1 = (r - 1)\, c - 1)$.
Denote

$$n_{i.} = \sum_{j=1}^{c} n_{ij}\,, \quad n_{.j} = \sum_{i=1}^{r} n_{ij} \tag{6.7}$$

then it is seen that the maximum likelihood estimates are given by

$$\hat{p_{i.}} = \frac{n_{i.}}{n} \tag{6.8}$$

Putting the values of the estimates in (6.6) we get that

$$\chi^2 = \sum_{i=1}^{r} \sum_{j=1}^{c} \frac{\left(n_{ij} - \dfrac{n_{i.}\, n_{.j}}{n}\right)^2}{\dfrac{(n_{i.})\,(n_{.j})}{n}} \tag{6.9}$$

has, for large n, approximately χ^2 distribution with $\nu = (r - 1)\,(c - 1)$ d.f.

Thus χ^2 test can be applied to test H_0 that the variables are independent. If the calculated value of χ^2 does not exceed the critical values (at a desired levels of significance) then the null hypothesis is accepted, otherwise it is rejected. We consider some examples below.

Example 6(a). In order to test the effectiveness of a new drug for common cold, a random sample of persons suffering from common cold was taken. About half the number were given tablets with the specific drug and about half were given similar tablets containing only sugar. The results of the experiment are noted below. Test whether the specific drug has proved effective in curing patients.

Result / Treatment	Cured	Not cured	Total
Tablet with specific Drug	56	64	120
Tablet with Sugar only	52	78	130
Total	108	142	250

This is a 2 × 2 contingency table with $n_{11} = 56, n_{12} = 64, n_{21} = 52, n_{22} = 78$.

The expected frequencies are

$$e_{11} = \frac{(n_{1\cdot})(n_{\cdot 1})}{n} = \frac{120 \times 108}{250} = 51.8,\ e_{12} = 120 - 51.8 = 68.2$$

$$e_{21} = \frac{(n_{2\cdot})(n_{\cdot 1})}{n} = \frac{130 \times 108}{250} = 56.2,\ e_{22} = 130 - 56.2 = 73.8$$

$$\chi^2 = \frac{(56-51.8)^2}{51.8} + \frac{(64-68.2)^2}{68.2} + \frac{(52-56.2)^2}{56.2} + \frac{(78-73.8)^2}{73.8} = 1.15 \text{ for } 1\ d.f.$$

From table, we get $\{\chi^2 > 1.15\} > .20$ for 1 d.f. Thus there is no reason to reject the null hypothesis that treatment and cure are independent. There is no evidence to show that the medicine was effective in curing common cold.

Example 6(d). Consider the data given in Table 11.2. Test whether the two classifications are independent.

Here $n_{11} = 52, n_{12} = 62, n_{1\cdot} = 114$

$n_{21} = 97, n_{22} = 22, n_{2\cdot} = 119$

$n_{31} = 36, n_{32} = 31, n_{3\cdot} = 67$

$n_{\cdot 1} = 185, n_{\cdot 2} = 115, n = 300$

The expected frequencies e_{ij}'s are

$$e_{11} = \frac{n_{1\cdot}\, n_{\cdot 1}}{n} = 70.3,\ e_{12} = 114 - 70.3 = 43.7$$

$$e_{21} = \frac{n_{2\cdot}\, n_{\cdot 1}}{n} = 73.4,\ e_{22} = 119 - 73.4 = 45.6$$

$$e_{31} = \frac{n_{3\cdot}\, n_{\cdot 1}}{n} = 41.3,\ e_{32} = 67 - 41.3 = 25.7$$

Writing the expected frequencies within parenthesis, we can rewrite the table as below :

Table 11.3

	Location		
Nature of occupancy	**Big City**	**Small Town**	**Total**
Owner occupied	52 (70.3)	62 (43.7)	114
Rented to office etc.	97 (73.4)	22 (45.6)	119
Rented to Private parties	36 (41.3)	31 (25.7)	67
Total	185	115	300

The calculated value of

$$\chi^2 = \frac{(52-70.3)}{70.3} + \frac{(62-43.7)^2}{43.7}$$

$$+ \frac{(97-73.4)^2}{73.4} + \frac{(22-45.6)^2}{45.6}$$

$$+ \frac{(36-41.3)^2}{41.3} + \frac{(31-25.7)^2}{25.7}$$

$$= 4.76 + 7.66 + 7.59 + 12.21 + 0.68 + 1.09$$

$$= 33.99 ;$$

and the d.f. = (2 – 1) (3 – 1) = 2.

The critical value of χ^2 for d.f. 2 is 5.99 at 5% level of significance and 9.21 at 1% level of significance. The calculated value being greater than 1% value is highly significant. We reject the null hypothesis of independence; in other words, there is no evidence to believe that the sample is from a population in which the two classifications are independent.

EXERCISES - 11

Sections 11.1 – 11.2

1. State the properties of *t*-distribution. How does it differ from a standard normal distribution?
2. Mention some applications of *t*-distribution, indicating clearly the situations where a test with this distribution is applicable.
3. What are the basic assumptions for student's *t* test ? How does non-normality of the population affect the test ?
4. What is meant by a small sample test ? How does such a test differ from a large sample test?
5. Can a small sample test be used for large samples ?
   Discuss the possible effect of applying a large sample (normal) test used for a small sample in a situation where *t* test is appropriate.
6. It is intended to produce wall pins of length 2.54 cm. A sample of 12 pins has mean length of 2.52 with a s.d. of 0.16. Test whether the sample mean differs significantly from the intended mean length.

7. A random sample $X_1, X_2, \ldots, X_{16}$ taken from a normal population gives the following data. You are to test whether it is reasonable to assume that the sample comes from a normal population with mean 50.
[Given $\overline{X} = 48.5 \;\; \Sigma\,(X_i - \overline{X})^2 = 240.$]
8. It is known from past experience that the mean weight of the products from a filling machine is 10 gms. A random sample of 10 shows that the weight (in gms.) of the sample products are as follows :
9.8, 10.3, 10.4, 9.6, 9.5, 9.7, 9.8, 10.4, 10.2, 10.3.
Test the significance of the difference of the sample mean from the population mean.
9. With a view to test whether a certain additive will increase the strength of a certain type of synthetic yarn, two random samples, one without the additive and the other with the additive were taken. The results noted are as follows :

Type	Sample size (n)	$\overline{X}$	$\Sigma\,(X_i - \overline{X})^2$
With Additive	10	82.5	324
Without Additive	15	76.5	570

Can we conclude that the strength has increased due to the use of the additive ? State the assumptions you make.
10. It is desired to test the effectiveness of an advertising campaign on the sale of a commodity. The records of daily sales before the campaign (X) and after the campaign (Y) are as given below :

(Sales in 1000 Rupees)

X: 50.5, 60.5, 45.5, 49.5, 52.5, 55.0, 56.0
Y: 52.5, 56.0, 59.5, 48.5, 56.5, 58.0, 59.5

Could you infer that the advertising campaign has been effective ? State the assumption(s) you make.

Section 11.3

11. Describe F distribution, clearly indicating the underlying assumption. State some of its chief properties. Mention some of its uses.
12. If F has F distribution with d.f. n_1, n_2, then what is the distribution of $\frac{1}{F}$?
13. The variate F has F distribution with 15 and 12 d.f. Find from the table
(i) $F_{0.05}$ (5 per cent critical value of F)
(ii) $P\,(F \geq 2)$.
14. In Exercise 9, verify the assumption of equality of variances at 5% level of significance.
15. Test the hypothesis $H_0 : \sigma_x = \sigma_y$ against the alternative hypothesis $H_1 : \sigma_x \neq \sigma_y$ for the data of Exercise 10.

Sections 11.4 – 11.6

16. Describe the χ^2 distribution, clearly indicating the underlying assumptions. State some of its important properties.
17. Mention some of the applications of the χ^2 distribution as a large sample test.
18. Mention some of the limitations as well as of the general nature of the χ^2 test.
19. Describe a contingency table. What is the purpose behind construction of such a table ? How can χ^2 test be used in this connection ?
20. What is a goodness of fit test ? Describe how χ^2 can be used as a goodness of fit test. What precautions would you take in applying this test ?
21. A random sample of size 10 from a normal population gives $s^2 = 90$. Test at the 5% level of significance the hypothesis :
$$H_0 : \sigma^2 = \sigma_0^2 \,(= 80) \text{ against } H_1 : \sigma^2 \neq \sigma_0^2 \,(= 80)$$
22. Consider the data of Exercise 7. Test whether the sample can be considered to be one from a normal distribution with s.d. 4.5.

23. Consider the data of Exercise 6. Test whether the sample standard deviation differs significantly from the intended (desired level of) s.d equal to 0.15.
24. A die is thrown 60 times, in which an odd face (with 1, 3 or 5) shows 36 times. Test whether the die shows odd faces more often than even faces.
25. The percentage distribution of household sizes in a certain district A was obtained on the basis of a large scale survey. A random sample of 292 households in another district B was taken and the frequency distribution was obtained. Test whether the structure of household size of B differs significantly from that of A.

Family size	Percentage dist.(A)	Observed frequency (B)
1–3	15.4	42
4–6	42.5	127
7–9	30.8	93
10–12	9.6	25
13–15	1.4	4
16–18	0.3	1

26. The number of times floods occurred in a year during the last 80 years in a state was as follows :

No. of occurrences (k)	Frequency (n_k)
0	24
1	30
2	16
3	8
4	2
5 or more	0
	80

Fit a Poisson distribution to the above data and test its goodness of fit ($e^{-1.2} = 0.301$).

27. The following table gives the frequency of sales of radio sets in a store for 100 days.

Sales per day (k)	No. of days with sales (n_k)
0	11
1	31
2	26
3	17
4	10
5	4
6	1

Fit a Poisson distributon and test its goodness of fit ($e^{-2} = 0.135$).

28. A group of 200 individuals were classified according to marital status and smoking habit as follows :

	Smoker	Non-smoker
Married	25	95
Unmarried	45	35

Test whether the two criteria of classification are independent.

29. Use χ^2 test to ascertain the effectiveness of inoculation against certain ailment on the basis of the following data.

	Inoculated	Not inoculated
Attacked	25	70
Not attcked	85	20

30. Use χ^2 test to test the effectiveness of using a lubricant in the manufacture of a product on the basis of the following data.

	Defective	Non defective
With lubricant	8	137
Without lubricant	32	203

31. Show that the chi-square for a 2 × 2 contingency table having cell frequencies *a*, *b*, *c*, *d*, respectively is given by

$$\chi^2 = \frac{(a+b+c+d)(ad-bc)^2}{(a+b)(c+d)(a+c)(b+d)}$$

32. Consider the large sample method for testing the difference of percentages (Ch. 10). Show that the method is equivalent to the χ^2- test as applied in the case of a 2 × 2 contingency table of successes and failures.

PART III

Special Topics : Applied Statistics

"The time has come
the Walrus said,
To talk of many things
of shoes – and ships –
and sealing wax
Of cabbages – and kings – ."

Lewis Carroll in *Alice in Wonderland*

Lewis Carroll is the pseudonym of Charles L. Dodgson, a Victorian mathematician at Cambridge, England. Queen Victoria being greatly amused by reading *Alice in Wonderland* asked for other books, if any, of the author — then the mathematician sent a book on Determinants to the Queen!

Chapter 12

Time Series Analysis

> *"Time present and time past,*
> *Are both present in time future*
> *And time future contained in time-past'* — T.S. Eliot

12.1. INTRODUCTION

A series of observations recorded over time is known as a *time series*. The data on the population of a country over equidistant time points constitute a time series, e.g. the population of India recorded at the ten-yearly censuses. Some other examples of time series are : annual production of a crop, say, rice over a number of years, the wholesale price index over a number of months, the turn-over of a firm over a number of months, the sales of a business establishment over a number of weeks, the daily maximum temperature of a place over a number of days, and so on. In fact, economic data are, in general, recorded over time and are released at regular intervals. These constitute economic time series.

The objectives of analysis of a time series are (1) to give a general description of the past behaviour of the series, (2) to analyse the past behaviour, and (3) to attempt to forecast the future behaviour on the basis of the past behaviour; however, great caution is needed to do this.

A *forecast* does not tell what will happen but indicates what *would* happen if the past behaviour (as reflected in trends etc.) continues.

The analysis of time series is of interest in several areas, such as economics, commerce, business, sociology, geography, meteorology, demography, public health, biology, and so on. The techniques of time series analysis have largely been developed by economists. Empirical investigations dealing with economic theory are largely dependent on time series analysists. Social scientists, in general, do not have the privilege of conducting studies through laboratory experimentation. Studies are to be based on time series data collected over time in such cases. For example, trade cycles are important to economists and others in business and commerce. The exact behaviour of the cycles and their causes are of interest to them. Various theories explaining the phenomena are put forward. Analysis of time series provides an important tool for testing the theories and the explanations. Consumer behaviour is studied mainly with the help of time series data.

The analysis of time series plays an important role in empirical investigations, leading to quantitative revolution, in economics and in several other areas of social sciences and even in biological sciences. Thus political economy (usually known as economics) has been described as 'the oldest of the arts, the newest of the sciences-indeed the queen of the social sciences.'

We shall now discuss the techniques used in the analysis of time series. We begin with the main components or characteristic movements in a time series.

12.2. CHARACTERISTIC MOVEMENTS IN A TIME SERIES

Long time series, in general, reveal the following types of characteristics :

1. Secular Trend
2. Periodic Movements
3. Cyclical Movements
4. Irregular Movements.

Ordinarily, all the four types will be found in a time series. These characteristics are also called *components* of a time series.

By *secular trend* is meant the general long-term movement of a series. A time series taken at chosen time points over a period is likely to show a tendency to increase or to decrease. For example, a time series on population shows a tendency to increase; time series of sales of products shows a tendency to increase, and so on. On the other hand the time series on crude death rate shows a tendency to decrease (see Fig. 3.6).

Sometimes, a movement recurs, with some degree of regularity, within a definite period. Such a movement, called periodic movement, is associated with seasonal variation which occurs within a period. As such it is also known as *seasonal movement* or *seasonal fluctuation* or simply *seasonal*. The period during which it appears may be a year, a quarter of a year, a month or a week or even a day, whenever data are collected at appropriate time points. Seasonal fluctuations do not appear in a time series giving annual total figures.

In certain cases annual figures may themselves manifest periodic movements. Such movements are of longer duration than a year and these do not show the regularity as in the case of periodic movements. Movements of this type are called *cyclical movements*. Trade cycles are movements of this nature. Apart from such cycles, there occur also other types of cycles, known as long cycles.

Irregular movements in a time series may be of two types (i) random or chance and (ii) episodic. Random or chance factor is present in everything that is real. It does effect a series in a random way, sometimes in one direction, sometimes in the opposite direction, but in general, the effect of chance on a series is small. Episodic movements arise due to specific events or episodes like epidemic, strike, or fire or political upheaval or natural calamities like flood, drought, earthquake or even some more common phenomena like early onset or late arrival of monsoon etc.

Apart from the four prominent movements that are found in time series, evidence of cyclic movement with a very *long cycle* is also found by researchers in economics. These long movements of 50 years or over are found to be mixed up with secular trends.

12.3. TIME SERIES MODELS

One of the objectives of time series analysis is to give a general description of the behaviour of the series. In order to achieve this objective, it is necessary to break down the series into its main components – the secular trend, the periodic or seasonal movement, the cyclic movement and the irregular movement, and to estimate the magnitudes of each of these components separately.

The method of analysis would depend to a large extent, on the hypothesis as to how the components interact. The simplest hypothesis is to assume that the effects of the distinct components are independent and are additive. This leads us to the time series model

$$Y_t = T_t + S_t + C_t + I_t \tag{3.1}$$

where Y_t is the value of the variable at time t, T_t is the trend value, S_t is the seasonal variation, C_t is the cyclic variation and I_t is the irreguaIr variation. The component S_t will be positive or negative according to the season of the year; so also will be C_t according to the phase 'above normal' or 'below normal' of the cycle. I_t will also have positive or negative values, and for a long series the total ΣI_t is assumed to have the value zero. For simplicity, (3.1) is also written as

$$Y = T + S + C + I \tag{3.2}$$

If annual figures are taken, the seasonal variation will not appear and S will be zero.

The model considered above is known as *additive model*. An alternative model, which is considered to be more useful, is the *multiplicative model*, which can be obtained as follows. The relation (3.2) can be written as :

$$\begin{aligned} Y &= T + C + S + I \\ &= T\left(\frac{T+C}{T}\right)\left(\frac{T+C+S}{T+C}\right)\left(\frac{T+C+S+I}{T+C+S}\right) \\ &= T C_1 S_1 I_1 \end{aligned} \tag{3.3}$$

where $\quad C_1 = \dfrac{T+C}{T}, S_1 = \dfrac{T+C+S}{T+C}, I_1 = \dfrac{T+C+S+I}{T+C+S}$

Here C_1, S_1, T_1, instead of having positive or negative values will have indexing values greater than or less than unity. The movements here are considered in relative terms rather than in absolute terms. This model is considered more useful because the cyclic, seasonal and irregular variations C, S, I tend to remain more nearly constant in magnitude relative to the trend T, rather than in absolute terms.

Again, by a taking logarithms of (3.3), we find

$$\log Y = \log T + \log C_1 + \log S_1 + \log I_1 \tag{3.4}$$

which reduces to an additive model, indicating that there is no difference, in principle, between the two models.

12.4. MEASUREMENT OF TREND

The immediate objectives behind the study of the trend of a time series are (1) to describe the general underlying movement of the series, and (2) to eliminate the trend in order to bring into focus the other movements in the series. It is necessary therefore to estimate the trend by statistical methods. Some methods of estimation or determination of trend are described below.

12.4.1. Inspection Method

The simplest way to describe a trend graphically is first to graph the series and then to draw a free hand curve through the points on the graph. The method, though simple, is very subjective and can be adopted only to have a general idea of the nature of the trend.

12.4.2. Method of Moving Averages

This method attempts to smooth out the bumps or irregularities in a series by a process of averaging, known as *moving averages*. Suppose that the successive observations are taken at equal intervals of time, say, monthly or yearly. Denote the successive values by $Y_1, Y_2, Y_3, \ldots$. If the data are annual figures, then by a three-yearly moving averages, we shall obtain average of first three consecutive years and place it against time $t = 2$; then the average of the next three consecutive years (beginning with the second year) and place it against time $t = 3$, and so on. This is illustrated below :

Time t	Observation Y_t	Moving Total (3 year)	Moving Average (3 Year)
1	Y_1		
2	Y_2	$Y_1+Y_2+Y_3$	$\frac{1}{3}(Y_1+Y_2+Y_3)$
3	Y_3	$Y_2+Y_3+Y_4$	$\frac{1}{3}(Y_2+Y_3+Y_4)$
4	Y_4	$Y_3+Y_4+Y_5$	$\frac{1}{3}(Y_3+Y_4+Y_5)$
5	Y_5	$Y_4+Y_5+Y_6$	$\frac{1}{3}(Y_4+Y_5+Y_6)$
.	.	.	.
.	.	.	.
.	.	.	.

For an even order moving average, two averaging processes are necessary in order to centre the moving average against periods rather than between periods. For example, for a four-yearly moving average we shall first obtain the average $\bar{Y}_1 = \frac{1}{4}(Y_1 + Y_2 + Y_3 + Y_4)$ of the first four years and place it in *between* $t = 2$ and $t = 3$ then the average $\bar{Y}_2 = \frac{1}{4}(Y_2 + Y_3 + Y_4 + Y_5)$ of the next four years and place it in between $t = 3$ and $t = 4$, and finally obtain the average $\frac{1}{2}(\bar{Y}_1 + \bar{Y}_2)$ of the two averages and place it *against* time $t = 3$. Thus the moving average is brought against time or period rather than between periods. The procedure is continued to get further results. This method is considered in Example 4(a).

Example 4(a). The following table gives the production of Cotton in India in million bales (1 bale = 174 kg.) from 1975-76 to 1985-86.

Year	1975-76	1976-77	1977-78	1978-1979	1979-80	1980-81
Production	4.76	5.95	7.24	7.96	7.65	7.01
Year	1981-82	1982-83	1983-84	1984-85	1985-86	
Production	7.88	7.53	6.39	8.51	8.61	

Fit a trend line by a four-year moving average.

We construct the table as follows :

Year	Production	4 yr. moving total	4 yr. moving average	2-term moving total	2-term moving average (4 yr. centred moving average)
[1]	[2]	[3]	[4]	[5]	[6]
			([3]÷4)		([5]÷2)
1975-76	4.76				
1976-77	5.95				
		25.91	6.48		
1977-78	7.24			13.68	6.84
		28.80	7.20		
1978-79	7.96			14.67	7.34
		29.86	7.47		
1979-80	7.65			15.10	7.55
		30.50	7.63		
1980-81	7.01			14.15	7.08
		30.07	7.52		
1981-82	7.88			14.72	7.36
		28.81	7.20		
1982-83	7.53			14.78	7.39
		30.31	7.58		
1983-84	6.39			15.34	7.67
		31.04	7.76		
1984-85	8.51				
1985-86	8.61				

The last column gives the 4-year centred moving average, that is, the trend fitted by 4-year cycle moving average. A figure drawing the graphs of the observed values and the moving average values, side by side, is useful for comparison.

The moving average method tends to produce a smoother curve by smoothing the irregular fluctuations while preserving the main characteristics of the time series. Larger the number of terms included in the moving average procedure the smoother is the resulting series or curve. Inclusion of more terms for averaging means that more information is lost at the beginning and at the end of the time points. For example, by taking four-yearly moving average, we do not get the trend values for $t = 1$ and $t = 2$ and similarly for the last two values of t. A moving average of $(2k + 1)$ terms will mean losing k terms each at the beginning and at the end. Exactly how many terms should be included in the moving average is largely a matter of judgment in each particular case.

It is easy to employ the moving average method in practice using a calculator.

We have described the simple moving average method here. More complicated moving averages are sometimes used in special types of problems.

Spencer's 15-point and 21-point formulas are some such moving average formulas.

A moving average may sometimes generate an irregular oscillatory movement though there was none in the original series. This is known as *Slutzky-Yule Effect*.

When a moving average is applied to a series, a new series is obtained. The new series may show spurious oscillatory movements not present in the original series. Though such movements may not be regular, these may show striking resemblance to the kind of movements actually observed in some real economic time series.

Merits and Demerits

The method of moving average is quite simple to apply. It is suitable for trend elimination when a linear trend is present. If there are erratic fluctuations around a trend in a series, a moving average would tend to smooth and reduce them. If there are periodic/cyclical movements in a series with period, say m, then an m-year moving average method is also useful for measuring seasonal variation.

A moving average method leads to sacrifice of some terms (expected observations) at the beginning and at the end. This may be serious in case of a short series. If a series consists of a non-linear trend, a moving average will not reveal the trend present in the original series.

Further, if a series consists of completely random fluctuations (without any trend or periodic movement), then a moving average will tend to show a somewhat irregular periodic movement, known as Slutzky-Yule Effect.

12.4.3. Fitting of Mathematical Curves

There is a third method for obtaining the trend of a time series; this is through a mathematical equation, which would give a concise form on the trend. The method is objective – once the form of the equation has been decided upon. The simplest type of trend equation is the linear equation, a first degree equation of the form

$$y = a + bx$$

where x represents time and y the values of the variable. The curve represented by the above equation will have no bend. It would generally provide a reasonably good description of the trend of a time series for short periods of time. The trend represented by the above is called a *linear trend*.

A curve of the second degree of the form

$$y = a + bx + cx^2$$

will have one bend. A curve of the third degree will have two bends, and so on. These will give *non-linear* trends. A rough idea about the form of the mathematical equation, that is, the degree of the equation may be had by first drawing the trend by inspection method. The number of distinct bends would give an idea of the degree of the polynomial curve to be fitted. The most convenient method of fitting a trend equation is the method of least squares, described in Chapter 9. In case of time series, the variable x represents time. The method described in section 9.5.1 could be used for a linear trend and the one described in section 9.9.1. for a non-linear (polynomial) trend. In fact, there will be considerable simplification, as x is given in equidistant intervals of time, such as in yearly intervals, monthly intervals and so on. We describe the method below.

12.5. SECULAR TREND

12.5.1. Linear Trend

Let x represent time and y the value of the variable and let the equation be

$$y = a + bx \tag{5.1}$$

If n sets of annual values (x_i, y_i) are given, then the normal equations for finding the constants a and b are

$$\Sigma y_i = na + b\ \Sigma x_i \tag{5.2}$$

$$\Sigma x_i y_i = a\,\Sigma x_i + b\,\Sigma x_i^2 \tag{5.3}$$

(see section 9.5.1 equations (5.5 and 5.5a)).

Two cases may arise. The number of years n covered by the time series may be odd or even.

(i) n is odd :

We can then take the origin at the middle year. Then $x = 0$ will correspond to the middle year, $x = 1$ the next year, $x = -1$ the preceding year and so on. We shall then have

$$\Sigma x_i = 0$$

(ii) n is even ($n = 2m$ say) :

Then there will be two middle years, the mth and the $(m+1)$th year. We take the origin at the middle of these two years and take *a half year as the unit*. Then the mth year will correspond to $x = -1$ and $(m + 1)$th year to $x = +1$, the $(m-1)$th year to $x = -3$ and $(m + 2)$th year to $x = +3$ and so on. We shall have then

$$\Sigma x_i = 0$$

Thus, whether n is odd or even, we can, by properly choosing the origin and the unit of time make $\Sigma x_i = 0$.

The normal equations (5.2) and (5.3) will then reduce to

$$\Sigma y_i = n\,a \tag{5.4}$$

$$\Sigma x_i y_i = b\,\Sigma x_i^2 \tag{5.5}$$

These give the following estimates of a and b :

$$\hat{a} = \frac{1}{n} \Sigma y_i = \bar{y} \tag{5.6}$$

$$\hat{b} = \frac{\Sigma x_i y_i}{\Sigma x_i^2}. \tag{5.7}$$

The trend equation (5.1) is thus completely determined.

When writing the trend equation, the origin and the units of time must be clearly specified, as an equation without them will be quite meaningless.

We illustrate the method with the help of two examples.

Note : When we choose the origin and the unit of time (for n odd or even) as indicated above, sums of all odd powers of x_i will be zero. Thus

$$\Sigma x_i = \Sigma x_i^3 = \Sigma x_i^5 = \ldots = 0.$$

Example 5(a). India's imports for 1980-81 to 1986-87 are as given below (in 100 crores of Rupees)

Year	1980-81	1981-82	1982-83	1983-84	1984-85	1985-1986	1986-87
Imports	125.5	136.1	142.9	158.3	171.3	197.7	200.8

To fit a linear trend by the method of least squares.

Here n = no. of years covered = 7(odd). We take the middle year, here 4th year $\left(4 = \frac{7+1}{2}\right)$, that is, 1983-84 as origin and unit as one year and get the table as follows :
(except the last column).

Year	x	y	x^2	xy	Trend values $\hat{y}$
1980-81	−3	125.5	9	−376.5	121.4
1981-82	−2	136.1	4	−272.2	134.8
1982-83	−1	142.9	1	−142.9	148.3
1983-84	0	158.3	0	0	161.8
1984-85	1	171.3	1	171.3	175.3
1985-86	2	197.7	4	395.4	188.8
1986-87	3	200.8	9	602.4	202.2
Total	0	1132.6	28	377.5	1132.6

We have

$$\hat{a} = \frac{1132.6}{7} = 161.80$$

$$\hat{b} = \frac{377.5}{28} = 13.48$$

and the trend equation is

$$y = 161.80 + 13.48\,x \tag{5.8}$$

with origin 1983-84; x units, 1 year.

We can find the trend values $\hat{y}$ by putting the values of x. For example, for $x = -3, \hat{y} = 161.80 + (13.48)(-3) = 121.36 \cong 121.4$. Thus other trend values are obtained and given in the last column of the above table.

We can compare the observed values y with the trend values $\hat{y}$ obtained as above (and given in the last column). A figure (Fig.12.1) showing the trend line (5.8) and the curve represented by the given values (x_i, y_i) side by side on the same graph paper is very useful.

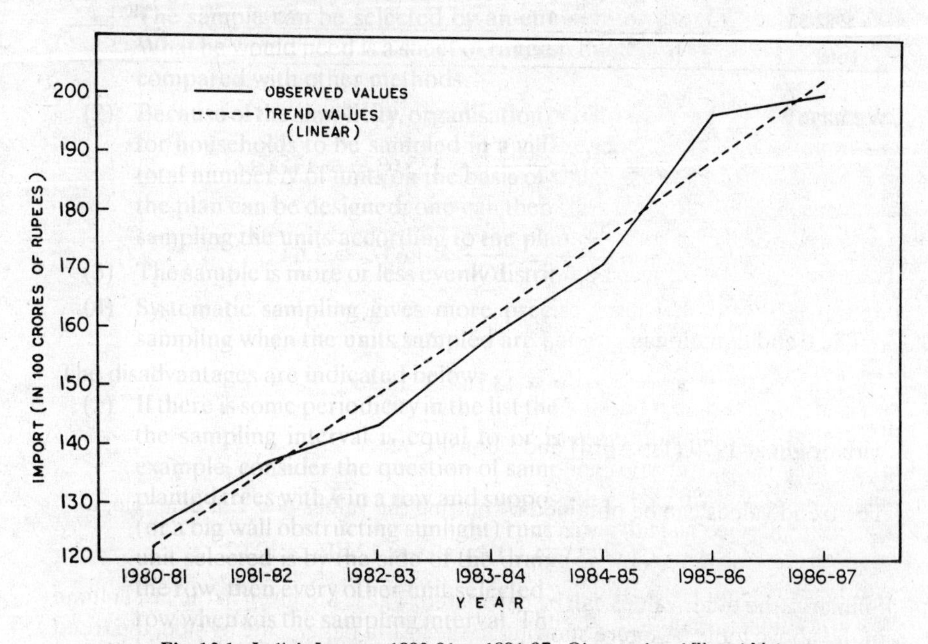

Fig. 12.1. India's Imports 1980-81 to 1986-87 : Observed and Trend Values.

Example 5 (b). The following table gives the index numbers of agricultural production for India with the triennium ending 1969-70 = 100 as base year.

Year	1975-76	1976-77	1977-78	1978-79	1979-80
Index number	125.1	132.9	138.0	117.0	135.3
Year	1980-81	1981-82	1982-83		
Index Number	142.9	137.5	156.4		

To fit a trend line.

Here $n = 8$ is even. We take as origin at the middle of the two middle years and as unit a half year.

A table can be constructed as follows (except the last column).

Year	x	y	x^2	xy	Trend values $\hat{y}$
1975-76	–7	125.1	49	– 875.7	124.2
1976-77	–5	132.9	25	– 664.5	126.8
1977-78	–3	138.0	9	– 414.0	130.7
1978-79	–1	117.0	1	– 117.0	134.0
1979-80	1	135.3	1	135.3	137.3
1980-81	3	142.9	9	428.7	140.6
1981-82	5	137.5	25	687.5	143.8
1982-83	7	156.4	49	1094.8	147.1
Total	0	1085.1	168	275.1	1084.5

We have

$$\hat{a} = \frac{\Sigma y_i}{n} = \frac{1085.1}{8} = 135.6375 \cong 135.64$$

$$\hat{b} = \frac{\Sigma x_i y_i}{\Sigma x_i^2} = \frac{275.1}{168} = 1.6375 \cong 1.64$$

The trend equation is

$$y = 135.64 + 1.64x \tag{5.9}$$

with origin at 1979 (1st April) and x units, $\frac{1}{2}$ year.

The trend values can be obtained by putting the values of x. For example, for

$$x = -7, y = 135.64 + 1.64(-7) = 124.16 \cong 124.2.$$

Similarly, the other values can be calculated. These are given in the last column of the above table. A figure showing in the graph paper, side by side, the trend line and the curve represented by joining the adjacent points (x_i, y_i) is very effective for comparison.

Example 5 (c). The projected number of women of childbearing age (15-49) for India from 1978 to 1985 are as follows :

Year	1978	1979	1980	1981	1982	1983	1984	1985
No.of women (in millions)	152.6	156.4	160.3	164.4	168.5	172.7	176.9	181.2

To fit a trend line.

Here n = 8 (even). We take the origin at the middle between two middle years 1981 and 1982 and as unit a half year. A table can be constructed as follows :

Year	x	y	x^2	xy	Trend value $\hat{y}$
1978	–7	152.6	49	–1068.2	152.3
1979	–5	156.4	25	– 782.0	156.4
1980	–3	160.3	9	– 480.9	160.5
1981	–1	164.4	1	– 164.4	164.6
1982	+1	168.5	1	168.5	168.7
1983	3	172.7	9	518.1	172.8
1984	5	176.9	25	884.5	176.9
1985	7	181.2	49	126.4	181.0
Total :	0	1333.0	168	344.0	1333.2

We have

$$\hat{a} = \frac{1330.0}{8} = 166.625 \cong 166.63$$

$$\hat{b} = \frac{344.0}{168} = 2.047 \cong 2.05$$

The trend equation is

$$\hat{y} = 166.63 + 2.05x \qquad (5.9)$$

with origin 1981-82 and x units, $\frac{1}{2}$ years.

The trend values can be obtained by putting the values of x. For example, for $x = -7$

$$\hat{y} = 166.63 + (2.05)(-7) = 152.28.$$

Similarly, the other values can be obtained. These are given in the last column. A graph showing the trend line and the curve represented by (x_i, y_i) is very effective for comparison.
Here the trend values $\hat{y}$ are very close to the observed values y. This is because y values are not actual observed values but are projected values based on some projection formula.

Example 5(d). The following table gives the index number of industrial production (1958 = 100) of U.K from 1962 to 1964.

Quarter	Ist	2nd	3rd	4th
1962	116	117	109	118
1963	115	119	114	127
1964	128	130	121	134

To fit a trend line and to find the trend values. Here $n = 3 \times 4 = 12$. We take the origin midway between the 2nd and 3rd quarters of 1963 and as unit, a half year. A table is constructed as below (without the last column).

Quarter		x	y	x^2	xy	Trend Values $\hat{y}$
	I	-11	116	121	-1276	111.76
1962	II	-9	117	81	-1053	113.38
	III	-7	109	49	-763	115.00
	IV	-5	118	25	-590	116.62
	I	-3	115	9	-345	118.24
1963	II	-1	119	1	-119	119.86
	III	1	114	1	114	121.48
	IV	3	127	9	381	123.10
	I	5	128	25	640	124.72
1964	II	7	130	49	910	126.34
	III	9	121	81	1089	127.96
	IV	11	134	121	1474	129.58
Total		0	1448	572	462	1448.04

We have

$$\hat{a} = \frac{1448}{12} = 120.67$$

$$\hat{b} = \frac{462}{572} = 0.80769 \cong 0.81$$

The trend equation is

$$y = 120.67 + 0.81\,x$$

with origin midway between the second and third Quarters of 1963 and x units, $\frac{1}{2}$ year.

The trend values $\hat{y}$ can be obtained by putting the values of x in the above equation. For example, when $x = -11$,

$$y = 120.67 + (0.81)(-11) = 120.67 - 8.91 = 111.76.$$

Similarly, other values of $\hat{y}$ can be found. These are given in the last column of the above table.

12.5.2. Non-linear Trend

12.5.2.1. Polynomial Fitting

While a straight line may be suitable for short periods of time or for a short time series, a curved line may provide a better description for long periods of time or for long time series, specially when bends appear. Then we may consider, instead of a straight line which is given by a first degree equation, a curved line given by a higher degree equation, say, second degree or third degree and so on.

Consider a second degree polynomial of the form

$$y = a + bx + cx^2 \tag{5.10}$$

We have described the method of fitting such a polynomial in section 9.9.1. The normal equations for finding the constants are given by the equations (9.2a), (9.3a), (9.4a).

For simplifying the calculations, we can choose the origin and the unit of time as indicated in the preceding section. Then we shall have (see Note in that section) :

$$\Sigma x_i = \Sigma x_i^3 = \Sigma x_i^5 = \ldots = 0$$

and the normal equations (9.2a), (9.3a), (9.4a) (of Chapter 9) will reduce to

$$\Sigma y_i = na + c\,\Sigma x_i^2 \tag{5.11}$$

$$\Sigma x_i y_i = b\,\Sigma x_i^2 \tag{5.12}$$

$$\Sigma x_i^2 y_i = a\,\Sigma x_i^2 + c\,\Sigma x_i^4 \tag{5.13}$$

The above equations can be easily solved and the trend equation (5.10) can then be written.

Example 5 (e). The following table gives the total expenditure of the Govt. of India during 1978-79 to 1984-85.

(in Rs. 100 Crores)

Year	78-79	79-80	80-81	81-82	82-83	83-84	84-85
Expenditure	177.2	185.0	224.9	254.0	304.9	359.9	438.8

To fit a quadratic trend to the above data. Here $n = 7$, we take 1981-82 as origin and x units as 1 year. We can construct the table as follows : (except the last column)

Year	x	y	xy	x^2	x^2y	x^4	Trend values $\hat{y}$
78-79	–3	177.2	–531.6	9	1594.8	81	175.92
79-80	–2	185.0	–370.0	4	740.0	16	191.05
80-81	–1	224.9	–224.9	1	224.9	1	217.48
81-82	0	254.0	0	0	0	0	255.21
82-83	1	304.9	304.9	1	304.9	1	304.24
83-84	2	359.9	719.8	4	1439.6	16	364.57
84-85	3	438.8	1316.4	9	3949.2	81	436.20
Total	0	1944.7	1214.6	28	8253.4	196	1944.67

For a quadratic trend

$$y = a + bx + cx^2 \quad (1)$$

the normal equations are

$$1944.7 = 7a + 28c \quad (2)$$

$$1214.6 = 28b \quad (3)$$

$$8253.4 = 28a + 196c \quad (4)$$

From (3), we get $\hat{b} = 43.38$.

Multiplying both sides of (2) by 4 and subtracting from (4) we get $474.6 = 84\hat{c}$, so that $\hat{c} = 5.65$. Putting this value of $\hat{c}$ in (2) we get $7\hat{a} = 1944.7 - 28 \times 5.65 = 1786.5$ so that $\hat{a} = 255.21$. Thus we get

$$\hat{y} = 255.21 + 43.38x + 5.65x^2 \quad (5)$$

with origin at 1981-82, x units as 1 year.

Putting the values of x in (5) we get the trend values, which are given in the last column of the above table. A graph with observed and trend values against x is helpful.

Example 5 (f). Last column of Table 3.1 of Chapter 3 gives the average annual rate of increase (in per cent) in population of India during the decades from 1921-31 to 1971-81. To find a linear trend and a quadratic trend and to compare the two.

Here $n = 6$. we can take origin at the middle of 1941-51 and 1951-61 and x units as half a decade and construct the table as follows :

Year	x	y	xy	x^2	x^2y	x^4
1921-31	–5	1.05	–5.25	25	26.25	625
1931-41	–3	1.34	–4.02	9	12.06	81
1941-51	–1	1.26	–1.26	1	1.26	1
1951-61	1	1.98	1.98	1	1.98	1
1961-71	3	2.24	6.72	9	20.16	9
1971-81	5	2.28	11.40	25	57.00	81
Total	0	10.15	9.57	70	118.71	1414

For linear trend $y = a + bx$, we get

$$\hat{a} = \frac{\Sigma y}{n} = \frac{10.15}{6} = 1.69167 \cong 1.6917$$

$$\hat{b} = \frac{\Sigma xy}{\Sigma x^2} = \frac{9.57}{70} = 0.13671 \cong 0.1367$$

$$y = 1.6917 + 0.1367x\,;$$

for a (polynomial) quadratic trend

$$y = a_1 + b_1 x + c_1 x^2.$$

We get the normal equations :

$$10.15 = 6\,a_1 + 70\,c_1$$

$$9.57 = 70\,b_1$$

$$118.71 = 70\,a_1 + 1414\,c_1$$

From the second equation, we get

$$\hat{b}_1 = \frac{9.57}{70} = 0.1367$$

and from the 1st and 3rd equations, we get

$$\hat{a}_1 = 1.6858, \; \hat{c}_1 = 0.0005$$

so that

$$\hat{y} = 1.6858 + 0.1367\,x + 0.0005\,x^2$$

with origin midway between 1941-51 and 1951-61 and x units, $\frac{1}{2}$ years.

Since the value of c_1 is very small, there is not much difference between the linear and quadratic trends.

Higher Degree Polynomial

We can likewise apply the method of least squares to fit a polynomial of any specified degree

$$y = a_0 + a_1 x + a_2 x^2 + \ldots + a_k x^k \tag{5.14}$$

where the a's are constants. These constants can be obtained by solving the normal equations.

12.5.2.2. *Other Mathematical Curves : Exponential Curves*

There are other types of mathematical equations, which are appropriate in some cases for fitting non-linear trends. One such equation is of the exponential type

$$y = ab^x \tag{5.15}$$

where a, b are constants (to determine). The trend given by the above equation is called an *exponential trend*.

Taking logarithms of both sides of (5.15), we get

$$\log y = \log a + x \log b$$

This is of the linear form

$$Y = A + Bx \tag{5.16}$$

where $Y = \log y, \quad A = \log a, \quad B = \log b.$

We can employ the method described in the preceding section to fit the linear trend (5.16). By suitable choice of origin and unit of time we can get the estimates as follows :

$$\hat{A} = \frac{1}{n}\Sigma y_i \tag{5.17}$$

$$\hat{B} = \frac{\Sigma x_i y_i}{\Sigma x_i^2}$$

By taking antilogarithms we can then get $\hat{a}$ and $\hat{b}$ and the exponential trend (5.15).

Example 5(g). To fit an exponential trend curve to the data given in Example 5(e) on Central Govt. expenditure.
Let the curve be

$$y = ab^x$$

Taking logarithms, we get

$$Y = A + Bx$$

where $Y = \log y$, $A = \log a$, $B = \log b$.
We are thus to fit the straight line $Y = A + Bx$.
Here $n = 7$, we take as origin 1981-82 and x unit, 1 year. We construct the following table (without the last two columns).

Year	x	y	$Y = \log y$	xY	$\hat{Y}$	$\hat{y}$ = antilog $\hat{Y}$
78-79	–3	177.2	2.2485	– 6.7455	2.218	165.2
79-80	–2	185.0	2.2672	– 4.5344	2.286	193.2
80-81	–1	224.9	2.3520	– 2.3520	2.354	225.9
81-82	0	254.0	2.4048	0	2.422	264.3
82-83	1	304.9	2.4842	2.4842	2.490	309.0
83-84	2	359.9	2.5562	5.1124	2.558	361.4
84-85	3	438.8	2.6423	7.9269	2.626	422.7
Total :	0	1944.7	16.9552	1.8916	16.954	1941.7

We have

$$\hat{A} = \frac{\Sigma Y}{n} = \frac{16.9552}{7} = 2.4222 \cong 2.422$$

$$\hat{B} = \frac{\Sigma xY}{\Sigma x^2} = \frac{1.8916}{28} = 0.0676 \cong 0.068$$

Thus

$$\hat{Y} = \hat{A} + \hat{B}x = 2.4222 + 0.0676x \cong 2.422 + 0.068x$$

with origin at 1981-82 and x unit, 1 year.

Thus we can find the trend values $\hat{Y}$, which are given in column (6). Taking antilogarithms we get the trend values $\hat{y}$ which are given in column (7).
Again,

$$\hat{a} = \text{antilog}\,\hat{A} = \text{antilog}\ 2.4222 = 264.3$$

$$\hat{b} = \text{antilog}\,\hat{B} = \text{antilog}\ 0.0676 = 1.168$$

so that the exponential curve fitted is given by

$$y = (264.3)\,(1.168)^x$$

From a comparison with the results of Example 5(e), it appears that the quadratic gives a slightly better fit. However, it may be noted that more than one equation can be fitted to the same data.

12.5.2.3 *Growth Curves*

There are more complicated types of equations which are suitable in certain cases. For example, the *logistic and Gompertz* curves are useful in fitting time series data relating to growth. Such curves, called *asymptotic growth curves*, are suitable for spatially limited universe, in which a population grows; these are also useful in describing growth of an industry, where initially the growth is slow during the period of experimentation, then the growth is rapid during the period of development and then the growth is again slow and stable when a period of stability is reached.
The logistic curve has the equation

$$\frac{1}{y} = k + ab^x \tag{5.18}$$

where k, a, b are constants. It is merely a modified exponential in terms of $\frac{1}{y}$.
There are several methods of fitting this curve, but we shall not discuss the matter further here.

We have described the various methods as well as the mathematical curves generally used in determining the trend in a time series. The choice of a particular curve is largely a subjective matter : it depends on the person who tries to fit the curve, when there is no hard and fast rule available to him. It is useful to fit a trend at first by inspection and to examine the same with a view to finding out the most appropriate curve to fit. *A priori* knowledge of the forces behind the movement provides a useful guidance.

12.6. SEASONAL MOVEMENTS

As already indicated earlier, there are some movements which occur with some degree of regularity within a definite period, where the period may be a year, a month, a week or even a day. An idea may be had from Fig. 3.8(b), Chapter 3. Climatic conditions affect production in agriculture and in some industries which reflect in the yield and production figures. For example, sale of cold drinks goes up during summer. Apart from natural causes, there are other man-made factors which cause variations in demand. For example, the custom of giving gifts during certain periods of the year results in increasing sales in those periods. The demands for consumer products generally go up during the early part of the month, so also small saving deposits rise during the early part of the month. The traffic in a city is high during the rush hours. All these show seasonal variations.

The objectives in analysing seasonal movement are as follows :

First, one may be interested to know the seasonal movement itself. One may be interested in the forces behind the seasonal movement.

Secondly, one may wish to eliminate the seasonal from the time series so as to bring into focus the other movements in the series. Statistical procedures are adopted to *de-seasonalise* (or to *adjust seasonally*) time series data.

To deseasonalise a series, we express the observed values as percentages of the indices of seasonal variation for the corresponding period. The new series of deseasonalised values becomes free from seasonal variation, and can be used as indicators of the trend present in the series.

When annual figures are given, there is no seasonal variation. Seasonal variation may be present only when data are given for specific periods of the year i.e. for data such as quarterly, monthly, weekly, daily or hourly. The study of seasonal variations which are generally present in data relating to agriculture, industry, commerce, business etc., are of importance to planners, industrialists, economists, businessmen, traders and so on. Analysis of seasonal variations would provide guidelines for scheduling production, for inventory and personnel management, for launching effective and timely sales promotion and advertising programmes and so on.

For analysing seasonal variation, it is necessary to assume that the seasonal pattern is superimposed on the values; a quarter (for example, winter) may exert a particular effect this way or the other on the values of a series. The multiplicative model is often used in analysing seasonal variations, which are usually expressed in terms of ratios and percentages. The methods usually adopted to measure seasonal variation are the following :

(i) Method of Simple Averages
(ii) Ratio to Trend Method (or Percent of Trend Method)
(iii) Ratio to Moving Averages (Percent of Moving Average Method)
(iv) Link Relative Method.

Before proceeding to determine seasonal variations, it is useful to ascertain, at first, by means of a rough graph, that a seasonal movement is actually (or appears to be) present.

12.6.1. Method of Simple Averages

Suppose that monthly data are given by years and months. The figures are added up for each month and averages are obtained by dividing the monthly totals by the number of years covered. Suppose that $\bar{x}_1$ is the average for January, $\bar{x}_2$ is the average for February, and so on. Then obtain the overall average $\bar{x}$ of these montly averages. We have $\bar{x} = \frac{1}{12}(\bar{x}_1 + \bar{x}_2 + \ldots + \bar{x}_{12})$.

Seasonal indices for different months are obtained by expressing the monthly averages as percentages of the overall average $\bar{x}$. Thus,

$$\text{Seasonal index for January} = \frac{\bar{x}_1}{\bar{x}} \times 100$$

$$\ldots \quad \ldots \quad \ldots$$

$$\text{Seasonal index for December} = \frac{\bar{x}_{12}}{\bar{x}} \times 100$$

If quarterly data are given we first compute the average for each quarter and then express this average as a percentage of the overall average.

It may be noted that the average of the indices will always be 100, i.e. sum of the indices will be 1200 for 12 monthly data, and 400 for 4 quarterly data. We consider an example below :

Example 6(a). The price of a certain commodity during 1980 to 1983 were as follows :

(Price in Rs. per Quintal)

Year	Jan-Mar (I)	April-June (II)	July-Sept (III)	Oct-Dec. (IV)
1980	321	348	348	348
1981	327	351	354	348
1982	342	359	381	345
1983	364	390	401	385

To compute the seasonal indices by the method of simple average and to find the deseasonalised values.

We construct the following table

	I	II	III	IV	
1980	321	348	348	348	
1981	327	351	354	348	
1982	342	359	381	345	
1983	364	390	401	385	
Total	1354	1448	1484	1426	Overall average
Quarterly Average	338.50	362.00	371.00	356.50	357.0
Seasonal Index	94.82	101.40	103.92	99.86	
	$\left(\frac{338.5}{357} \times 100\right)$	$\left(\frac{362}{357} \times 100\right)$	$\left(\frac{371}{357} \times 100\right)$	$\left(\frac{356.5}{357} \times 100\right)$	100

The deseasonalised values are

	I	II	III	IV
1980	338.54	343.20	334.87	348.49
1981	344.86	346.15	340.65	348.49
1982	360.68	350.04	366.63	345.48
1983	383.89	384.62	385.87	385.54

12.6.2. Ratio to Trend Method or Percent of Trend Method

This method is designed to overcome the difficulty of the simple average method when trend is present in the time series data. The steps involved in this method for measuring seasonal variations are as follows :

First, determine an appropriate trend equation for the time series data by using the principle of least squares. Then compute the trend values based on the trend equation (which may be linear or parabolic or exponential or of some other form).

Secondly, express the original time series values as percentages of the corresponding trend values.

Thirdly, arrange these percentages according to the years and months for monthly data (or according to years and quarters for quarterly data or likewise as the case may be).

Then find the monthly (or quarterly, in case of quarterly data) averages by taking the arithmetic means or the medians as may be considered suitable.

Fourthly, find the overall average of these monthly (or quarterly) averages. If the over-all average is 100, then these monthly (or quarterly) average will give the seasonal indices.

In case the overall average is different from 100, then express the monthly (or quarterly) averages as percentages of the overall average to obtain the seasonal indices.

The logic of this procedure is explained below.

Assume the multiplicative model

$$Y = TCSI \tag{6.1}$$

described in section 12.3.

By the first step, trend values T are sought to be determined. The second step which gives

$$\frac{Y}{T} \times 100 = \frac{TCSI}{T} \times 100 = (CSI) \times 100 \tag{6.2}$$

implies that the components C, S and I are left.

The third step seeks to eliminate the cyclic (C) and the irregular (I) components through averaging. The averages (monthly or quarterly as the case may be) give preliminary indices of seasonal variations. The final step is aimed at adjusting the average seasonal indices to an overall average of 100.

Merits and Demerits

This method is an improvement over the simple average method in that the trend is eliminated. But the method does not effectively tackle the cyclical movement that may be present; whereas averaging, as indicated in the third step, may be effective in smoothing out irregular fluctuations, the same may not altogether do away with cyclic fluctuation. Moreover, the method depends on the averaging process (the median or the arithmetic mean) to eliminate cyclical highs and lows. It can be recommended only if the cyclic movements are either absent or when present, are unimportant or insignificant relative to the seasonal movements. When the data exhibit strong cyclic movements, then the method of 'Ratio to Moving Average' is considered to be more suitable and is more often employed.

Inspite of the distinct advantage that the trend values corresponding to all observed values can be had, the method of Ratio-to-Trend is not widely used.

We consider an example below :

Example 6 (b). To find the seasonal indices for the data, given in Example 5(d) by the 'Ratio-to-Trend Method'. We construct the following table with observed (y) and trend values $\hat{y}$ as obtained in Example 5(d).

	I		II		III		IV	
	y	$\hat{y}$	y	$\hat{y}$	y	$\hat{y}$	y	$\hat{y}$
1962	116	111.76	117	113.38	109	115.00	118	116.62
1963	115	118.24	119	119.86	114	121.48	127	123.10
1964	128	124.72	130	126.34	121	127.96	134	129.58

Then we proceed as follows ;

Second Step : Calculate $(y/\hat{y}) \times 100$

1962	103.79	103.19	94.78	101.18
1963	97.20	99.28	93.84	103.17
1964	102.63	102.90	94.56	103.41

Third Step :

Average (A.M.)

101.23	101.79	94.39	102.59

Fourth Step :

Overall average = 100.00.

Since the overall average is 100.00, the seasonal indices are as obtained in the third step. That is, seasonal indices are as follows :

	I	II	III	IV
Seasonal Index	101.23.	101.79	94.39	102.59

In the third quarter, the index is 5.41% below and in the fourth quarter, the index is 2.59% above than these would have been otherwise. These indicate the extent by which industrial production goes below or above in the corresponding quarters.

12.6.3. Ratio-to-Moving Average Method or Percent-of-Moving Average Method

This method differs from the 'Ratio-to-Trend Method' in that the original data are expressed as percentages of moving averages instead of as percentages of trend values.

The steps, excepting the first two, are essentially the same. The steps are indicated below. Consider that the data given are monthly data.

First, find the centred 12-month-moving average values from the observed time series data.

Secondly, express the original time series values as percentages of the corresponding centred moving average values.

Thirdly, arrange these percentages according to year and month. Then find the averages over the years for all the 12 months.

Fourthly, find the overall average of these 12 monthly averages.

If the overall average is 100, then the 12 monthly averages will give the seasonal indices for the 12 months. If it is different from 100, then the seasonal indices will be the monthly averages expressed as percentages of the overall average.

If the data are quarterly, then we take 4-quarter moving average; if the data are given for different days of the week, then we find 7-day moving average and proceed in the same manner.

The logic behind the procedure is explained below : Consider that monthly data are given. The 12-month centred moving average will smooth out the seasonal and to a large extent the irregular component also. The first step of taking 12-month centred moving average will thus lead us to a rough estimate of the other two components i.e. to an estimate of $T \times C$ in the multiplicative model; it may also introduce a bias as explained by Slutzky-Yule effect. The second step which is then equivalent to

$$\frac{Y}{T \times C} \times 100 = \frac{T \times C \times S \times I}{T \times C} \times 100 = (S \times I) \times 100 \qquad (6.3)$$

leads us to the components S and I, with a possible bias. The third step of averaging seeks to eliminate the irregular component I and gives us preliminary indices of seasonal variation with the possible bias, if any. The final step aimed at adjusting the preliminary indices to an overall average of 100 removes the bias present, if any. We then get the indices of seasonal variations.

The computation, though a little lengthy, is straight forward and can be easily done by using a pocket calculator. An example is given.

Merits and Demerits

This method is widely used. It is because the method seeks to smooth out both the trend and cyclical components through the centred moving average procedure. However, it would smooth out cyclical fluctuations only if the cyclical components are regular in amplitude and periodicity. One disadvantage of the method is that while taking the 12-month centred moving average, 6 values at the beginning of the series and 6 values at the end are lost.

Example 6(c). The index numbers of Wholesale Prices (WPI) for India (base 1970-71 = 100) from 1985-86 to 1987-88 are as given in Table 12.1.
(the figures for 1987-88 are provisional that for March 1988 being for the week ending March 5).
To find the seasonal indices and the deseasonalized values for the year 1987. We proceed to find the 12-month centred moving average. Column [5] of **Table 12.1** gives these average, while column [6] gives the original values expressed as percentages of the moving averages

Table 12.1 (WPI).

	1985-86	1986-87	1987-88
April	350.5	363.0	381.2
May	353.7	368.6	390.3
June	357.5	373.1	394.0
July	362.3	378.3	400.6
August	363.1	381.6	409.6
September	357.6	381.5	408.9
October	360.0	384.8	408.4
November	357.9	380.6	411.1
December	356.4	377.9	410.3
January	357.5	377.7	413.5
February	358.9	376.6	414.5
March	359.8	378.1	413.8

Table 12.2

Month [1]	Original Value [2]	12-months moving total [3]	12-months moving average [3] ÷ 12 [4]	12-month centred moving average [5]	Original Values as percentage of centred moving average [[2] ÷ [5]] × 100 [6]
1985-86					
April	350.5				
May	353.7				
June	357.5				
July	362.3				
August	363.1				
Sept.	357.6				
		4295.2	357.93		
Oct.	360.0			358.46	100.42
		4307.7	358.98		
Nov.	357.9			359.60	99.53
		4322.6	360.22		

Table 12.2 (continued)

[1]	[2]	[3]	[4]	[5]	[6]
Dec.	356.4			360.87	98.77
		4338.2	361.52		
Jan.	357.5			362.18	98.71
		4354.2	362.85		
Feb.	358.9			363.62	98.70
		4372.7	364.39		
March	359.8			365.38	98.47
		4396.6	366.38		
1986-87					
April	363.0			367.42	98.80
		4421.4	368.45		
May	368.6			369.40	99.78
		4444.1	370.34		
June	373.1			371.24	100.50
		4465.6	372.13		
July	378.3			372.98	101.43
		4485.8	373.82		
August	381.6			374.56	101.88
		4503.5	375.29		
Sept.	381.5			376.06	101.45
		4521.8	376.82		
Oct.	384.8			377.58	101.91
		4540.0	378.33		
Nov.	380.6			379.24	100.36
		4561.7	380.14		
Dec.	377.9			381.01	99.18
		4582.6	381.88		
Jan.	377.7			382.81	98.67
		4604.9	383.74		
Feb.	376.6			384.91	97.84
		4632.9	386.08		
March	378.1			387.21	97.65
		4660.3	388.36		
1987-88					
April	381.2			389.33	97.91
		4683.9	390.33		
May	390.3			391.60	99.67
		4714.4	392.87		
June	394.0			394.22	99.94
		4746.8	395.57		
July	400.6			397.06	100.89
		4782.6	398.55		
Aug.	409.6			400.13	102.37
		4820.5	401.72		
Sept.	408.9			403.20	101.41
		4856.2	404.68		

Table 12.2 (continued)

[1]	[2]	[3]	[4]	[5]	[6]
Oct.	408.4				
Nov.	411.1				
Dec.	410.3				
Jan.	413.5				
Feb.	414.5				
March	413.8				

Then we arrange these percentages given in column[6] monthwise for the years 1985, 86, 87 in the first three rows of Table 12.3. Then find the averages over the years for each month. These averages are given in row [4]. Then find the overall average of these 12 monthly averages. This is given in the last column of row [4]. Since the overall average is different from 100, the averages given in row [4] are only preliminary indices. Expressing the preliminary indices as percentages of the overall average 99.845, we find the 12 seasonal indices as given in row [5]. Expressing the original values as percentages of the seasonal indices we find the deseasonalised (or seasonally adjusted) figures for the corresponding months. These are given in row [7].

Interpretation of Seasonal Indices

Seasonal pressure on prices in India usually begins in April and lasts till September. Due to purely seasonal influences, the average price for February tends to be about 1.58% lower and that for August tends to be about 2.28% higher than these would have been otherwise.
Seasonal Indices are shown in Fig. 12.2.

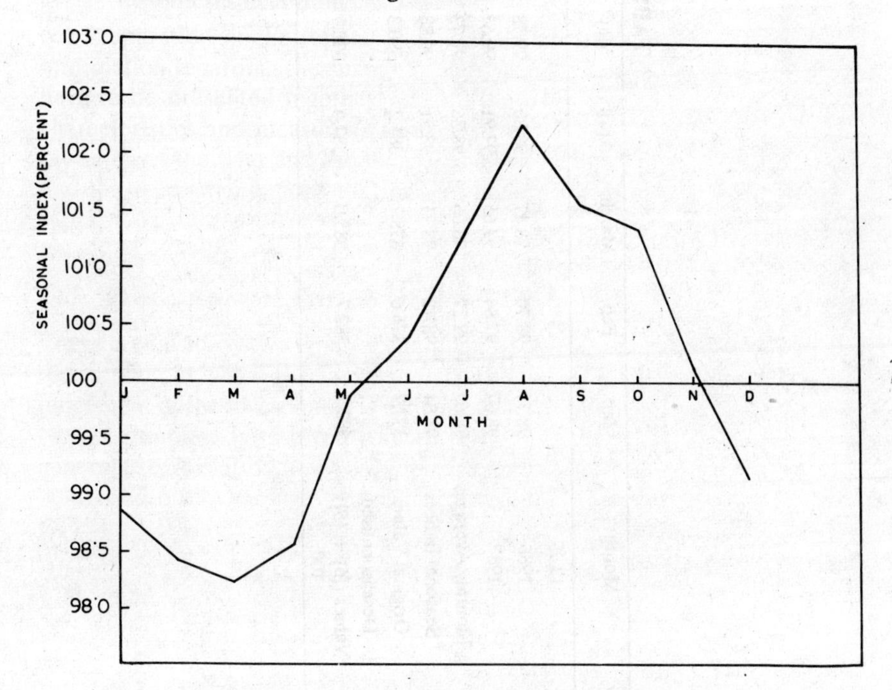

Fig. 12.2 Seasonal Index of Wholesale Price Index

TABLE 12.3

Row	Month	Jan	Feb	March	April	May	June	July	Aug.	Sept.	Oct.	Nov.	Dec.	Overall average
1	19.95										100.42	99.53	98.77	
2	1986	98.71	98.70	98.47	98.80	99.78	100.50	101.43	101.88	101.45	101.91	100.36	99.18	
3	1987	98.67	97.84	97.65	97.91	99.67	99.94	100.89	102.37	101.41				
4	Monthly Averages	98.69	98.27	98.06	98.36	99.73	100.22	101.16	102.12	101.43	101.17	99.95	98.96	99.845
5	Seasonal Indices	98.84	98.42	98.21	98.51	99.88	100.38	101.32	102.28	101.59	101.33	100.11	99.13	100.00
6	Original Values	377.7	376.6	378.1	381.2	390.3	394.0	400.6	409.6	408.9	408.4	411.1	410.3	
7	Deseasonalised Values ([6] ÷ [5]) × 100	382.1	382.7	385.0	387.0	390.7	392.6	395.4	400.5	402.5	403.0	410.6	413.9	

Deseasonalised data which are free from seasonal fluctuations, show the trend present in the series. The increasing trend in wholesale prices for 1987 is evident from the deseasonalised data given in row [7].
This is shown in Fig. 12.3.

Deseasonalised Data

One objective of studying seasonal variation is to deseasonalise the data. This is done by expressing the observed time series data as percentages of the corresponding seasonal indices. The deseasonalised data can be used as indicators of the current trends in time series, seasonal influences having been eliminated. Most economic time series are often presented as deseasonalised or seasonally adjusted data with a view to bring into focus the trend present in the series.

Consider the time series of WPI of *Example* 6(b). As can be seen seasonal pressure on prices usually begins in April and lasts till September. Deseasonalised data clearly show the increasing trend in Prices (Fig. 12.3).

An estimate of the annual rate of inflation can be obtained from a comparison of deseasonalised data as the percentage change in the deseasonalised data. Deseasonalised WPI (old series with 1970-71 as base year) are as follows :

$$\text{For March 1986 : } \frac{359.8}{98.21} = 366.4$$

$$\text{For March 1987 : } \frac{378.1}{98.21} = 385.0.$$

Estimated annual rate of inflation for 1986-87 equals

$$\frac{385.0 - 366.4}{366.4} \times 100 = 5.08\%.$$

The WPI actually rose from 377.8 at the *end* of March 1987 to 417.7 at the *end* of March 1988. The point to point annual rate of inflation for 1987-88 equals (with seasonal index 98.21).

$$\frac{417.7 - 377.8}{377.8} \times 100 = 10.56\%.$$

Sometimes it is necessary to estimate the annual rate of inflation on the basis of indices on two points of time within a year. The annual rate is then estimated on the basis of the proportionate increase in the period to one year.
The percentage change in deseasonalised indices from the end of March 1986 to 24-1-87 (43 weeks) is 5.95. The estimated *annual* rate of inflation (as computed from the data for 43 weeks) equals

$$\frac{5.95}{43} \times 52 = 7.20\%$$

as against 5.08% calculated for one year (March 1986 – March 1987) basis.

Better and more accurate estimates of annual rate of inflation call for more accurate estimates of seasonal indices. In order to have them, data for a fairly large number of years are to be considered. The Reserve Bank of India compute

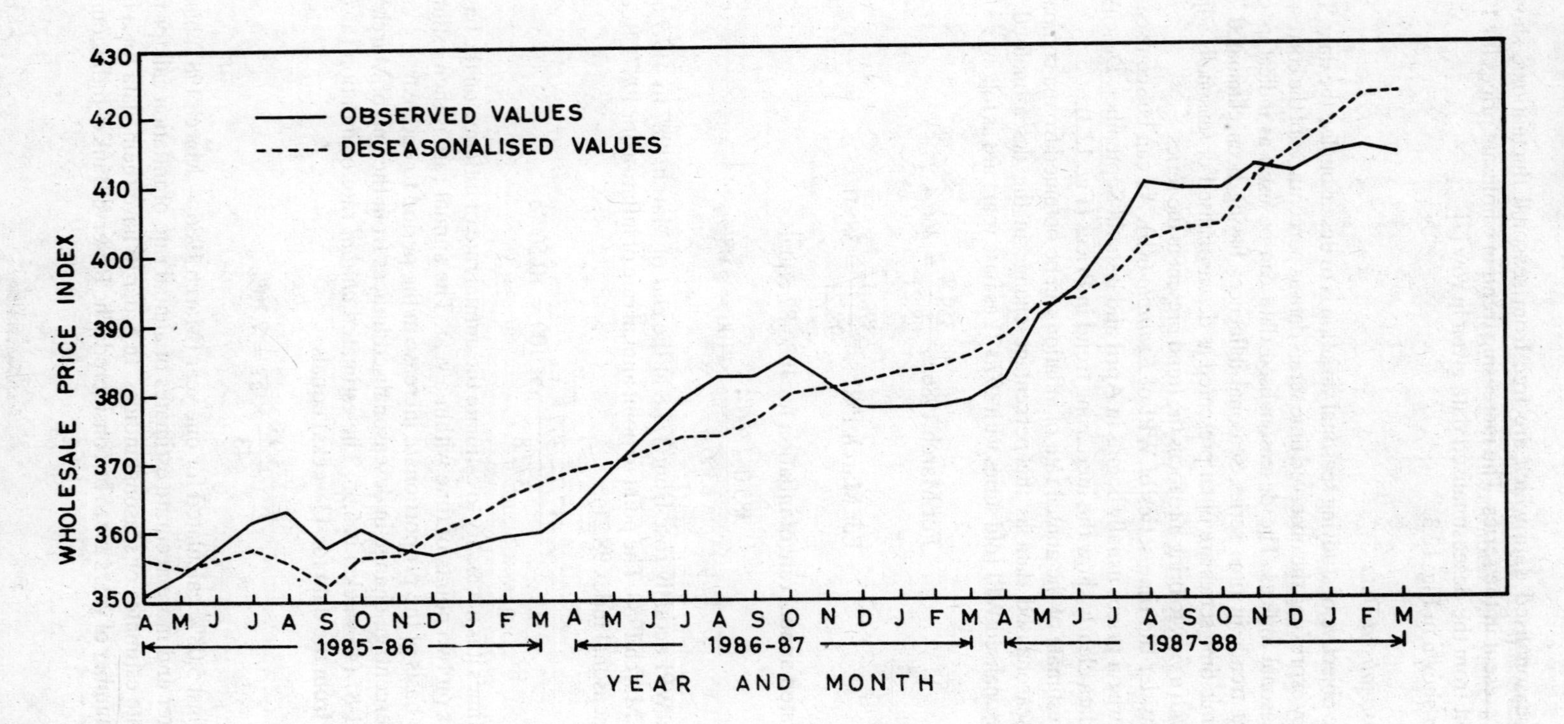

Fig.12.3 Wholesale Price Index 1985-86 to 1986-87 : Observed and Deseasonalised Values

the seasonal factors for WPI by the method of moving average based on data for 10 years.

Note : A new series of WPI with 1981-82 as base year has been released in July 1989 to replace the old series with base 1970-71 (see Chapter 5). The WPI (under the new series) which stood at 180.7 points on 1-9-90 rose to 208.3 points on 31-8-91 showing that the current rate of inflation (between September 90 to August 91) is 15.27%.

12.6.4. Method of Link Relatives

This method, which was at one time the most commonly used method for obtaining a seasonal index involves less extensive computations than required in the method of moving average. *Link relative* (L.R.) is the observed value of one period expressed as a percentage of the value of the preceding period : the period is a quarter for quarterly data, it is a month for monthly data and so on. The steps in the construction of a seasonal index by the method of link relatives are as indicated below.

Suppose that monthly data are given.

First, convert the monthly data in terms of link relatives. Thus

Link relative (L.P) for any particular month

$$= \frac{\text{observed value of the particular month}}{\text{observed value of the preceding month}} \times 100 \qquad (6.4)$$

Secondly, arrange the link relatives according to year and month. Then find the averages over the years for all the 12 months. Median is preferred to A.M. for averaging.

Thirdly, convert these L.R.'s into chain relatives (C.R) by using the formula:

Chain relative (C.R.) for any particular month

$$= \frac{\text{Average L.R. of the particular month} \times \text{C.R. of preceding month}}{100} \qquad (6.5)$$

By this we do not get the C.R. of the first month, which has to be assumed at first. Assuming that the C.R. of the first month is 100, C.R.'s of the other months can be obtained by using the formula (6.5).

Fourthly, compute the ratio

$$c = \frac{\text{L.R. of the first month} \times \text{C.R. of the last month}}{100} \qquad (6.6)$$

This ratio, which gives the *new* C.R. for the first month, is usually different from 100. Then find the correction factor $d = \frac{1}{12}(c - 100)$.

Fifthly, find the *adjusted* C.R.'s of the second and subsequent months by using the formula :

Adjusted C.R. for rth month = original C.R. of rth month $- (r - 1)d$, (6.8)
taking $r = 1$ for the 1st month, $r = 2$ for the 2nd month, and so on.

Note that the adjusted C.R.'s of the subsequent months are so adjusted, the adjsuted C.R. for the first month being fixed at 100.

Finally, find the overall average of the adjusted C.R.'s. If it is 100, then the adjusted C.R.'s give the seasonal indices. If it is different from 100, then the seasonal indices are given by the adjusted C.R.'s expressed as percentages of the overall average of adjusted C.R.'s.

The above is a *modified* link relative method. There is also a simpler link relative method. Here the first step is the same as the first step of the modified method described above (where L.R.'s are obtained by formula (6.4)). From this point, the procedure is the same as the third and the fourth steps in the moving average method.

The procedure described above will also hold good for quarterly data with necessary modification.

Merits and Demerits

Though the logic behind this method is not as easy to understand, the computations involved in this method (whether the simpler or the modified method is used) are much less extensive than in the 'Ratio-to-Trend' or in the 'Ratio-to-Moving Average Method'. In this method only one L.R., that of the first period is lost whereas in the Ratio-to-Moving Average Method, a number of values at the beginning and at the end are lost. The method would give good estimates of the seasonal indices provided the trend present in the time series is linear. However, this may not be the case with many economic time series.

Example 6(d). Quarterly average of wholesale Price Index (India) (1970-71 = 100) for the period 1985-86 to 1987-88 are as given below :

Wholesale Price Index

Quarters

Year	I (April-June)	II (July-Sept.)	III (Oct.-Dec)	IV (Jan-Mar.)
1985-86	353.9	361.0	358.1	358.7
1986-87	368.2	380.5	381.1	377.5
1987-88	388.5	406.4	409.9	413.9

To find the seasonal indices by the Link Relative Method.

The computations are as given below :

Table 12.4

First step	Link Relatives (L.R.)			
	I	II	III	IV
1985-86	–	102.01	99.20	100.17
1986-87	102.65	103.34	100.16	99.06
1987-88	102.91	104.61	100.86	100.98
Second step Average of L.R's (A.M.)	102.78	103.32	100.07	100.07

Third step Chain Relatives (C.R.)	100.00	$\frac{100 \times 103.32}{100}$ = 103.32	$\frac{103.32 \times 100.07}{100}$ = 103.39	$\frac{103.39 \times 100.07}{100}$ = 103.46

Fourth step : Calculation of correction factor

$$c = \frac{(\text{Av.L.R. for Qr I}) \times (\text{C.R. for Qr IV})}{100} = \frac{102.78 \times 103.46}{100} = 106.336$$

$$d = \frac{1}{4}(c - 100) = 1.58$$

Fifth step Adjusted C.R.	100.00	103.32 – 1.58 = 101.74	103.39 – 3.16 = 100.23	103.46 – 4.74 = 98.72
Final step	Average Adjusted C.R. = 100.1725			
Seasonal Indices	99.83	101.56	100.06	98.55
				Average = 100.00

The data can be deseasonalised by expressing the observed values as percentages of the seasonal indices for the 4 quarters.

Note :- The data considered in the above example are the quarterly averages of Wholesale Price Indices given in Example 6(b).

The quarterly averages of the seasonal indices are found to be :

Quarterly Average of	April-June	July-Sept.	Oct.-Dec.	Jan-March
	I	II	III	IV
Seasonal Indices	99.59	101.73	100.19	98.49

The seasonal indices found by the Link Relative method are

99.83	101.56	100.06	98.55

The two sets of seasonal indices are close enough. The agreement is quite satisfactory.

12.7. CYCLICAL MOVEMENT

As already discussed, a typical time series has four characteristic movements or components, secular trend (T), seasonal variation (S), cyclical movement (C) and irregular fluctuations (I). We have indicated some methods for determining trend and seasonal variation. We now discuss a method of determination of cyclical movement. There are two distinct view points regarding the cause of cyclical movement.

The first is the "self-generative theory" according to which a cyclical movement is self generated and the other is that cycles are caused by "external factors". Whatever the causes, cyclical fluctuations are common in business and economic time series, though cycles may not be of the same amplitude and periodicity.

12.7.1. Residual Method

This is the most commonly used method for determining cyclical movement. It consists of deseasonalising the data by eliminating the seasonal variation, and then of eliminating the trend.

For the multiplicative model, this would lead to

$$(T \times S \times C \times I) \div S = T \times C \times I$$

and then

$$(T \times C \times I) \div T = C \times I$$

so that one would be then left with cyclical and irregular component. The irregular component can then be smoothed so that only the cyclical component is left out. This is the residual method. Thus it consists of the following steps. First, estimate the seasonal indices and obtain the deseasonalised data by expressing the observed values as percentages of seasonal indices.

Secondly, obtain the trend values from the deseasonalised data by an appropriate method of determination of trend, such as fitting of a suitable mathematical curve. Then express the deseasonalised values as percentages of the trend values.

Thirdly, smooth out the irregular movements by using a short-term moving average. For monthly data, a centred-two-month moving average (which is a weighted three month moving average) may be used. The final series will then show the cyclical component.

As can be seen the determination of truly representative cyclic components would depend on the accurate determination of seasonal indices, trend values and the period of moving average (for smoothing irregular components).

12.7.2. Other Methods

Besides the residual method, which is the most widely used procedure, there are other procedures for estimation of cyclical components. These are mentioned below :

Direct Method: The steps are the following :

First, express the observed value of each month as a percentage of the corresponding month of the preceding year. Thereby the secular trend and seasonal variations are roughly eliminated. The resulting series will then comprise of the cyclical and irregular components; some residual trend may also remain.

Then eliminate the residual trend and the irregular component by a moving average. From the resulting series a rough estimate of the cyclical fluctuations can be obtained.

Periodogram Analysis : This method is effective when the cyclical movements are roughly of the same periodicity and amplitude. Then a sine-cosine curve, having regular periodic movement can be fitted to the observed data. The period λ of a sine series of the form

$$y_t = a \sin \frac{2\pi t}{\lambda} + \varepsilon_t$$

(where ε_t is the error component) may be estimated by periodogram or harmonic analysis.

This method involves extensive computations. Further, time series encountered in business, in economics and other social sciences do not, in general, possess cyclical movements with regular periodicity and amplitude. This method will not be discussed here.

12.8. IRREGULAR MOVEMENTS

Because of their very nature, an appropriate formula for determining irregular movements is not available. The residual method may be used to estimate the irregular components. This consists of determining the seasonal variations (S), the trend (T) and the cyclical components (C) and then by dividing the observed values by ($S \times T \times C$). We shall have for the muliplicative model

$$\frac{Y}{S \times T \times C} = \frac{T \times C \times S \times I}{S \times T \times C} = I$$

This amounts to obtaining the cyclical-irregular components at first through the first two steps indicated in the residual method (section 12.7.1) and then dividing these by the cyclical components obtained through the three steps indicated in the same section.

However an estimate of the variance of the irregular component can be obtained by the Variate Difference Method. We do not consider it here.

12.9. LONG CYCLES : KONDRATIEV WAVES

Evidences of occurrences of cycles of long duration of 50 years or more have been observed by some economists. Today this matter has been receiving considerable attention the world over because now is about the time that a certain phenomenon may recur. The work by Nikolai Kondratiev, a Russian economist of astonishing foresight, done in the 1920's is only now coming back to fashion. Kondratiev's (Kondratieff's) work was taken up by another well known economist Joseph Schumpeter in the late 1930's.

In his writings published in the 1920's Kondratiev put forward the theory that all economic activities move in long predictable waves, each wave lasts just over half a century and that there have been four such waves since the Industrial Revolution (see Fig. 12.4). Reference is to the Industrialised nations.

His reference was to world economy in general. Each such wave, known as *Kondratiev Wave*, has all the four characteristics of a business cycle. It rises on the back of new technologies and new forms of enterprise; the expansion leads to a peak, then topples over into a crash and reaches the trough. But at that very moment a new breed of entrepreneurs starts to create the basis for the new technology for the next wave to commence. The propellent that pushes a wave to its peak and collapse is a devastating war. To each wave can be associated the name of a nation(s).

The first Kondratiev wave of the Industrial revolution rode on the technologies of the cotton and wrought-iron industries; it reached its peak after the

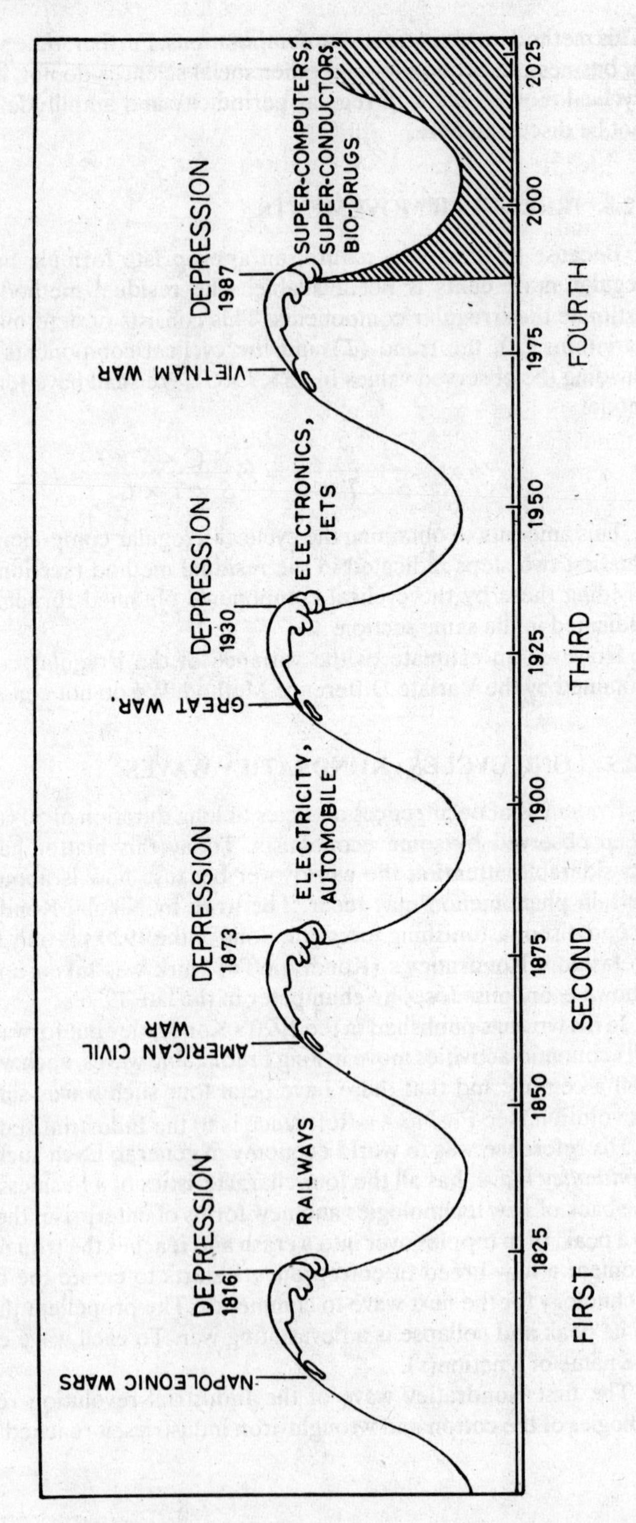

Fig. 12.4 Kondratiev Waves

Napoleonic wars; then began the slump leading to depression which began around 1816. Britannia ruled the first wave.

The second wave rode on the technology of Railways; its peak followed the American Civil War, then the slump began around 1873 after 57 years of the earlier one. Britain along with Germany ruled the second wave.

Electricity and automobile were the new technologies on which the third wave rode. Its peak followed the first Great War, then the slump began and depression set in around 1930 - after 57 years of the earlier one. Germany and USA ruled this wave.

The fourth wave rode on the new technologies like Electronics, Television, Jets and Computers. The nations associated with this wave are USA and Japan. The Vietnam war came in and it is believed that the peak of this wave has now been reached. It was feared that a depression might set in around 1987 again after 57 years. In fact, a great fall in the Dow Jones Average* (of 30 Industrial stocks) occurred in October 1987 : in the course of a single day on October 19,1987, the Dow Jones Average nosedived a record 508 points (22.6%) to 1,738.74 (this dwarfs the great crash of 1929). There has been considerable concern the world over now over the impending depression. In India also the share Price Index of *The Economic Times* fell and reached a trough in December 1987.

Each Kondratiev wave has (approximately) a 57 year cycle. On the basis of this, it is argued by some noted economists that the depression is about to commence and the trough will come around 2000.

For the fifth wave, new technology like super-conductors, super computers, bio-drugs will provide the basis of new expansion; it may reach its peak and then slump may occur, again after 57 years, in around 2044. The Pacific rim, in particular, Japan and perhaps China would, according to many, rule this wave.

This type of fluctuations need not necessarily happen according to many renowned economists, like Paul A. Samuelson, N.L., who thinks that business cycles have been tamed. We close this Chapter with his observation. He asserts that democratic countries having mixed economics are unlikely to experience ever again prolonged depressions as happened earlier and that recessions and periods of relative stagnation will no doubt still occur; however, through fiscal arid monetary policies, their frequency, intensity and duration can be moderated to a great extent. *(Economics* : P.A. Samuelson, 10th edition p. 267).

EXERCISES-12

Sections 12.1–12.3

1. What is a time series? Explain the objectives of analysis of a time series. Why is analysis of time series of importance in social sciences?
2. Give some examples of economic, business and demographic time series. Indicate the importance of analysis of such series in the respective fields.

---

*The Dow Jones Industrial Average (first published in 1896), is an indicator of the health of the economy of USA. Weathering the 'Black Monday' crash of October 19, 1987 (when this Index stood at 1738 points) and the lacklustre years that followed, it rose from 3000 level on April 17, 1991 and to 10,000 level on March 16, 1999, with the help from a growing economy and the increasing enthusiasm of investors in investing their money in stocks in USA.

3. Describe the components of a time series. Illustrate them with suitable examples and graphs. What is meant by time series analysis ? What does one seek to achieve through such an analysis?
4. Describe the models of a time series. How do they differ ? How can an additive model be considered as a particular type of multiplicative model ? Which of the two models is considered to be more useful and why ?

Section 12.4

5. Define trend (secular trend) in a time series. Give some illustrations of increasing as well as decreasing trends in a time series. What are the objectives behind the study of trend in a time series ?
6. What are the methods for estimation of trend in a time series ? How would you get the trend values from an observed time series ?
7. Explain the method of moving averages for determination of trend. How do you get a centred moving average in case of an even order moving average ? Describe the merits of this method over other methods.
8. The per capita Net National Products (in Rs.) of India from 1971-72 to 1985-86 are given below.

Year	At Current Prices	At 1970-71 Prices	Year	At Current Prices	At 1971-72 Prices
71-72	660.3	626.6	79-80	1337.5	664.7
72-73	711.1	603.4	80-81	1557.3	698.3
73-74	870.1	621.3	81-82	1743.0	719.5
74-75	1003.5	717.0	82-83	1887.3	721.5
75-76	1026.4	663.5	83-84	2186.0	763.8
76-77	1079.4	652.1	84-85	2354.8	774.6
77-78	1194.1	694.7	85-86	2596.6	797.7
78-79	1253.0	717.0			

(Figures for 79-80. . . . are provisional). Estimate the trend values by taking 3-year moving averages.

9. Estimates of birth and death rates in India. based on Sample Registration System 1970-79. are as follows :

Year	Birth rate	Death rate
1970	36.8	15.7
1971	36.9	14.9
1972	36.6	16.9
1973	34.6	15.5
1974	34.5	14.5
1975	35.2	15.9
1976	34.4	15.0
1977	33.0	14.7
1978	33.3	14.2
1979	33.0	12.8

Obtain the trend values by taking 3-year moving averages. Graph the observed values and the trend values.

10. The following table gives the total fertility rates (TFR) for two low fertility countries in Asia.

TFR

Year	Japan	Singapore
1969	2.13	3.14
1970	2.13	3.09
1971	2.16	3.04
1972	2.14	3.05
1973	2.14	2.08
1974	2.05	2.36
1975	1.91	2.08
1976	1.85	2.11
1977	1.80	1.82
1978	1.79	1.80
1979	1.77	1.79
1980	1.74	1.74

Estimate the trend values by taking 4-year centred moving averages.

11. What is Slutzky-Yule effect ? Explain.

Section 12.5

12. Describe the method of least squares. Explain how you would obtain estimates of the constants of a straight line fitted for time series data over an (i) odd number of years. and (ii) even number of years.
13. Fit a linear trend for per capita Net National Product (in Rs.) of India at 1970-71 prices from the data given in Exercise 8. Draw a graph of the observed values and the trend values.
14. Fit a linear trend for birth and death rates for India from the data given in Exercise 9. Draw a graph for the observed and the fitted values. From the graph estimate the birth and death rates for 1980.
15. Obtain the trend line for the total fertility rate for Singapore from the data given in Exercise 10.
16. The annual sales for a store for a number of years are as given below :

Year	1980	1981	1982	1983	1984	1985	1986
Sale (in 0000 rupees)	28	34	32	36	40	43	44
Profit (in 0000 rupees)	1.5	2.0	3.0	2.8	3.6	3.9	3.8

Obtain the trend values by fitting a linear trend for (1) sales (2) profits.

17. The production of sugarcane in India from 1981-82 to 1985-86 was as follows :

Year	1981-82	1982-83	1983-84	1984-85	1985-86
Production of sugarcane (in million tonnes)	186.4	189.5	174.1	170.3	171.7

Fit a linear trend and graph the trend and observed values.

18. The estimated total fertility rates (TFR) of India and Bangladesh are as given below :

Year	1960-65	1965-70	1970-75	1975-80	1980-85
India	6.0	5.8	5.6	4.9	4.2
Bengladesh	6.7	6.9	7.0	6.7	6.2

Fit a linear trend for TFR for each of the countries. Graph the observed and trend values.

19. What is meant by a non-linear trend ? How would you proceed to fit a quadratic trend by the method of least squares ?
20. Explain how you would proceed to fit an exponential trend of the type $y = ab^x$.

21. The population of India during the census years 1921-81 is given in Table 3.1 (Chapter 3). Taking population figures to nearest million, fit a parabola of the second degree to the data.
22. The decennial growth rates of population during 1921-81 are given in Table 3.1 (Chapter 3). Fit a linear and a parabolic trend and graph the trend values along with the observed values in the same graph.
    Estimate the percentage increase for 1981-91 on the basis of the linear trend on the assumption that the same trend would continue during the decade.
23. How would you decide whether to fit a straight line, a polynomial of second degree or a polynomial of third degree to some observed data ? Discuss.
    What are the mathematical curves generally considered for study of growth of population or growth of an industry during a certain period ?
24. Describe the merits of fitting of mathematical curves over other methods.
    What is the main difficulty of considering mathematical curves for trend elimination ?
25. Compare the method of moving average and the method of polynomial fitting for determination of trend.
    Show how the former method can be considered as a special case of the latter method.

Section 12.6

26. What is meant by 'seasonal variation' in a time series ? Illustrate your answer with suitable examples.
    Mention the objectives behind the analysis of seasonal variation in a time series.
27. What are the characteristic fluctuations observed in a time series of a large number of years ? Distinguish between the seasonal and cyclical variations.
28. Discuss the different methods for determining seasonal variation in a time series. Which one is the most commonly used method for computing seasonal indices? Explain how you would interpret the values of seasonal indices.
29. Describe the method of simple averages for determination of seasonal fluctuations in a time series.
30. Estimate the seasonal indices by the method of simple averages for the data given in Example 6(c).
31. Describe the 'Ratio-to-Trend Method' for measurement of seasonal fluctuation in a time series. When would you recommend this method ? Mention its merits and demerits.
32. Fit a linear trend to the data given in Example 6(c). Then compute the seasonal indices by the 'Ratio-to-Trend Method'.
33. Explain the logic of the procedure adopted in 'Ratio-to-Trend Method' with the help of the multiplicative model.
34. Describe the 'Ratio-to-Moving Average method' for measurement of seasonal fluctuations in a time series. How does his method differ from the 'Ratio-to-Trend Method'?
    What are the merits and demerits of this method ? Discuss.
35. Explain the justification of the procedure adopted in 'Ratio-to-Moving Average Method' for finding seasonal indices.
    What kind of bias may creep in while taking the moving averages ? How is the bias. if any, sought to be eliminated? Explain.
36. Compute the seasonal indices for the data given in Example 6(c) by the 'Ratio-to-Moving Average Method.'
37. The following table gives the number of patients visiting an outpatient department of a hospital during a course of 4 weeks (from Monday to Saturday) :

Week	M	T	W	T	F	S
I	352	307	285	270	275	314
II	325	318	300	280	275	310
III	335	324	290	276	282	318
IV	324	315	285	274	288	322

Find the seasonal indices by the Ratio-to-Moving Average Method.
Interpret the values of the indices.

38. What do yo mean by deseasonalisation of data ?
    How are deseasonalised data obtained from observed data ? What is the purpose behind deseasonalisation ? Mention some economic time series usually given as seasonally adjusted series.
39. Explain how the annual rate of inflation is estimated from deseasonalised wholesale price index.
40. Explain the method of Link Relatives for measuring seasonal fluctuations.
    Mention the merits and demerits of the method.
41. Find the seasonal indices for the data given in Exercise 37 by the method of Link Relatives.
42. Find the seasonal indices for the data given in Example 6(a) by the method of Link relatives.
43. Discuss the relative merits and demerits of the four methods of measurement of seasonal indices. Indicate the situations where you would recommend the use of a particualr method.

Sections 12.7 – 12.8

44. What are cyclical fluctuations?
    Describe the methods used for measurement of cyclical fluctuations in a time series.
45. What are irregular movements ? What are the main types ? How are their occurrences explained?
    Explain the residual method of estimating irregular components in a time series.

Miscellaneous

46. Draw a line graph of the Dow Jones Average (as it stands at the beginning of each month) for as many years as you can collect your data.
    Try to fit a trend line.
47. Fit a curve for the population of your country as enumerated in the last 10 censuses.

Chapter 13

Demography

13.1. INTRODUCTION

By population is meant the totality of all human beings living at a certain point of time within a territory demarcated by geographical, political or cultural boundaries. Demographic statistics deal with the quantitative aspects of the population of a community. Demography relates to the study of the measurements of characteristics related to populations and is mainly concerned with the growth of populations and the techniques of measurement of population growth. The study of demography is of great importance in economics and other social sciences. The balance between growths of population and of resources has been a topic of great interest to economists since the days of Thomas Malthus (1766-1834), who held that man tends to increase faster than his means of subsistence.

The primary statistical unit of observation in demography is the individual. Demographic statistics are acquired or gathered by enumeration of a population, periodically through censuses, continuously through registrations of births, deaths, marriages etc. and occasionally through sample surveys of cross sections of the population.

We shall give an account of Censuses and also an account of the Indian Census in section 13.7.

13.2. DEFINITIONS

Vital events refer to such events of human life as birth, death, sickness, marriage, divorce, adoption etc.

The term vital statistics refers to the basic or derived data relating to vital events. It also refers to the methods applied for the analysis of vital events.

The basic data of vital statistics are usually recorded at the time of occurrence of the event: this is usually done through established registration system and registration of vital events is a legal requirement for many countries. In India Registration of Births and Deaths Act was passed in 1969, whereby registration of Births and Deaths is compulsory in India and violations are punishable.

Cohort. All persons born during a given year form a cohort or birth cohort of that year. A population is thus the aggregate of the survivors of cohorts of different years.

The term cohort is also used in different contexts; by marriage cohort is meant the totality of persons married during a given year.

Vital statistics rates and ratios are measures which are computed for meaningful comparisons of vital events between different populations or communities or for ascertaining the trends in the same community over a time period. If N represents the number of occurrences of a certain vital event E (such as birth, death etc) during a stated interval of time, and P represents the size of the population within which the vital occurrence took place, then N/P is regarded as the *rate* of that vital statistics. The period of time is usually taken as one year. The size P of the population is exact or approximate; usually mid-year population size is taken. The rate is a number which lies between 0 and 1. This definition of rate is similar to classical definition of probability.

A multiplier called *radix* is used to round off the decimals. Usually 1000 is taken as radix so that the rate is per 1000 of population; thus, using the radix k, the rate is expressed as

$$\frac{N}{P} \times k$$

with k usually taken as 1000.

Various rates, such as birth rate, death rate, fertility rate etc are considered as measures of the vital events concerned; these are discussed later.

If the denominator represents the number of occurrences N_1 during the same period of time, of another vital event E_1 which may or may not include the vital event E then the quotient N_1/N is usually described as a vital statistics *ratio*. The two classes of events need not be non-overlapping. Using the radix k, the ratio is expressed as

$$\frac{N_1}{N} \times k.$$

For example, if N is the total number of deaths in a year and N_1 is the number of deaths due to a particular cause, then

$$\frac{N_1}{N} \times 1000$$

is the ratio of the death due to that cause per 1000 deaths.

The relative number of males and females in a population is measured by the sex-ratio, which is defined as

$$\text{sex-ratio} = \frac{N_1}{N} \times k$$

where N_1 is the number of females in a population at time t and N is the number of males in the same population at time t. With k usually taken as 1000, the ratio gives the number of females per 1000 males.

The sex-ratio is also defined as $\frac{N}{N_1} \times 1000$, which gives the number of males per 1000 females.

Note:

1. The sex-ratio (number of females per 1,000 males) in India as recorded in the successive censuses from 1901 to 2001 are as follows:

   972, 964, 955, 950, 945, 946, 941, 930, 934, 927 and 933.

   The declining sex-ratio has been a matter of serious concern as well as a subject of deep investigation.

2. For analysis in regard to certain economic and social purposes the population is divided into 3 broad categories : young, adult work force population and old. While division of the three categories by age-groups 0-14, 15-64, 65 and above is more common, some demographers in India consider the age-groups as 0-14, 15-60, 60 + (61 and above) (because of lower expectation of life and shorter working life). In Chapter 3 (Tables 3.16, 3.17 and Fig. 3.21) we have considered some specific measures of interest–the index of ageing and dependency ratios.

13.3. BIRTH AND DEATH RATES

13.3.1. Crude Rates

Crude Birth Rate (CBR) is defined as follows:

$$\text{Crude birth rate} = \frac{\text{No. of live births during a year}}{\text{Mid–year population}} \times 1000$$

$$= \frac{B}{P} \times 1000$$

The CBR gives the annual birth rate per 1000 of population. The CBR for the countries of the world lies between 10 and 55. The CBR in India for 1986 was 29.6 (as estimated by ESCAP population division). It came to 27.2 in 1997 (SRS estimate).

The level of CBR depends on the following:

(i) the age and sex distribution of the population, and

(ii) the average rate of child bearing of the female population.

Countries with relatively high proportion of female population in the fertile age group 15-49 years will record a relatively high CBR. Similarly in the case of countries with higher average rate of child bearing, CBR will be higher.

Crude Death Rate (CDR) is defined as follows:

$$\text{Crude death rate} = \frac{\text{No. of deaths in a year}}{\text{Mid–year population}} \times 1000$$

$$= \frac{D}{P} \times 1000$$

The CDR gives the annual death rate per 1000 of population. The CDR for countries of the world lies between 8 and 30. The CDR in India for 1986 was 11.5 (as estimated by ESCAP population division). In 1997 it was 8.9 (SRS).

The level of CDR depends on the following:

(i) the age and sex distribution of the population,

(ii) the average longevity of the population.

Countries with high proportion of old population will exhibit a relatively high CDR. Similarly, in the case of countries with low longevity, CDR will be high.

Crude rate of natural increase

The annual crude rate of natural increase is defined as the difference (CBR – CDR) between CBR and CDR. This gives the crude rate of natural increase for a given year. For India this rate equals 29.6 – 11.5 = 18.1 for 1986. This implies that the population has registered a natural increase of 18.1 per 1000 of population during 1986. Table 3.11 and the corresponding Fig. 3.6 give the CBR, CDR and the crude rate of natural growth during the decades 1901-11 to 1971-81.

13.3.2. Specific Death Rates

The CDR is a rough measure, and may be used to compare mortality situation of the same region at different periods, not too far apart. To get a better idea of mortality situation in a population, sex and age specific rates are considered. Annual CDR for males (females) is defined by

$$\frac{\text{No. of male (female) deaths in a year}}{\text{Mid–year male (female) population}} \times 1000$$

Similarly age-specific rates are defined.

The annual age-specific death rate (per 1,000) of age x is defined as

$$M_x = \frac{D_x}{P_x} \times 1000 \tag{1.1}$$

where D_x is the number of deaths of persons aged x and P_x is the mid-year population aged x during a certain year.

Note that by a person aged x is meant a person who, at last birth day, was aged x (or who has lived x full years, ignoring the part, if any, of another year lived).

We have $\sum_x P_x$ = total mid year population

and $\sum_x D_x$ = total number of deaths

so that
$$\frac{\sum M_x P_x}{\sum P_x} = CDR. \tag{1.2}$$

Thus CDR is a weighted average of M_x with weights $\frac{P_x}{\sum_x P_x}$, which is the proportion of population aged x.

Instead of considering distinct age x, we consider age group x to $(x + n)$; then we get the age-specific death rate for that age group. If ${}_nP_x$ denotes the mid-year population of persons aged between x and $(x + n)$ and ${}_nD_x$ denotes the number of deaths of persons aged between x and $(x + n)$ during a certain year, then

$$_nM_x = \frac{_nD_x}{_nP_x} \times 1000 \tag{1.3}$$

gives the age-specific death rate for the age group x to $(x + n)$ during that particular year.

The CDR is then given by

$$\frac{\sum_x {}_nM_x \cdot {}_nP_x}{\sum_x {}_nP_x} \tag{1.4}$$

Restricting to male (female) populations we can likewise define age-specific death rate for male (female) populations separately. A specific death rate can likewise be computed for segments of the community differentiated by characteristics such as race, region, marital status etc.

Cause-specific death-rate

When the total number of deaths D can be subdivided to show the number of deaths D_c from each specific cause c, then the cause-specific death rate due to the cause c is

$$\frac{D_c}{P} \times 1000.$$

It gives the number of deaths per 1000 of population. The quantity

$$\frac{D_c}{D} \times 1000$$

gives the proportion of deaths due to cause c per 1000 deaths. This is age-specific death ratio.

The specific death rates are useful measures of mortality conditions of a community. Rates for one community (or one geographical region) are often compared with those of another community (or other geographical region). Similarly rates for components of population such as race, religion and also of population divided on the basis of locality of dwelling (rural and urban) provide useful measures for comparison of mortality conditions.

The age specific death rates show a striking behaviour for communities in general. The age specific death rates start at a rather high level in the first year of life, then gradually diminish and reach a minimum around the age group 10-14, then again rise slowly until midlife and then rise rapidly at the older ages. The curve with specific death rate along the .y-axis and age along the x-axis is roughly U-shaped (or bath-tub shaped).

We give below CDR and age specific death rates for males and females and for rural and urban population of India for 1997.

Table 13.1 Age–specific death rates & CDR, India 1997
(Rates are on as per-thousand basis)

Age Group	Rural		Urban		Total		
	Male	Female	Male	Female	Male	Female	Combined
0-4	24.2	27.2	12.5	13.8	21.8	24.5	23.1
5-9	2.0	2.7	1.0	1.2	1.8	2.4	2.1
10-14	1.2	1.3	0.9	0.7	1.1	1.2	1.2
15-19	1.4	2.4	1.0	1.5	1.4	2.1	1.7
20-24	2.3	1.3	1.8	2.0	2.2	2.8	2.5
25-29	2.8	2.9	2.1	2.1	2.6	2.7	2.7
30-34	3.5	3.2	2.9	1.9	3.4	2.8	3.1
35-39	4.0	3.2	3.5	2.0	3.9	2.9	3.4
40-44	6.6	4.2	4.4	3.6	6.0	4.0	5.0
45-49	9.6	5.8	7.5	5.5	9.0	5.8	7.5
50-54	14.7	11.0	12.5	7.3	14.2	10.2	12.2
55-59	20.1	14.9	18.2	11.2	19.7	14.1	16.9
60-64	33.4	23.1	28.2	18.2	32.3	22.1	27.1
65-69	46.4	35.3	46.2	30.8	46.3	34.4	40.2
70+	90.0	78.6	88.4	73.7	89.7	77.6	83.5
CDR	9.8	9.4	7.0	6.0	9.2	8.6	8.9

Source: SRS, Registrar General, India, 1997

13.3.3 Infant Mortality Rate (IMR)

If D_0 is the number of deaths between birth and age one (i.e. the number of infants dying before attaining the age one) during a certain year and B is the total number of live births during the same year, then

$$IMR = \frac{D_0}{B} \times 1000$$

Note that fetal deaths and still births are excluded in this computation.

The IMR as computed above does not give an accurate measure of the risk of death of infants during the first year of life. Here D_0, and B refer to the corresponding numbers during a *certain year.* In fact many of the deaths D_0 during the year may be of infants born in the preceding year and many of those born B during the year may die in the following year while still under one year of age. When the number of births does not show rapid change from year to year, the IMR computed as above (from figures during a certain year) gives a reliable measure of infant mortality.

Specific IMR can be computed according to sex, and other characteristics and also according to cause of death.

The infant mortality rate in India in 1986 was 95 (according to ESCAP population estimate). There has been a decline in IMR since 1950 (when IMR' was about 134) due to Government's efforts to extend health services to the villages. The IMR of some countries are given in Table 13.2. The IMR for India in 1997 came to 72 (while for Kerala it was 12 (SRS)).

Table 13.2 IMR

Country	IMR	Country	IMR
India	95	Indonesia	77
Bangladesh	133	Malaysia	27
Nepal	136	Japan	5
Pakistan	126	Australia	9
Sri Lanka	30	China	35

Source: 1986 Escap Population Data Sheet

13.3.4 Adjusted Measures of Mortality : Standardised Death Rate

In order to get a clear insight into the differences in the mortality situation of two communities or even of the same community over two different time periods, it is necessary to make comparisons of death rates according to a number of different characteristics such as age, sex, marital status, socio-economic class, locality of dwelling (rural and urban) etc. Such an investigation would involve a huge mass of data, which may be necessary only for some very specific purpose. Ordinarily this would not be necessary nor could such a large amount of information be readily assimilated. This brings out the necessity of computing a single index of mortality which is some sort of an average of the death rates for various segments of the population. One such rate is the CDR. However, the CDR is greatly influenced by the distribution of the population according to age, sex etc. The CDR may be a useful index of comparison of mortality situation of a single community over two different periods, provided these periods are close enough so that the distributions according to age, sex etc., do not show marked changes. But for comparison of the mortality situation of two communities, whose distributions according to age, sex etc. are very much different, the CDR will not provide a reliable index. From (1.2), it can be seen that CDR is the weighted average of age specific death rates with weights equal to the proportion of population with corresponding ages. The weights reflect the age distribution of the population. Thus for two communities, the age distributions may not be identical, in which case, their CDR's will differ though the age specific death rates for different ages may be comparable for the two communities. Differences in crude death rates for two communities may be due to the differences of their age and sex distributions and also to the differences in the age-specific death rates for each sex. To obviate the defect in using CDR as an index for comparison of mortality situation of two communities, an index is constructed with some standard weights (the same set of weights for the two communities). The index thus constructed is called the standardized or adjusted or corrected death rates. The population, to whose age distribution the standard weights correspond is called the standard population. We have considered age distribution as the basis of finding standard weights : the standardised death rates will be called age-standardized.

Similarly, standard weights may correspond to a distribution of the population according to some other characteristic; then the standardisation will refer to that characteristic. However, age standardisation is the common method.

Direct method of standardisation

In this method the weights used in computing standardized death rates are taken from some *actual* population which we call standard population. Thus for computation of age-adjusted death rate for each state of India, the All India population may be taken as the standard population, and the proportion of population in different age groups may be taken as the standard weights. Adjusted or standardized death rates may be computed by suitably modifying formula (1.4). Thus if ${}_nP^s_x$ denotes the number of persons between ages x and $(x + n)$ in the standard population, then the age-adjusted (or age-standardised) death rate is given by

$$\frac{\sum_x {}_nM_x \cdot {}_nP^s_x}{\sum_x {}_nP^s_x} \tag{1.5}$$

Here the weights are $\dfrac{{}_nP^s_x}{\sum_x {}_nP^s_x}$ (for age groups x to $(x + n)$). (1.6)

The same weights are used for comparing age-adjusted rates for males and females. Replacing ${}_nM_x$ by the age-specific death rate for males (females) in the age group x to $(x + n)$ and using the same weights (1.6) for each sex, the age-adjusted death rate for males (females) can be computed from formula (1.5). With a break up according to both age and sex, the age-sex-adjusted death rates can be computed.

One advantage of the adjusted death rate is that it is easy to compute and is also easy to interpret. If the age specific death rates of one community are higher than those of another, the fact would get reflected in the adjusted death rate. One drawback of considering adjusted death rate as a reliable index is the subjective consideration in the choice of the standard population. If the standard population does not vary too much in its age-distribution from those of the communities compared, then the difficulty may not be great. If the age-distributions of the two communities differ widely, then one of them may be chosen as the standard. It is convenient to choose as standard population, the population whose mortality measure is of greater immediate concern.

Indirect method of standardisation

The direct formula for standardisation as given by (1.5) involves ${}_nM_x$, the age-specific death rates for the community as well as ${}_nP^s_x$, the age distribution of the standard population. When ${}_nM_x$ are not available, the direct method cannot be used and an indirect method is used. The indirect formula, given in (1.7) below requires the age distribution ${}_nP_x$ of the given population as well as the age-specific death rates and age distribution of the standard population. The indirect age-standardised death rate is given by

$$\frac{D}{\sum_x {}_nM^s_x \cdot {}_nP_x} \left\{ \frac{\sum_x {}_nM^s_x \cdot {}_nP^s_x}{\sum_x {}_nP^s_x} \right\} \tag{1.7}$$

where D is the total number of deaths in the given population. Comparing with (1.5) we get that the second factor is the (CDR)s for the standard population, so that (1.7) reduces to

$$\frac{D}{\sum_x {}_nM_x^S \cdot {}_nP_x} \cdot \text{(CDR)s.} \qquad (1.7a)$$

13.4. LIFE TABLES

It is often of interest to study the mortality experienced in a particular community or country during a given period of time. A method of expressing the pattern of mortality is to be found. The simplest way of finding this is through a life table. Suppose we start with a hypothetical group of 1,00,000 births at a certain instant of time (a cohort of 1,00,000) and estimate the number of persons which will survive to age 1, 2, 3, . . . , if they are subjected to the given mortality conditions throughout their lives. The given mortality conditions refer to the age-specific death rates prevalent during a certain year or a period of time. Subjected to the mortality conditions of a certain year (or a certain period), the number of 1,00,000 will get depleted through deaths at each age till finally the number living will become 0. A life table presents a life history of the cohort of persons. This enables us to measure the longevity of persons. It also serves as an essential tool for more detailed studies relating to mortality, longevity, fertility and population growth.

13.4.1. Basic Assumptions in construction of a life table

There are a few basic assumptions, which are discussed below.

(1) A start is made with a hypothetical cohort of births or new borns (each exactly aged 0).

(2) The cohort cannot be increased or decrased due to other factors of population change. In other words, the life table population is closed to migration and emigration.

(3) The distribution of deaths over the year of age is uniform. In other words, deaths occur uniformly over the age-interval, specially when it is one year.

(4) Deaths due to severe famine, epidemic or other calamities are not anticipated.

13.4.2. Description of Various Columns of a Life Table

Life tables are based upon mortality rates. The following are the notations commonly used.

x = the age at last birthday (or number of complete years lived) (x takes non-negative integral values)

l_x = the number of persons who attain the exact age x

l_0 = cohort or radix of the table : the assumed number of births or new borns (at a time point) with which we start (l_0 is usually taken as 1,00,000)

d_x = the number of persons of age x who die before attaining the age $x + 1$ $= l_x - l_{x+1}$

q_x = the probability that a person of exact age x would die before attaining the age $(x + 1) = \dfrac{d_x}{l_x}$

(q_x is called the mortality rate with radix $k = 1$; $k \times q_0$ being the infant mortality rate per k live births)

p_x = the probability that a person of exact age x would survive till his next birthday (would not die before attaining the age $x + 1$) $= 1 - q_x$

${}_nq_x$ = the probability that a person of exact age x would die before attaining the age $(x + n)$ $({}_1q_x \equiv q_x)$, ${}_np_x = 1 - {}_nq_x$

L_x = the number of person-years lived by the l_x persons during the age interval $(x, x + 1)$. We have

$$L_x = \int_0^1 l_{x+t}\, dt \tag{4.1}$$

If it is assumed that d_x deaths which occur in the age-interval $(x, x + 1)$ are approximately uniformly distributed over this interval, then

$$L_x = \frac{l_x + l_{x+1}}{2} \tag{4.2}$$

$$= \frac{l_x + l_x - d_x}{2} = l_x - \frac{1}{2} d_x$$

L_x is the mid-year population (for $x = 3, 4, \ldots$)

L_x can be interpreted as the average *size* of the cohort between the ages x and $x + 1$.

T_x = the total number of person-years lived after the age x (it is the total person-years lived by l_x persons of age x).

$$T_x = L_x + L_{x+1} + \ldots$$

$$= \sum_{r=0} L_{x+r} \tag{4.3}$$

e_x^0 = the average number of years lived *after* age x (it is the additional number of years a person attaining age x can expect to live)

$$e_x^0 = \frac{T_x}{l_x} \tag{4.4}$$

e_0^0 = the average number of years a new born child can expect to live (it is the expectation of life at birth).

13.4.3. Construction of Life Tables

The mortality rates q_x are fundamental in the construction of a life table. Once these are known then starting with a suitable cohort, of say, l_0 = 1,00,000, the quantities for the other columns of the life table can be computed. We get $d_0 = l_0 q_0$; then $l_1 = l_0 - d_0$. Again $d_1 = l_1 q_1$ and $l_2 = l_1 - d_1$, and so on. Thus we get the columns for l_x, d_x. Then with these values, we can proceed to find L_x, T_x and e_x^0, and the corresponding columns.

The basic question is : how to estimate the mortality rates q_x. The procedure for estimating q_x is discussed below.

Suppose that the annual (observed) age-specific death rates M_x for the population under consideration are known (calculated). Taking the radix k = 1, we have

$$M_x = \frac{D_x}{P_x} \tag{4.5}$$

where D_x is the number of deaths of persons aged x and P_x is the mid-year population aged x during the year under consideration. The probability m_x that a person of the age group $(x, x + 1)$ would die in that age-interval is

$$m_x = \frac{d_x}{L_x} \tag{4.6}$$

Assume that deaths occur uniformly in the age interval so that $L_x \cong P_x$. Thus we can estimate m_x by M_x . Now

$$m_x = \frac{d_x}{L_x}$$

$$= \frac{d_x}{l_x - \frac{1}{2} d_x} \quad \text{from (4.2)}$$

$$= \frac{d_x / l_x}{1 - \frac{1}{2} d_x / l_x}$$

$$= \frac{q_x}{1 - \frac{1}{2} q_x}$$

or

$$q_x = m_x - \frac{1}{2} q_x m_x$$

i.e.,

$$q_x = \frac{m_x}{1 + \frac{1}{2} m_x}$$

$$= \frac{2 m_x}{2 + m_x} \tag{4.7}$$

Thus q_x can be estimated from (4.7) by replacing m_x by its estimate M_x. While this formula (4.7) could be used for q_x for $x = 3, 4, \ldots$, an alternative formula is suggested for q_x for $x = 0, 1, 2$ (even upto $x = 5$) because the age-specific death rates m_x, for $x = 0, 1, 2, 3, 4$ obtained from census data as such are not considered reliable. Further the assumption that deaths occur uniformly in an age-interval is not valid for the early ages, specially for age 0. It is found that mortality which is generally high immediately after birth and during the first few weeks after birth shows a rapid decline thereafter. For the same reasons, the formula (4.2) for L_x is not used for $x = 0, 1, 2, 3, 4$ (at least upto $x = 2$).

There are elaborate methods for estimating mortality rates at ages under five years, i.e. for estimating q_x for $x = 0$ upto $x = 4$. So also for the calculation of L_x for $x = 0$ upto $x = 4$. We shall not discuss them here.

Note: The mortality rates q_x play a very pivotal role in the construction of a life table. As such the validity and utility of a life table depend, to a large extent, on reliable estimates of q_x. Now q_x is defined as a probability, whereas from empirical data one can get proportions or rates. Another principal difficulty is that whereas a life table deals with a cohort with exact ages (i.e. age-year), actual empirical data are available on yearly basis (in general, calendar-year) and rates are calculated for events occurring during a year.

13.4.4 Complete and Abridged Life Tables

By a complete life table is meant a life table where values of the functions l_x, q_x etc are presented for every integral value of x, i.e. for every completed age. Construction of such a life table is quite laborious. An abridged life table is one with values for the functions computed at every 5th or 10th age. Such an abridged life table meets the requirements of considerably large group of workers, such as those involved with vital statistics, population problems, public health, etc. Several short-cut methods have been put forward for constructing abridged life tables. Mention may be made of King's method, Greville's method and Reed-Merrell method.

For construction of an abridged life table, estimates of ${}_nq_x$ are needed. Reed and Merrell have arrived at such estimates as functions of ${}_nm_x$. They have also constructed tables of ${}_nq_x$ for observed values of ${}_nm_x$ for $n = 3, 5$ and 10. With the help of these tables it is possible to compute an abridged life table very easily and rapidly. The method of Reed-Merrell has been widely used for construction of abridged life table by public health workers and others.

Indian Life Tables

In the absence of accurate life and death data, life tables in India have been prepared on the basis of age distribution from each census since 1872–81 (except for 1911-21 and 1931-41). As reliable registration data on mortality rates by age have not been available, the census actuaries had to adopt improvised methods for construction of life tables.

In the table given below we note the expectation of life in India.

Table 13.4 Expectation of Life (in India) at Ages 0, 20, 40, 60

Period	At Ages							
	0		20		40		60	
	Male	Female	Male	Female	Male	Female	Male	Female
1921-30	26.9	26.6	29.6	27.1	18.6	18.2	10.3	10.8
1941-50	32.5	31.7	33.0	32.9	20.5	21.1	10.1	11.3
1961-70	46.4	44.7	41.1	39.9	25.9	25.4	13.6	13.8
1996	62.4 (combined)							

Source: Census Actuarial Tables

It can be seen that there is a steady increase in the expectation of life at all ages. Estimates of expectation of life at birth for a few countries are given in the Table below.

Table 13.5 Expectation of Life at Birth (1986 Estimates)

Country	Male	Female	Country	Male	Female
India	56.2	57.0	Japan	75.1	80.8
Bangladesh	50.2	49.2	China	67.6	70.3
Nepal	48.0	46.5	Singapore	69.9	76.2
Pakistan	52.4	50.6	Indonesia	53.9	56.7
Sri Lanka	68.0	71.2	Malaysia	67.0	71.2
			Australia	72.4	79.4

Source: 1986 ESCAP Population Data Sheet

13.4.5. Applications of the Life Tables

It is possible to make further analysis of the life table by cause of death. Suppose that the total number of deaths d_x can be broken up into components according to the cause of death, i.e.

$$d_x = d_x^{(a)} + d_x^{(b)} + \dots$$

where $d_x^{(a)}$ is the number of deaths of age x who die in the age-interval $(x, x + 1)$ due to cause (a) and so on. When this additional detail is available, one can compute the probability of ultimately dying to any specified cause and also estimate the increase in the expectation of life by eliminating completely the effect of any specific cause of death.

From the life table it is possible to find probabilities of some more complex events. We can find the probability at *birth* of dying in any specified age-interval $(x, x + 1)$ by dividing d_x by the radix (or cohort). Similarly, when a break up of d_x into specific causes is available, the probability at *birth* of dying in the age-interval $(x, x+1)$ due to cause (a) is obtained by dividing $d_x^{(a)}$ by the radix (or cohort). One can compare the probability at birth of dying due to different causes. Relative mortality by age and cause as also by race and community can be studied.

The information that can be extracted from a life table are useful to health workers, and those dealing with vital statistics and population studies, in general. These are useful to the Government for planning purposes and for determination of superannuation benefits.

The life table is of very great importance to life insurance agencies; it was originally developed to meet their needs. The calculation of premiums to be paid for a sum assured is based on life table data. The insurance and census actuaries have played a vital role in the development of life table studies.

13.5. FERTILITY AND ITS MEASUREMENT

The crude birth rate is a measure which indicates at what rate births have increased the population over the course of the year considered. There is a more precise concept in demography, called *fertility* which refers to the *actual production* of children. *Fertility* is to be distinguished from *fecundity* which refers to the *capacity to produce* or to bear children, irrespective of whether or not children have been produced; fecundity refers to the physiological potential and there is no apparent way of measuring fecundity. The capacity to bear children is possessed by only a section of the population, namely the female population and that too only in the reproductive age-group. The reproductive age-group is usually taken to range from 15 years of age to 50 years of age.

13.5.1. Measures of Fertility

General Fertility Rate

Crude birth rate gives annual birth rate per 1000 of total population P, and its level depends on the composition of the total population. A more effective measure of fertility is obtained by relating birth to the total female population in the reproductive age-group instead of the total population.

Thus general fertility rate (GFR) is defined by

$$\text{GFR} = \frac{\text{Number of births during a year}}{\text{Mid–year female population in the age group (15–49)}} \times k \qquad (5.1)$$

where the radix k is usually taken to be 1000. Births include only live births.

Specific Fertility Rate

The GFR takes into account the sex-composition and also the age-composition in a very broad way, covering the entire reproductive age-group 15-49. The reproductive age-group 15-49 can be divided into further subgroups of age interval of 1, 5 or more years. One then gets age-specific fertility rates which would give a better idea of the fertility situation of a community. The age-specific fertility rate f_x for women aged x is defined as

$$f_x = \frac{\text{Number of births to women aged } x \text{ during a year}}{\text{Mid}-\text{year population of women aged } x} \times k \qquad (5.2)$$

Women aged x refer to women in the age-interval $(x, x + 1)$ of 1 year. If $(n + 1)$ –year age-interval $(x, x + n)$ is considered, one can likewise define ${}_nf_x$ for women aged x to $x + n$. Thus

$$ {}_nf_x = \frac{{}_nB_x}{{}_nW_x} \times k \tag{5.3} $$

where ${}_nB_x$ is the number of births to women aged x to $x + n$ during a year, and ${}_nW_x$ is the mid-year female population, i.e., the number of women aged x to $x+ n$. The radix k is usually taken as 100.

When the numbers both in the denominator and numerator in (5.3) are restricted to married women and to legitimate births respectively, then one gets 'age specific marital fertility rates'. When restricted to any particular segment like race, religion etc., one gets age specific rates corresponding to that segment.

The study of age pattern of fertility throws considerable light on several aspects of reproduction for that community or segment of population. It gives indication of the age when women start participating in the reproductive process, of the tempo of child bearing and of the average family size etc.

The curve of f_x against x (or of ${}_nf_x$ against x), called the fertility curve, is more or less similar in shape for different communities, races etc. It is positively skewed. In India the curve starts from a low point, rises rapidly to a peak around the age groups 20-29 and then declines steadily thereafter. This can be seen from the following data obtained from a survey.

Table 13.6 Age specific fertility rates: india, 1997

Age Group x to $x + n$, $n = 4$	Age specific fertility rate (per woman, $k = 1$)		
	Rural	Urban	Combined
15-19	0.061	0.032	0.054
20-24	0.242	0.178	0.226
25-29	0.200	0.152	0.188
30-34	0.122	0.071	0.109
35-39	0.063	0.029	0.055
40-44	0.030	0.012	0.026
45-49	0.009	0.003	0.008
Total	0.727	0.477	0.666
TFR (15-49)	3.63	2.38	3.33
CBR (Per 1000 population)	28.9	21.5	27.2

Source: SRS, Registrar General, India

There are several factors which influence the age-specific fertility rates and thus the shape of the fertility curve. The important factors are : female age at marriage, incidence of widowhood (also divorce, separation) among women, and the extent of the adoption of birth control and family planning methods etc.,

Total Fertility Rate

Though age-specific fertility rates reflect the fertility experience of a community in a precise manner, these cannot be readily used in comparing the fertility experience of two regions, or of two communities nor of the same region at two different periods. For spatial and temporal comparison, the age specific fertility rates are combined into a single index or measure. This is called the *total fertility rate* (TFR). The TFR is the sum of the age-specific fertility rates, when the rates are available for each single age. Thus

$$TFR = \sum_{x=15}^{49} f_x \tag{5.4}$$

per 1000 women (taking $k = 1000$).

If the age-specific fertility rates are available only for five year age-groups (15-19,20-24,...; $n = 4$), then the TFR is the sum of these rates multiplied by 5. Thus

$$TFR = (n+1)\sum_{x} {}_nf_x \tag{5.5}$$

per 1000 women (taking $k = 1000$).

From Table 13.6, we find that TFR for rural and urban components would be about 5 × 0.727 = 3.63 and 5 × 0.477 = 2.38 respectively per woman (k = 1). Combined TFR is 3.33. A replacement level, at which, on an average, each woman is replaced by one daughter, occurs at TFR 2.1 (appro.). India is long away from reaching this.

The TFR for some countries are given below.

Table 13.7 TFR for some Countries (1986)

Country	TFR	Country	TFR
India	3.9	Japan	1.8
Bangladesh	5.7	China	2.2
Nepal	6.1	Singapore	1.6
Pakistan	5.5	Indonesia	3.7
Sri Lanka	2.8	Malaysia	3.8
		Australia	1.9

Source: 1986 ESCAP Population Data Sheet

The TFR is a hypothetical figure. The figure indicates the number of children that would be born per woman if the females in the reproductive age span 15-49 were subjected to the observed age-specific fertility rates and none died in the reproductive period 15-49. The TFR is independent of the age distribution of the female population within this period.

13.6. REPRODUCTION RATES

Gross Reproduction Rate

The TFR indicates the number of children, male or female, which a female can expect to produce during her life time, provided she is not subject to mortality conditions over her child-bearing period. We can obtain another measure by restricting the births in the specific fertility rates to female births only, i.e. a modified TFR in terms of female births only. This modified measure is called *Gross Reproduction Rate* (GRR). The GRR gives the number of *daughters*, which a female can expect to produce during her life time if she is subject to the observed fertility rates during her child bearing period and if she lives right through this period. The GRR is a measure of the mean number of female children which would be born to a new-born female who is subject to the given fertility conditions but *not* subject to mortality. The GRR, like the TFR, is a hypothetical figure.

With $k = 1$, the GRR is

$$\sum_x \frac{F_x}{W_x} \qquad (6.1)$$

where F_x is the number of female births to women aged x during a year and W_x is the mid-year population of women aged x.

$\frac{F_x}{W_x} = r_x$ gives the age-specific gross reproduction rate, for women aged x (r_x is the number of female children born per woman aged x).

When age intervals $(x, x + n), \ldots$ are considered, then with $k = 1$,

$$\text{GRR} = (n + 1) \sum_x \frac{{}_nF_x}{{}_nW_x} \qquad (6.2)$$

$$= (n + 1) \sum {}_nr_x , \qquad (6.2a)$$

${}_nr_x$ being the age-specific gross reproduction rate for women aged x to $x + n$.

Even when ${}_nF_x$ are not available but only ${}_nB_x$ are known, i.e. break up of births in the age group $(x, x + n)$ according to male and female components are not available, but the sex-ratio at birth is known, the GRR can be estimated fairly accurately by multiplying the TFR by the proportion of female births to all births (male and female). Thus

$$\text{GRR} = \text{TFR} \times \frac{\text{Number of female births}}{\text{Total number of births}} \qquad (6.3)$$

It is assumed that the sex-ratio at birth is more or less the same over all ages of mothers.

For India, sex-ratio at birth (number of male babies per 100 female babies) is approximately 107 according to 1981 census. It is slightly high for urban population. Taking the sex-ratio to be 107 : 100, we get the estimated GRR's

$$4.56 \times \frac{100}{100 + 107} = 4.56 \times 0.483 = 2.20$$

and
$$3.29 \times \frac{100}{100 + 107} = 3.29 \times 0.483 = 1.59$$

for rural and urban population respectively. These, however, appear to be underestimates of GRR.

Net Reproduction Rate

The GRR does not take into account the mortality situations. As such the GRR appears to present overestimates of population replacement and population growth; the population growth depends on the balance between fertility and mortality when the effect of migration is reglected. A more appropriate measure can be obtained by combining both mortality and fertility situations. A modification of GRR with mortality component built into it serves such a purpose : the corresponding measure is called *Net Reproduction Rate* (NRR).

Suppose that we have a life table for females only, constructed on the basis of mortality rates of females only (obtained as estimates from observed age-specific female death rates). Let the cohort be l_0. The column L_x gives the mean cohort size at age x. Then $r_x \cdot L_x$ gives the number of female children that would be born to the (mean) cohort L_x at age x (i.e. to the total women living at age x). Thus $\sum_x r_x L_x$ is the total number of female children that would be born to the total cohort l_0 during their life time. The mean number of female children that would be born per newly born female during her life time is

$$\frac{\Sigma \, r_x L_x}{l_0} \tag{6.4}$$

which is the net reproduction rate.
Thus

$$\text{NRR} = \sum_x r_x \frac{L_x}{l_0}$$

$$= \sum_x \frac{F_x}{W_x} \cdot \frac{L_x}{l_o} \tag{6.4a}$$

where x ranges from 15 to 49 (i.e. reproductive age span for women). Now $\frac{L_x}{l_0} < 1$ for each x and thus

$$\text{NRR} = \sum_x \frac{F_x}{W_x} \cdot \frac{L_x}{l_0}$$

$$< \sum_x \frac{F_x}{W_x} = \text{GRR} \tag{6.5}$$

In other words the NRR cannot exceed the GRR. This is evident also from the fact that NRR is modified GRR with mortality situation built into it. The difference between the GRR and the NRR indicates the extent to which mortality affects fertility in the replacement of population.

When female life table is available for age groups x to $(x + n)$, the formula for NRR given in (6.4) can be suitably adjusted as in the case of GRR.

The NRR gives the mean number of female children which would be born to a newly born female during her life time under the given fertility and mortality conditions throughout her life time. The NRR is also a hypothetical figure. It gives the replacement potential of the population.

If the NRR = 1 for a community, then each female, during her life time, would on the average, produce one female to replace herself. In such a situation, the population would be said to possess a tendency to remain constant in size. That is, the population would ultimately become stationary. If NRR > 1, the population will ultimately increase, because each female on her death will have left more than one daughter to replace her. If NRR < 1, the population will ultimately decrease, and will ultimately die out (unless the fertility and mortality change). The NRR for countries of the world ranges from 0 to about 5.

The GRR and the NRR are hypothetical figures, the NRR being a measure of replacement potential of the population. These are hypothetical because these indicate what *would* happen if certain conditions continue to hold. These cannot indicate what *will* actually happen, since the conditions may actually change over time.

13.7 SOURCES OF DEMOGRAPHIC DATA IN INDIA

The main sources of demographic data are (A) population censuses, (B) civil registration, (C) sample surveys, and (D) family welfare programme data. Of these population censuses are the most comprehensive sources of data.

A. Population Census

The first systematic census on all India basis, referred to as the 1872 census, was spread over the period 1867-1872 in different parts of India. The first census was conducted in 1881 on a uniform basis over the entire country. Since then decennial censuses have been undertaken regularly at 10–year intervals. The reference date for 1981 was the sunrise of 1st March 1981. There are two methods of census enumeration : *de juré* method and *de facto* method. According to *de juré* method, persons are enumerated on the basis of their usual place of residence, whereas according to *de facto* method persons are enumerated according to their presence in the place of enumeration. In India *de juré* method is used. Since 1961, a uniform method of house listing has been adopted, the purpose being to assist the enumerator in covering the entire jurisdiction without omission or overlapping.

The coverages of items in the census are as follows:

(a) Population Census : This covers

(i) Demographic and personal characteristics,

(ii) Educational characteristics,

(iii) Migration characteristics, and (iv) Economic characteristics.

(b) Housing Census: This covers

(i) Type of house with material used, (ii) Purpase for which used, (iii) Number of persons residing in the household, and (iv) Other relevant information.

Further there are ancillary studies in censuses covering such areas as (i) village surveys, (ii) urban and town studies, (iii) ethnographic notes, (iv) scientific and technical personnel. See Section 2.4 for 1981 Census details.

The Census of India 1991

Some of the salient features of the Census of India, 1991 are as noted below:

1. House list. Besides name, sex, age, religion, etc., of the inmates of the house, information is also collected on such aspects as number of rooms in the house, material used for construction of the building, whether the house is rented or owned, facility for drinking water, toilet, electricity etc.
2. Enterprise list. Information is collected on the nature of the enterprise, ownership, number of persons employed, whether fulltime or parttime.
3. PG Degreeholders list. This is for information on PG degreeholders and technical personnel.
4. Individual list. This list contains 23 questions on name, sex, religion, mother tongue, economic activity, place of birth, reasons for migration, if applicable, etc. of every individual of the household. For married women there are questions such as age at marriage, number of children, sexwise, ever born and living and whether any child was born to a recently married woman during one year.

The Census of India 2001

In place of the Individual slip and the Household Schedule canvassed for population enumeration in the Censuses of 1961–1991, only the Household schedule was adopted for the Census of 2001; thus there were two schedules–the **Houselist Schedule** and the **Household Schedule.** While Houselistings were done during April–September, 2000, the referral time of population enumeration (through Household Schedule) was 00.00 hours on 1st March, 2001.

As for modifications in the Census of 2001, apart from expanding the scope of the questions canvassed during 1991, some new questions were added in both the schedules. As for the Houselist Schedule, new questions put included : (1) availability of certain assets (such as radio, TV, telephone, bicycle, scooter/ motor cycle/ car etc.,) and (2) whether any banking service was being availed of by the household. As for the Household Schedule, modifications questions in the Census 2001 included: (1) age at marriage for males also, (2) expansion of scope of work, (3) inclusion of tea, coffee, rubber, coconut and betelnut under 'plantation' and the other crops under 'cultivation', (4) a new response category 'moved after birth' included on 'reasons for migration, to bring out additional migration patterns, and (5) number of children born alive, separately for male and female children, to currently married women during the last one year.

The new questions in the Household Schedule in the Census 2001, included: (1) travel to place of work mainly for non agricultural activities (2) for land under cultivation/plantation: net area sown, net area of land irrigated, and tenure status, (3) questions on partial or total disability; further (4) the signature or thumb impression of the respondent of the Household Schedule taken for the first time in the history of Indian Census. Another new feature of the 2001 Census is to bring forth data on socio-economic parameters like literacy and sex ratios for religious groups while in earlier censuses information on religious groups was limited to their population size alone.

The Provisional Population Totals (Census Paper I of 2001) were made available in March, 2001 itself, while the final figures and more details (including breakup regionwise) were made available in 2004. The final figures are as given below:

		2001	1991
Population of India	Total	1,028,610,328	846,302,688
	Males	532,156,772	439,230,458
	Females	496,453,556	407,072,230
Growth of Population (Percent)	Annual	1.96	2.16
(1991-2001)	Decadal	21.5	23.85
Literacy Rate (aged 7+)	Total	64.8	52.21
	Males	75.2	64.18
	Females	53.6	39.29
Sex Ratio (Females per 1000 males)			
Overall		933	927
Child Pop. (0-6 age group)		927	945

Scheduled Caste and Scheduled Tribe Population Constitute 16.2% and 8.2% respectively of the total population. As against the global population in 2000 of 605.5 crore, India's population of 100.0 crore (May 2000) was the second to China's (127.76 crore). While the annual growth rate (1991–2001) for India was 1.96%, China registered a much lower growth rate of 1.00% (1990-2000). India's population would touch 146 crore by 2035 (that would exceed that of China and India would then be the most populous country on the earth).

B. Civil Registration

India has been collecting data on vital statistics for over a century. The Registration of Births and Deaths Act, 1969 provides for compulsory registration of vital events and issue of certificates, which are now being insisted upon for a number of purposes. At the urban centres the municipal authorities are responsible for registration, while in rural areas the same is done through health, police, revenue and panchayat.

C. Sample Surveys

Data collected through census and civil registration do not provide enough detail for estimation of vital rates. Periodic surveys are undertaken for reliable estimation of such rates. A survey by NSS on population, morbidity, family planning introduced in 1973-74 is being repeated every five years.

Sample Registration Scheme (SRS) is an important source of information which provides data for fertility and mortality. This has for basis a dual reporting system; a continuous enumeration of births and deaths by a local part-time enumerator and an independent survey by a full-time supervisor. The data obtained through these two systems are compared and unmatched data are re-verified in the field. Suitable sample designs are adopted for the purpose.

See also section 14.10.

D. Family Welfare Programmes

There are various forms and registers for compiling data on service statistics generated by family welfare programmes. The data also cover demographic aspects of target couples. The demographic data collected through these processes are subject to coverage and response errors. To a more or less extent this is bound to happen everywhere. A country like India, with so much diversity, with such low educational level and with so many other problems, is naturally no exception. However, in recent times several techniques have been developed and adopted to test the accuracy of the data and to take corrective measures. See National Family Health Survey (NFHS-II), 1998-99, IIPS, Mumbai (for interesting sample population data).

13.8. POPULATION GROWTH

13.8.1. Factors of Population Growth

Population growth is analysed by demographers in terms of four factors: fertility, mortality, immigration and emigration. Immigration did play and to some extent still plays an important role in the growth of population in U.S.A. Immigration across the international border in Eastern India (to Assam and other parts of N.E. India) is an important factor to be reckoned with. However, one would presume that the factors, immigration and emigration (except as the result of partition), have not played much significant role in population growth of India as a whole—the two major contributing factors being fertility and mortality.

As simple measures of fertility and mortality one can take crude birth and crude death rates respectively; their difference, the rate of natural increase can be taken as the composite measure.

Now the death rate the world over has declined significantly due to improved sanitation, better health and medical care (including techniques of epidemic control and invention of life saving drugs etc.); while the decline is very marked in developed countries, it is appreciable even in the rural areas of developing countries. As for birth rate, there has been very sharp decline in developed countries—the main contributing factor behind the significant decline being overall social and economic development following the industrial revolution. However, in developing countries birth rate has not been falling very much, because of various reasons including low level of development. The technology of death control is much easier to administer. Thus, while the rate of natural increase is very low or near zero in developed countries, it is significantly high in developing countries.

13.8.1.1 *Curves of population growth*. In section 12.5 we have discussed some curves suitable for fitting population growth. The logistic curve is one which would give a good fit for populations growing under constraints of food and space (spatially limited), but without any biological restrictions on reproduction.

The shape of the logistic curve is characterized by three phases : the first phase shows little change, the second phase shows rapid upward climb and the third phase shows slowing down of change to reach a stationary stage.

13.8.2. Theory of Demographic Transition

Based on the experience of the countries of Northern and Western Europe (countries which were able to reap great dividends from Industrialisation) an interesting theory of social science was developed, the main exponents of the theory being Landry, Thompson and Notestein. It is to be noted that there is a sharp distinction between theories of social science and those of pure experimental science. In case of the latter, laboratory experiments, under perfectly similar and controlled conditions, could be carried out any number of times (over time and space) to establish and verify theories of experimental science. Thus scientific truths and technology have universal validity. Situations are entirely different in case of investigations relating to a social science. Scholars of social sciences, while making efforts to enunciate a theory or principle, do not have that kind of advantage of being able to repeat an observation under similar conditions. Their analysis is based on historical processes involving demographic, economic as well as cultural changes. Though such analyses or theories, based on experiences of some countries at some length of the continum of time, cannot be expected to have universal or all-pervasive validity, attempts are made to explain social and economic phenomenal change on the basis of such theories. The results are often very encouraging and satisfying.

We now discuss below an interesting theory of social science.

Theory of Demographic Transition

Growths of population of countries can be observed to pass through some more or less well defined phases as described below :

1. *The pre-transitional phase.* During this phase, both birth and death rates remain high and are nearly equal; the rate of natural increase is negligible and so is the growth of population. The population remains more or less stationary.

2. *The transitional phase.* During this phase, the birth rates remain on the high side or decline slowly, while the death rates decline more sharply. As a result, the rates of natural increase are steady and over a period of time the growth of population is steady but substantial. The population grows fast, which leads to what is known as '*population explosion*'.

3. *The post-transitional phase.* During this phase, both the birth and death rates decline to more or less to comparable levels. The rates of natural increase remain small or low and the growth of population is very low or near zero.

Population growths of many developed countries follow the above pattern and the growths can be explained or analysed on the basis of the above theory. Many eminent scholars have attempted to apply the above theory for the analysis of the growths in developing countries, including India. We discuss India's case later.

13.8.3. Demographic Scenario of India

Population of India which stood at 238 million in 1901, rose to 251 million in 1921 and finally grew to 843 million in 1991 (see Table 3.1 for population of other Census years). The high growth of population can be viewed as falling into three major phases or stages. The first phase covers the period upto 1921. During 1901-11 and 1911-21 both the birth and death rates were high and nearly equal. The annual rate of natural growth per cent was 0.56 during 1901-11 and −0.03 during 1911-21. The increase of population was small, 13 million during 1901-21.

The second phase covers the period 1921-51, when both the birth and death rates were declining. The annual rates of natural increase remained on the modest side, being 1.05 per cent during 1921-31, 1.34 during 1931-41 and 1.26 during 1941-51. The increase of population was 110 million during the period.

The third phase covers the period since 1951. Since 1951 there has been a sharp decline in death rates while the decline in the birth rates has been lower. The annual rate of natural increase per cent was 1.98 during 1951-61, 2.24 during 1961-71, 2.25 during 1971-81 and 2.11 during 1981-91. The increase in population, which is 482 million, is quite high. The year 1921 from which the growth turned from negligible to modest is called 'the Great Divide'; so also is the year 1951, from which the growth turned from modest to substantial.

(See Fig. 3.6 and Table 3.11).

13.8.4. Indian Situation vis-a-vis Theory of Demographic Transition

Some of the renowned scholars such as Kingsley Davis, Gunnar Myrdal and Dudley Kirk have attempted to analyse the demographic situations in India and other developing countries in the light of the theory of demographic transition. There has been criticism to the application of the theory of demographic transition, as it is, in the Indian situation. It is evident that the theories of social science do not apply quite strictly in different times and under different cultural milieu. However a broad analysis in the light of the theory is not quite out of place.

India may be said to have been in the first stage of demographic transition, that is, the pre-transition stage upto 1921 or maybe upto a period of two to three decades thereafter. The average annual rate of natural increase since 1951 has been 2 per cent or over. India may be said to have entered the second phase of demographic transition, that is, the transitional phase in the middle of this century. India is still at this stage of demographic transition, it would appear. With the rate of natural increase still on the high side and with little prospect of the same coming down (as born out by present trends), India seems to have been caught up in the second stage of demographic transition, in what may be termed, to have got into a demographic trap. Is the third stage of demographic transition, that is, the post-transitional stage still a longway off, that is the question.

It may be mentioned in this connection that some other Asian countries were moving faster from the second towards the third stage of demographic transition. Growth has been checked to very low level, for example, in the Republic of Korea.

Table 13.8. Trends of vital statistics, India

	1951	1981	1991	1997
CBR	39.9	33.9	29.5	7.2
TFR	6.0	4.5	3.6	3.3
IMR	146	110	80	72
CDR	27.4	12.5	9.8	8.9

Among the major states, Kerala has the highest life expectancy at birth (72.9) and literacy rate (86.17) and the lowest IMR (16), CDR (6.4) and CBR (18.2) in 1991.

13.9. CONCLUDING REMARKS

13.9.1. Malthusian Theory of Population

Thomas Robert Malthus (1766-1834), an English clergyman was the first to develop a theory of population and to draw pointed attention about population growth. His book *Essay on the Principle of Population,* published in 1798, has been influencing the thinking of people on this issue for quite sometime. His theory leans heavily on the law of diminishing returns. His thesis was that a population, if its growth is unchecked, would *tend* to grow in geometric rate (or geometric progression); as a result a population would double itself in not too long a time. However, the land and natural resources available, per capita, of the growing population would dwindle and would be less and less, further, because of the law of diminishing returns, food supply would *not tend* to keep up with the geometric rate of population growth. Population tends to increase faster than the means of subsistence and that its growth could only be checked by moral restraint or by disease and war. It may be noted that he mentioned only about *tendencies.*

Malthus' views are today considered to be gross oversimplifications. He never anticipated the miracles of the industrial revolution nor of the green revolution. The American wheat scientist Norman Ernest Borlaug, N.L. (b. 1914) is the person behind and is responsible for the green revolution which is transforming the agricultural prospects of the developing countries of the world; his invention has gone a long way to alleviate the world from the somewhat 'gloomy' prospect that Malthus envisioned. However, the theory of Malthus is very much relevant even today and is useful in understanding the pattern of population growth in several developing countries of the world.

13.9.2. World Population Scenario

Malthus and Mother Nature

The World population which was 3,632 million in 1970, rose to 4,000 million in 1975 and to about 5,000 million in 1985. The 'most likely projection' of population growth implies that the world population would grow to 5,480 million by mid-1992. According to forecast based on medium projection, the population would be 8,500 million by 2025 and to 10,000 million by 2050 (which is almost double the population as of today). On the high projection, population would grow to 1,25,000 million by 2050 and to 20700 million a century later.

Much depends on action to reduce family size during the next decade. Delay could make a difference upto a staggering 4,000 million, by 2050; this difference is almost equal to the whole world population in 1975. The world population growth is much more rapid now than ever before in the history of the world.

Apart from food, there are other serious problems posed by rapid growth of population. Consequences for development prospects and for the environment would be very serious, if not catastrophic. Large population growth would threaten the rain forests and would lead to ecological imbalance (by wiping out several species useful to living beings); it would lead to thinning of the ozone layer causing global warming. In fact, the quality of human life is intimately linked with the quality of the environment. Both are inseparable from the context of human numbers and concentrations.

The State of the World Population Report issued annually by the United Nations Fund For Population Activities (UNFPA) have constantly been sending warning signals. In the 1990 Report, it emphasises that developed or developing, the more people the more pollution. The top billion people of the industrialised countries have been utilising the largest share of the resources and have been creating the greatest amount of wastes. These countries are, the Report says, overwhelmingly responsible for the damage of the ozone layer, and acidification as well as for two-thirds of global warming. The bottom billion, in developing countries, are responsible for damaging the environment through deforestation and land degradation, the Report says. Action in three major areas are recommended : (1) A shift to cleaner technologies; (2) A direct and all out attack on poverty itself; (3) Reductions on the overall rates of population growth. The report argues that investment in human resources provides a firm base for rapid economic development. In the past, however, it has received a lower priority that agriculture, industry or military affairs.

The 1992 Report (released in March 1992), UNFPA reveals fresh information on population growth (some of which are indicated above in the projections), economic growth as well as on impact of population on environment. The Report calls for "a sustained and concerted programme, starting immediately" with the objective of reducing population growth. The emphasis is on Women's Education : Women's access to better education, health care and family planning and welfare programmes. A reduction in family size can, on the other hand, make direct contribution to better education, health care and so on.

Fixed or non-renewal resources have so far been able to meet the demand. New reserves have been found and new technologies have been developed. But the simultaneous effect of population explosion and increasing consumption level, per capita, might jeopardise the adjustment mechanisms. Rising population and consumption levels, it warns, will impose severe constraints not only on development but on the availability of the basic human need like water. The need of the hour is to act decisively to slow population growth and to protect the environment and the Mother Nature.

EXERCISES-13

Sections 13.1-13.3

1. Define : vital events, vital statistics rate, vital statistics ratio, dependency ratio.
2. What are crude birth and death rates? On what factors do these rates mainly depend?
3. What are specific death rates? Describe the pattern of behaviour that age-specific death rates generally show.
4. Why are basic measures of mortality adjusted to obtain standardised death rates? Explain clearly the purposes behind computation of standardised death rates.
5. Describe the direct and indirect measures of standardisation.
   Can the two methods lead to the same result? Under what conditions can these methods lead to the same result? Discuss.
6. Describe the merits of using standardised rates in place of crude rates. Explain why would you consider standardised death rates to give a better measure for comparison of mortality situation of two communities.

Section 13.4

7. The following table gives the relevant data for males of Australia, 1967 :

Table 13.8 Australian Males, 1967

Age Group	Mid-year Population	Deaths during 1967
0	117, 700	2, 421
1-4	474, 900	462
5-9	613, 300	268
10-14	568, 200	236
15-19	538, 300	698
20-24	477, 000	806
25-29	399, 400	609
30-39	756, 500	1504
40-49	755, 000	3792
50-59	609, 100	8139
60-69	387, 400	13282
70-79	195, 500	15393
80–	57, 000	10468

   Find the crude death rate and age specific death rates.
   Find estimates of age specific mortality rates. Taking l_0 = 1,00,000, find the number of survivors at subsequent age groups.
8. What is a life table? What are the basic assumptions in its construction? Explain them.
9. Explain the various columns of a life table and the relations between them.
10. Write a note on the construction of a life table.
11. Distinguish between abridged and complete life tables. Name some methods of construction of abridged life table. What are the merits of Reed-Merrell method?
12. Explain the application and uses of a life table. What are its limitations?
13. What do you mean by expectation of life? How is it calculated?
14. What are the basic data required for construction of a life table?
    Do you think that such data are available in India? If not, how are life tables prepared in India? Discuss.

Section 13.5

15. What is meant by fertility and how is it measured? Describe the various fertility rates commonly used. Discuss their relative merits.

16. Define total fertility rate. How is it calculated?

17. Enumerate the factors which influence the age-specific fertility rates.

Section 13.6

18. Define gross and net reproduction rates. How does gross reproduction rate differ from total fertility rate?

19. Show that the NRR cannot exceed the GRR.

20. Explain the importance of NRR as a replacement potential of the population. Interpret the results NRR < = > 1. What would ultimately happen if NRR < 1?

Section 13.7

21. What are the sources of demographic data in India? Enumerate them.

22. Write a note on population census of India. What are the objectives behind the decennial-censuses?

23. Describe the salient features of 1981 and 1991 population censuses in India. List the schedules used for the censuses.

24. Write a note on Civil Registration and sample surveys as sources of demographic data.

Miscellaneous

25. If those babies which survive to a year can expect 60 years more life, but one tenth die in the first year of life what was the expectation of life at birth?

26. The following table is an extract from a hypothetical cohort, with : x = age at last birth day; l_x = no. of persons who attain age x.

x	l_x	x	l_x	x	l_x	x	l_x
50	100	55	86	60	30	65	5
51	99	56	77	61	21	66	4
52	98	57	70	62	12	67	3
53	95	58	67	63	8	68	2
54	90	59	45	64	6	69	1
						70	0

Assume that deaths take place uniformly during each single year.
Estimate the age specific death rates at ages 58 and 60.

27. Estimate the expectation of life at ages 50 and 60 from the data of Exercise 26.

28. Discuss about theories of experimental and social sciences and bring out their essential differences.

29. Explain the theory of demographic transition, with a graph and an illustration.

30. Discuss the scope of analysis of population change in India (also in your country) in the light of the theory of demographic transition.

Examine the cases of the states of the Indian Union, such as Goa and Kerala.

31. PROJECTION OF POPULATION (INDIA)

Technical Group on Population Projections (Registrar General, India), in August 1996, made Population Projections of India for 1991-2016 on the estimated annual exponential growth rates of 1.98 during 1991-96, 1.58 during 1996-2006, and 1.44 during 2006-2016. Based on the above growth rates and the Census Population of 1991 find the projected population of India for 1996, 2006 and 2016.

Chapter 14

Sample Surveys

14.1. INTRODUCTION

In Chapter 8, we have considered some basic concepts related to sampling, such as population, sample, parameter, statistics, random sampling, sampling distribution etc. We have also explained the method of sampling with replacement (SWR) and sampling without replacement (SWOR). Here we shall consider methods of sampling and various questions associated with sampling and sample survey.

The basic idea behind sampling is to draw inference about the population from which the sample is drawn. This time old device is a very common practice. We take a handful of rice from a bag of rice and infer about the quality of rice in the bag. Few drops of blood are taken, then chemical and microscopic examinations are undertaken from which inference is drawn about the blood in the human body.

Sampling methods are extensively used in socio-economic, demographic and various other studies with a view to find out about the characteristics under consideration of the population from which the sample is taken. One of the most important applications of the theory of sampling is the use of sample surveys to elicit specific information or characteristics about the population. In a *sample survey* one observes only a representative fraction, properly selected, of the whole population; one then calculates or draws inference about the characteristics or parameters of the population. A sample survey is to be distinguished from a *census* (or *complete enumeration* or full count) in which every single unit of the population is surveyed and is enumerated. In a census, the whole population is observed whereas in a sample survey only a representative fraction of the population is observed. In most countries of the world, complete count or census of the human population is undertaken regularly. In India, we have the decennial population censuses. For most socio-economic, demographic and other enquiries, however, instead of complete count or census, sample surveys are undertaken to infer about the characteristics of the population.

14.2. ADVANTAGE OF SAMPLE SURVEYS OVER COMPLETE CENSUS

In Chapter 2 we described briefly the advantages of sample method over census method.

The considerations for which a sample survey is preferred to a complete census are discussed more elaborately below.

1. *Practicability*. In certain cases, a complete count may be impracticable, e.g. for examination of blood of an individual, for examination of quality of wheat in a big bag, for examination of quality of a product manufactured in a factory and so on. A sample study only would be practicable in such a situation.
2. *Scope and speed*. In general, a sample survey has a greater and wider scope than a census. There could be wider coverage in a sample survey. A survey could be done with greater speed. The data may be collected and analysed in a shorter duration of time. Sometimes some information may be urgently needed for some government policy decision or for market study by some commercial organisation in respect of market or brand choice of a product. A sample survey would be more helpful than a census in such a situation.
3. *Quality and accuracy*. For sample survey better trained personnel could be employed and greater care could be taken for collection of data as well as for supervision. An enumerator may be able to spend more time in filling up a schedule and thus he may be able to elicit more information; this would be helpful in arriving at the accurate situation. Thus data collected would be more accurate. Better job performance could be expected and ensured in a sample survey than in a census. Errors of certain types could be minimised in a sample survey.
4. *Cost*. A sample survey will entail reduction of cost since in a sample survey only a fraction of the whole population is covered. The costs associated with an inquiry can be broken down into
   (i) overhead cost of organisation,
   (ii) costs of data collection and supervision,
   (iii) costs of processing and analysing data, and
   (iv) cost of publication of material collected.

The costs under (ii) and (iii) are of more variable nature, and will be much smaller in a sample survey than in a census. Even though the cost of data collection per unit covered may be higher in a sample survey, the total cost of all units covered would be much lower in a sample survey. Thus it would be possible to carry an inquiry through sample survey with less cost.

In spite of the fact that a sample survey has some very real advantages over a complete census, it must not be taken that a census should never be undertaken. A sample survey must give reliable results : reliability aimed at cannot be sacrificed because of other advantages. Sometimes it may not be possible to select a satisfactory sample, then a census is to be taken. Apart from this, whenever complete information is required, census must be taken. Population censuses at regular intervals (e.g. every tenth year in India) are taken in most countries. When time and money are not important factors or there are more compelling reasons for obtaining complete and accurate information, a census is to be taken.

14.3. ERRORS IN SAMPLE SURVEYS

There are two types of errors which may creep up in a sample survey. These are (i) sampling errors, and (iii) non-sampling errors.

Sampling Errors : A sample covers only a fraction of the population from which it is drawn, and is a sort of substitute for the complete census. The information gathered through a sample survey is used for drawing certain inference about the characteristics of the population. For example, inference about the population *parameters* are drawn from the corresponding sample *statistics*; e.g. the population mean is estimated from the sample mean. While the population mean is fixed for a population, the sample mean will vary from sample to sample. A discrepancy will arise between the population parameter and the sample statistics. The error thus introduced by the discrepancy is called sampling error.

There are two factors which lead to sampling error : (i) sampling bias, and (ii) chance factor. Sampling bias is caused by the method of selection of the sample; when the method of selection is defective, discrepancy between population parameter and sample statistics will be there and will persist even when the sample is large. Suppose that we are interested in the mean number of children in a family in a community. If the sample is taken from families where females are educated (and also work) then the mean number of children per family obtained from the sample will not give a reliable estimate of the mean number of children per family in the community, even when the sample is large. The sampling bias is due to selection of a sample which does not truly represent the population from which it is drawn. The error due to sampling bias could be eliminated by the use of proper sampling method, i.e. by taking a representative sample. Even when sampling bias is absent, discrepancies will occur due to chance, as a sample will never be the same as the population from which it is drawn and thus will never reproduce exactly the characteristics of the population.

Measure of sampling error. The question arises as to how to measure the sampling error. Consider that a sample survey is taken to estimate certain characteristics of a population such as mean, aggregate, proportion and so on. An estimate of the characteristics can be obtained from the corresponding sample statistics. Suppose that we are interested in estimating the proportion of educated unemployed amongst youths in a district. A representative sample of youths is taken and the sample proportion of educated unemployed is found. This serves as an estimate of the corresponding population proportion (i.e. proportion calculated on the basis of the whole population of youths in the district). If another sample is taken, the sample proportion is likely to be different from the one obtained from the first sample. Thus if repeated samples are taken we shall get as many sample proportions as the number of samples taken. The sample proportions will have values scattered around the population proportion. The dispersion of the sample proportion around the population value will give a measure of the sampling error. In practice, the standard error of a sample statistic can be taken as a measure of the sampling error.

A comparison of two (or more) methods of selection of samples can be done on the basis of the magnitudes of their sampling errors. Smaller the sampling error, more tightly will the sample statistics cluster around the population value.

Of the two methods of sample selection, the one with smaller sampling error is to be preferred, other things remaining the same. One way of minimising sampling error lies with the choice of proper sampling scheme.

The sampling error will depend on the size of the sample, and will decrease with increasing sample size. The decrease which is rapid at the beginning becomes slow as the sample size increases. In most cases the sampling error varies inversely as the square root of the sample size. As can be seen the sampling error can be reduced by increasing the size of the sample and can be eliminated by taking a complete count or census.

Non-sampling errors. There are various other sources of error which arise in an inquiry. These occur due to procedural biases, which may be described as follows:

(i) Response error : It is due to response bias which has its origin in the vague, inaccurate or wrong answers given by a respondent to the interviewer. There may be a tendency on the part of the respondent to make an under-statement in certain cases (such as income) or to make an over-statement in certain other cases (such as expenditure incurred).

(ii) Observational error : It is due to bias in observation. For example, in making an eye estimate on the extent of damage caused to crops by flood or by a certain plant disease, an element of observational error is likely to occur.

(iii) Error due to non-response : An interviewer may not be able to gather information from a respondent as he may not be available even after repeated calls. Thus information from a section of the respondents may not be available; this section may thus be excluded from the inquiry. For example, working women may be completely left out in an inquiry into attitude towards family planning. Again, sometimes an interviewer gathers information from a neighbour being unable to find the respondent after repeated calls.

Non-sampling errors are likely to occur in any inquiry, be it a sample survey or a complete census. Non-sampling errors can be minimised by proper planning and execution of the inquiry; deployment of competent and properly trained staff as well as effective supervision and management of the various operations involved are some of the steps in this direction.

While the sampling errors tend to decrease with the increase of sample size, the non-sampling errors tend to increase with the increase of sample size.

14.4 VARIOUS STAGES IN A SAMPLE SURVEY

The four principal stages in a sample survey are (a) the planning stage, (b) the execution stage, (c) the analysis stage, and finally (d) the stage of preparation of the report.

(a) The planning stage involves the following steps :

(i) Statement of the objectives of the survey in clear terms.

(ii) Definition of the population to be sampled.

(iii) Definition of the sampling units.

The population should be divisible into *sampling units*, the ultimate unit to be sampled. It should be possible to define unambiguously the sampling units as distinct and non-overlapping units covering the entire population. For example, in a socio-economic survey the sampling unit may be an individual, a family, a household, an institution or a village. One would then need a *sampling frame*, i.e., a complete list of the sampling units in the population. The frame is used for drawing samples. Such a frame may sometimes be available (e.g. a list of households in a town, when a household is the sampling unit). If such a frame is not available, one must prepare a sampling frame.

(iv) Determination of the type of data to be collected. The data should be collected in accordance with the objectives of the survey. When the type or nature of data to be collected is decided upon, the next step would be to frame the *questionnaire* or *schedule* through which the data are to be collected.

A *questionnaire* contains a set of questions to be answered by the respondent himself whereas in a schedule answers are recorded by the investigator or enumerator on the basis of information gathered from the respondent. The points to be taken into consideration while framing a questionnaire or a schedule are discussed in section 2.3.

(v) Determination of the method of data collection. We have discussed the methods of data collection in section 2.2. The objectives of the survey are to be taken into account for determining the method to be adopted. Merits and demerits of different methods are also to be examined. Suppose that the objective of the survey is to find about the appropriateness of syllabus followed vis-a-vis job requirement and employment opportunities of Commerce and Economics Honours graduates passing out from a college over a number of years. Mailed questionnaires are then to be sent to the units (graduates) to be covered in the survey. This will be more practical, as locating the units and visiting them in person would be costly, difficult and time consuming. Further the respondents would be quite competent to answer the questions themselves.

(vi) Selection of an appropriate sampling design. This involves the procedure of selection of the sampling units, and of drawing the sample after having decided upon the size of the sample. Sometimes two or more sampling plans may be available for a sample survey. Then some criterion for choosing one from amongst them will have to be laid down. The criterion would be the margin of sampling error that can be tolerated as well as the cost involved in the survey. One is faced with a sort of 'uncertainty principle' of Heisenberg as in quantum physics. Two options are available. One can determine and fix the margin of sampling error that can be tolerated and then choose the plan which would involve minimum cost. Or one can determine the amount that one would be prepared to spend for the survey and then find the plan which will have minimum sampling error. There can be 'best' design only in relation to a certain criterion, for example, one which minimises cost for a given error or one which minimises error for a given cost. The time limit allowed and other practical considerations will also have to be taken into account in the selection of a sample design.

(vii) Organisation of field work. This involves laying down the plan of field work and training of field staff, such as investigators, supervisors etc. A *pilot survey* is sometimes conducted on a small number of units before actually launching the field work. This is helpful for removing ambiguities, if any, in the schedule as well as for organisation of field work. Some estimates of population parameters could also be had through a pilot survey.

(b) The execution stage involves actual field work, such as identification of the sampled units, and collection of information from them through questionnaires or schedules.

(c) The analysis stage involves the following steps:

(i) Scrutiny of questionnaires/schedules filled. One has to verify that the information collected is genuine and plausible and is free from apparent contradiction. Doubt or suspicion could be removed through 'resurvey' of the particular unit or units where these arise.

(ii) Tabulation. Data collected are to be arranged in tabular form. For small scale survey manual tabulation may be possible while for large scale survey machine tabulation have to be taken recourse to. Now a days computer programs can be drawn up for tabulation. Questionnaires and schedules are to be properly designed keeping in view computerised tabulation.

(iii) Statistical Analysis. On the basis of data available, necessary estimates of the population values are to be drawn up. Relevant characteristic measures are to be calculated from the data collected. Testing of hypotheses in certain cases will also form part of statistical analysis.

(d) Preparation of the report. This is the final stage of the survey. All the information regarding the survey, such as sample design, field work, analysis, have to be presented together with all the data collected as well as all the characteristics and measures calculated from the data collected. The interpretation of the data and the measures, the conclusions arrived at therefrom and the recommendations that can be made form an important part of the report.

14.5. TYPES AND METHODS OF SAMPLING

There are basically two types of sampling – subjective and objective. Any type of sampling in which the sample selected depends on personal discretion or judgement of the investigator is called a *subjective or judgement sampling*. This type of sampling is resorted to with certain definite purpose and is not for general use. It is also known as *purposive sampling*. For example, a survey may be carried out by an investigator interested in opinions and views on certain specific issues (such as family welfare programmes) by sending questionnaires to a sample of knowledgeable persons selected by the investigator on his personal judgement. Such a survey will be a subjective one and will have its own importance, though the sample may not be a representative one. Such an opinion survey is to be distinguished from certain opinion polls (e.g. Gallup poll named after G.H. Gallup (1901-)) conducted on a representative sample of public. Objective sampling may be subdivided as *probabilistic* and *mixed*. Any type of sampling in which every unit of the population sampled has a definite, preassigned probability of being selected is called probabilistic sampling. Any type of sampling which is partly probabilistic and partly non-probabilistic is called mixed sampling. *Probabilistic sampling* is also known as *random sampling*.

14.5.1. Sampling Methods

Some of the different methods of sampling are as given below :

(i) Simple random sampling
(ii) Stratified random sampling
(iii) Systematic sampling
(iv) Multistage sampling
(v) Multiphase sampling
(vi) Cluster sampling
(vii) Quota sampling

We shall confine here to the first three methods of sampling, while the first two are of probabilistic type, systematic sampling is of the mixed type.

14.6. SIMPLE RANDOM SAMPLING

In a *random sampling*, each unit has a definite probability of being selected. If, in particular, each unit of the population has an *equal* chance (or probability) of being selected and included in the sample, we get what is known as *unrestricted random sampling or simple random sampling*. Simple random sampling may be with or without replacement, according as a unit selected is not or is excluded from further selection. Simple random sampling is often loosely termed as random sampling. In sampling theory (as developed in Chapter 8) it is assumed that the population from which samples are drawn is infinite. Theorem 8.1 states that the sample mean $\bar{x}$ based on a random sample of size n from a population (assumed to be infinite) has variance $\frac{\sigma^2}{n}$, σ^2 being the population variance, i.e. has standard deviation $\frac{\sigma}{\sqrt{n}}$. For most situations in real life when we deal with investigations in socio-economic, demographic and other allied fields, the population considered is found to be finite, of size N, say. We shall assume the population to be finite, unless otherwise stated.

In simple random sampling with replacement, (*swswr*) the probability of selecting any unit is $1/N$, N being the population size. In simple random sampling without replacement (*srswor*), the probability of selecting the *first* unit is $1/N$, that of selecting the *second* unit (from the remaining $N-1$ units) is $1/(N-1)$, and so on, the probability of selecting the nth unit is $1/(N-(n-1))$. When N is large compared to n, then $1/(N-1)$, $1/(N-2)$, . . ., $1/\{N-(n-1)\}$ are all approximately equal to $1/N$. Thus when the size N of the population is large in comparison to the sample size n, then *srswor* will be more or less the same as *srswr*. In *srswor*, the number of possible samples of size n is $\binom{N}{n}$, when N is finite.

We shall now consider two aspects of simple random sampling. First, the method of selecting the sample and secondly, the criterion for determining the sample size. Finally, the question of estimation of the values of the population characteristics from the sample data will arise.

14.6.1 Method of Selection of a Simple Random Sample

It is quite easy in principle. All the units in the population are thoroughly mixed up and a sample is drawn at random. A crude way to do this is by using lottery method. In practice, one uses the *tables of random numbers*. Many such tables are available, (i) Tippet's Tables (ii) Fisher and Yates Tables, (iii) Rand Corporation Tables. Such a table can be used as follows. First, prepare a sampling frame, i.e. list the units in the population and number them from 1 to N, (N is the population size). Suppose $N = 5{,}000$ and we wish to take a sample of size $n = 1{,}000$. We take any random number table and choose any page from the table. Starting at any row or column, we write down the numbers in groups of 5 (as N is a five digit number). Excluding the numbers which exceed 5,000 we can get the set of numbers ≤ 5000. The units which correspond to these numbers will be the units to be sampled. Better methods are available for increasing the speed of selection.

14.6.2 Determination of the Sample Size

The next important question is the determination of the sample size for a simple random sample. The sample size is to be determined according to some preassigned degree of precision. The degree of precision can be specified in terms of two criteria as given below:

(i) The margin of permissible error between the estimated value and the population value. In other words, this is the amount of discrepancy that can be allowed between the sample statistics and the population value.

(ii) The confidence limit within which one would desire the estimate to lie.

Suppose, first, that the population is infinite with μ, σ as the population mean and s.d. respectively. Let d be the margin of permissible error and α the level of confidence. Assume that σ is known. Suppose that one is interested in a survey from which the population parameter μ can be estimated from the sample mean $\bar{x}$ with a margin of error $\pm\, d$ and with confidence level α. That is, the two criteria of the degree of precision can be specified by means of the relation

$$P[-d \leq \bar{x} - \mu \leq d] = 1 - \alpha \tag{6.1}$$

or,

$$P[\,|\bar{x} - \mu| \leq d] = 1 - \alpha \tag{6.1a}$$

The sample size n to be determined will be such that (6.1) is satisfied. From Theorem 8.4, we get that, for large n (sample size),

$$Z = \frac{\bar{x} - \mu}{s.e.(\bar{x})} = \frac{\bar{x} - \mu}{\dfrac{\sigma}{\sqrt{n}}}$$

is approximately normally distributed with mean 0 and s.d. 1. Thus given α, we can find from the table of standard normal distribution, the value c which satisfies

$$P[\,|Z| \le c] = 1 - \alpha \tag{6.3}$$

From (6.2), we can write the above as

$$P\left[\frac{|\bar{x} - \mu|}{\frac{\sigma}{\sqrt{n}}} \le c\right] = 1 - \alpha$$

or

$$P\left[|\bar{x} - \mu| \le c \cdot \frac{\sigma}{\sqrt{n}}\right] = 1 - \alpha \tag{6.4}$$

Comparing (6.1a) with (6.4), we get

$$d = c \cdot \frac{\sigma}{\sqrt{n}} \tag{6.5}$$

or

$$\sqrt{n} = \frac{c\sigma}{d}$$

or

$$n = \left(\frac{c\sigma}{d}\right)^2 \tag{6.5a}$$

This gives the sample size subject to given d and α; given α, c can be determined from (6.3) by using table of standard normal distribution.

For example, let d = 10 per cent = 0.10, i.e. a margin of 10 per cent error and the level of confidence α = 0.05. From (6.3), by using table of standard normal distribution, we get c = 1.96. Thus when σ is known,

$$n = \left(\frac{1.96\sigma}{0.10}\right)^2$$

Let α = 0.01, then from the table of normal distribution, we get c = 1.64. Thus

$$n = \left(\frac{1.64\sigma}{d}\right)^2$$

Suppose next that the population size N is finite. Then it can be shown that the standard error (s.e.) of the sample mean is given by

$$s.e.\ (\bar{x}) \equiv s.d.\ (\bar{x}) = \frac{\sigma}{\sqrt{n}}\sqrt{\left(\frac{N-n}{N-1}\right)} \tag{6.7}$$

When the population size N is not very small, we get $(N-1)$ close to N, so that from (6.7) we have

$$s.e.\ (\bar{x}) = \frac{\sigma}{\sqrt{n}}\sqrt{\left(\frac{N-n}{N}\right)}$$

$$= \frac{\sigma}{\sqrt{n}}\sqrt{1-F} = \sigma\sqrt{\left(\frac{1}{n}-\frac{1}{N}\right)} \tag{6.8}$$

where $F = n/N$ is the sampling fraction.

Comparing (6.8) with the value $\frac{\sigma}{\sqrt{n}}$ for population of infinite size, we see that the factor $\sqrt{1-F}$ is introduced when we consider population of finite size N. The factor $\sqrt{1-F}$ is known as the correction factor for finite population.

Thus using (6.8), in place of $\frac{\sigma}{\sqrt{n}}$ in (6.2) we get from (6.5)

$$d = c \cdot \frac{\sigma}{\sqrt{n}}\sqrt{1-F}$$

$$= c\,\sigma\sqrt{\left(\frac{1}{n}-\frac{1}{N}\right)}$$

or,
$$\left(\frac{d}{c\sigma}\right)^2 = \frac{1}{n} - \frac{1}{N}$$

or,
$$\frac{1}{n} = \left(\frac{d}{c\sigma}\right)^2 + \frac{1}{N} = \frac{Nd^2 + c^2\sigma^2}{N(c\sigma)^2}$$

or
$$n = \frac{Nc^2\sigma^2}{Nd^2 + c^2\sigma^2} \tag{6.9}$$

The formula (6.9) gives an estimate of the size n of the sample to be taken.

In particular, when $\alpha = 0.05$, $c = 1.96$ so that

$$n = \frac{N(1.96\sigma)^2}{Nd^2 + (1.96\sigma)^2} \tag{6.10}$$

and when $\alpha = 0.01$, $c = 1.64$ so that

$$n = \frac{N(1.64\sigma)^2}{Nd^2 + (1.64\sigma)^2} \tag{6.10b}$$

We have assumed so far that σ is known. In practice, however, σ will seldom be known and then one will have to use an estimate $\hat{\sigma}$ of σ. This may be found from a previous survey or census, if undertaken earlier. Another way of getting an estimate is from a pilot survey conducted before an actual survey is undertaken.

The above calculation of sample size is in relation to the mean. The method can be applied when parameters or values other than the mean are concerned. When more than one characteristic is involved in an investigation by means of a sample survey, the sample size can be obtained separately for each characteristic on the basis of the amount of error tolerated for each. Then the largest value for the sample sizes will meet the error requirement for all.

14.7. STRATIFIED RANDOM SAMPLING

The simplest type of sampling design is the simple random sampling. Sometimes supplementary information about the population from which the sample is to be drawn is available. The population to be sampled can then be divided into strata or layers on the basis of the supplementary information available from which one may be able to evolve certain criterion for classifying the sampling units into a number of different strata. The criterion, which is also known as 'stratifying factor', may be sex, age group, educational level, economic status, physical dimension, geographical area and so on. The stratifying factor will be effective if it enables one to divide the population into a number of distinct strata such that

(i) the units *within* each stratum are as more or less homogeneous, and

(ii) the units in *different* strata are as unlike or widely varying as possible.

Dividing the population into a number of distinct strata, simple random samples can be drawn from each stratum. The design is known as *stratified random sampling.*

Suppose that the population of size N can be divided into k distinct strata of sizes $N_1, N_2, \ldots, N_k$ such that $\sum_{i=1}^{k} N_i = N$. Simple random samples of sizes $n_1, n_2, \ldots, n_k$ are drawn respectively from the k distinct strata. The sample constitutes a stratified random sample of size $n = \sum_{i=1}^{k} n_i$.

The idea behind using a stratified random sampling design is that under certain conditions, a stratified random sample will give smaller sampling error than a simple random sample of the same size. Consider the case of estimation of population mean from a sample survey.

For a finite population of size N the standard error of the same mean $\bar{x}$ is given by

$$s.e.\ (\bar{x}) = \frac{\sigma}{\sqrt{n}}\sqrt{1 - \frac{n}{N}}$$

That is

$$\text{var}\,(\bar{x}) = \frac{\sigma^2}{n}\left(1 - \frac{n}{N}\right) \tag{7.1}$$

Precision of $\bar{x}$ increases as s.e. ($\bar{x}$) decreases; then values of $\bar{x}$ will cluster more closely. This would happen as n, or the sampling fraction n/N increases and also in case of a design in which one gets a factor which gives a value smaller than σ.

14.7.1 Variance and s.e. of Weighted Sample Mean in a Stratified Random Sample

Suppose that the population of size N is divided into k strata on the basis of certain stratifying criterion. Let

$N_i =$ the number of units in ith stratum so that $\sum_{i=1}^{k} N_i = N$

$n_i = \frac{N_i}{N}$, $i = 1, 2, \ldots, k$; $\Sigma p_i = 1$

$n_i =$ the number of sampling units selected from the ith stratum so that $\sum_{i=1}^{k} n_i = n$ is the total sample size

$\mu_i =$ the population mean of ith stratum

$\sigma_i^2 =$ the population variance of ith stratum

$\mu =$ the population mean in the total population

$\sigma^2 =$ the population variance in the total population

$\bar{x}_i =$ the sample mean of ith stratum

$\bar{x} =$ the sample mean of the total sample.

We have

$$\mu = \frac{\Sigma N_i \mu_i}{N} = \sum_{i=1}^{k} \left(\frac{N_i}{N}\right) \mu_i \tag{7.2}$$

and

$$\bar{x} = \frac{\Sigma n_i \bar{x}_i}{n} = \sum_{i=1}^{k} \left(\frac{n_i}{n}\right) \bar{x}_i \tag{7.3}$$

The sample mean $\bar{x}$ is a weighted average of the stratum sample means $\bar{x}_i$ with weights $\frac{n_i}{n}$. The quantity $\frac{n_i}{n}$ is the sample proportion i.e. the ratio of the ith sample size n_i to the total sample size $n = \Sigma n_i$.

The population mean μ can be estimated by $\bar{x} = \Sigma \left(\frac{n_i}{n}\right) \bar{x}_i$. However, by taking weights as $\frac{N_i}{N}$ in correct proportions according to population sizes, in place of weights $\frac{n_i}{n}$, we can arrive at another weighted average

$$\bar{x}^* = \sum_{i=1}^{k} \left(\frac{N_i}{N}\right) \bar{x}_i \tag{7.4}$$

as an estimate of μ. Of the two estimators (7.3) and (7.4), the estimator $\bar{x}^*$ is better (this can be seen from theoretical considerations, which we will not discuss here; $\bar{x}^*$ is the best linear unbiased estimator of μ). Thus we shall use $\bar{x}^*$ as an estimator of the population mean in a stratified random sample. We shall denote this by the notation

$$\bar{x}_{st} \equiv \bar{x}^* = \sum_{i=1}^{k} \left(\frac{N_i}{N}\right) \bar{x}_i \tag{7.5}$$

We then get

$$\text{var}\,(\bar{x}_{st}) = \sum_{i=1}^{k} \left(\frac{N_i}{N}\right)^2 \text{var}\,(\bar{x}_i) \tag{7.6}$$

Now var $(\bar{x}_i)$ can be obtained from (7.1) considering the stratum i of the population and replacing σ, n, N by σ_i, n_i, N_i respectively. Thus we get

$$\text{var}\,(\bar{x}_{st}) = \sum_{i=1}^{k} \left(\frac{N_i}{N}\right)^2 \cdot \frac{\sigma_i^2}{n_i}\left(1 - \frac{n_i}{N_i}\right) \tag{7.7a}$$

$$= \frac{1}{N^2} \sum_{i=1}^{k} N_i(N_i - n_i)\,\frac{\sigma_i^2}{n_i} \tag{7.7 b}$$

The standard error (*s.e*) is given by

$$s.e.\ (\bar{x}_{st}) = \frac{1}{N}\sqrt{\left[\Sigma(N_i(N_i - n_i)\sigma_i^2/n_i\right]} \tag{7.7c}$$

We have

$$\text{var}\,(\bar{x}_{ran}) = \frac{\sigma^2}{n}\left(1 - \frac{n}{N}\right)$$

for sample mean $\bar{x}_{ran}$ from a simple random sample of the same total size n. When N is large compared to n, also N_i large compared to n_i for each i, then we can write from (7.7a) and (7.1)

$$\text{var}\,(\bar{x}_{st}) = \sum_{i=1}^{k} \left(\frac{N_i}{N}\right)^2 \frac{\sigma_i^2}{n_i} \tag{7.8}$$

$$\text{var}\,(\bar{x}_{ran}) = \frac{\sigma^2}{n} \tag{7.9}$$

Thus

$$s.e.\ (\bar{x}_{st}) = \frac{1}{N}\sqrt{\left[\Sigma\{(N_i\sigma_i)^2/n_i\}\right]} \tag{7.8a}$$

Of the two sampling designs, simple random and stratified random, the one with smaller variance (i.e. with smaller s.e.) is to be preferred. The question arises under what conditions (7.8) will be smaller than (7.9), so that stratification can be taken recourse to.

14.7.2. Allocation of Sample Size between Strata

Given the total sample size n, there are many ways in which the sample size n_i may be allocated to strata i, $i = 1, 2, \ldots, k$, subject to $\sum_{i=1}^{k} n_i = n$. The variance of $(\bar{x}_{st})$ depends on n_i and so the allocation of sample sizes between different strata will be an important consideration in stratified random sampling. Two ways of allocation of sample size between strata generally considered are

(i) proportional allocation,

(ii) optimum allocation.

14.7.3. Proportional Allocation

Here the sample is allocated between strata in the same proportions as in the population. We take

$$\frac{n_i}{n} = \frac{N_i}{N} = p_i, i = 1, 2, \ldots, k \tag{7.10}$$

This implies that

$$\frac{n_i}{N_i} = \frac{n}{N} = \text{constant} = c, i = 1, 2, \ldots, k \text{ for given } n \text{ and } N$$

that is, $$n_i = \frac{n}{N} N_i$$

or $$n_i = cN_i \tag{7.11}$$

or $$n_i \propto N_i$$

In this case the two estimates given by (7.3) and (7.4) are identical. The estimator can then be directly calculated from the sample by using (7.3). Then we have from (7.7a)

$$\text{var}(\bar{x}_{st}) = \sum_{i=1}^{k} \left(\frac{n_i}{n}\right)^2 \cdot \frac{\sigma_i^2}{n_i}\left(1 - \frac{n}{N}\right)$$

$$= \left(1 - \frac{n}{N}\right) \sum_{i=1}^{k} \left(\frac{n_i}{n}\right) \left(\frac{n_i}{n}\right) \frac{\sigma_i^2}{n_i}$$

$$= \left(1 - \frac{n}{N}\right) \sum_{i=1}^{k} p_i \frac{\sigma_i^2}{n} \tag{7.12}$$

For N large compared to n, neglecting n/N, we get

$$\text{var}\ (\bar{x}_{st}) = \sum_{i=1}^{k} p_i \frac{\sigma_i^2}{n} = \left(\sum_{i=1}^{k} p_i\ \sigma_i^2 \right) \Big/ n \qquad (7.13)$$

Thus $$s.e.\ (\bar{x}_{st}) = \sqrt{\left(\frac{\Sigma\ p_i\ \sigma_i^2}{n} \right)} \qquad (7.14)$$

It can be shown that under certain conditions, stratified random sampling under proportional allocation gives a smaller standard error (and therefore greater precision) compared to simple random sampling for the total sample size n when estimation of population mean from the sample mean is concerned. (see Remarks noted below).

14.7.4. Optimum Allocation

We have indicated certain criteria for selection of an appropriate design (section 14.4) from amongst a number of alternative designs. The criteria could be minimisation of error for a given fixed cost or minimisation of cost given a total permissible error. Alternative procedure of allocation of sample size between strata could be looked into from some of the above criteria. The design selected will be optimum according to that particular criterion. Allocation of sample size between strata according to some of the criteria is called *optimum allocation*. Allocation of n_i may be such that

(a) error is minimum for fixed n,

(b) error is minimum for total cost, or

(c) total cost is minimum for fixed error.

We shall measure error by the *s.e.* of the estimate of population mean, i.e. error = $s.e.\ (\bar{x}_{st})$. To simplify the algebra we may take the square i.e. var $(\bar{x}_{st})$.

(a) Neyman Allocation

Optimum Allocation of n_i *under fixed total sample size*

We shall now consider the question of determining n_i given $n = \sum_{i=1}^{k} n_i$ (the total sample size) such that var $(\bar{x}_{st})$ is minimum.

Suppose that N is large compared to n, then from (7.8)

$$Z \equiv \text{var}\ (\bar{x}_{st}) = \sum_{i=1}^{k} p_i^2 \frac{\sigma_i^2}{n_i}, \quad p_i = \frac{N_i}{N}$$

We are to minimise Z subject to $n = \sum_{i=1}^{k} n_i$

Now $$n = \sum_{i=1}^{k} n_i$$

$$\Rightarrow n_1 = n - n_2 - n_3 - \ldots - n_k \qquad (7.31)$$

or n_1 can be considered fixed when $n_2, \ldots, n_k$ are found.

In other words we are to find $n_2, n_3, \ldots, n_k$ for which Z will be minimum. These will be given by the solutions of

$$\frac{\partial Z}{\partial n_i} = 0 \text{ for } i = 2, 3, \ldots, k \qquad (7.32)$$

(i.e. for all i except $i = 1$)

Now

$$Z = \sum_{i=1}^{k} p_i^2 \frac{\sigma_i^2}{n_i} = p_1^2 \frac{\sigma_1^2}{n_1} + \frac{p_2^2\sigma_2^2}{n_2} + \ldots + \frac{p_k^2\sigma_k^2}{n_k}$$

$$= p_1^2 \frac{\sigma_1^2}{(n - n_2 - \ldots - n_k)} + \frac{p_2^2\sigma_2^2}{n_2} + \ldots + \frac{p_k^2\sigma_k^2}{n_k}$$

$$\frac{\partial Z}{\partial n_2} = 0 => \frac{p_1^2\sigma_1^2}{(n - n_2 - \ldots - n_k)^2} - \frac{p_2^2\sigma_2^2}{n_2^2} = 0$$

or,

$$\frac{p_2^2\sigma_2^2}{n_2^2} = \frac{p_1^2\sigma_1^2}{(n - n_2 - \ldots - n_k)^2}$$

$$= \frac{p_1^2\sigma_1^2}{n_1^2} \qquad \text{from (7.31)}$$

Similarly,

$$\frac{\partial Z}{\partial n_3} = 0 \quad \text{gives} \quad \frac{p_3^2\sigma_3^2}{n_3^2} = \frac{p_1^2\sigma_1^2}{n_1^2}$$

$$\ldots$$

$$\frac{\partial Z}{\partial n_k} = 0 \quad \text{gives} \quad \frac{p_k^2\sigma_k^2}{n_k^2} = \frac{p_1^2\sigma_1^2}{n_1^2}$$

Thus we have

$$\frac{p_1^2\sigma_1^2}{n_1^2} = \frac{p_2^2\sigma_2^2}{n_2^2} = \ldots = \frac{p_k^2\sigma_k^2}{n_k^2} = \text{constant}$$

or,

$$\frac{n_1^2}{p_1^2\sigma_1^2} = \frac{n_2^2}{p_2^2\sigma_2^2} = \ldots = \frac{n_k^2}{p_k^2\sigma_k^2} = \text{constant}$$

$$= \lambda^2 \text{ (say)} \qquad (7.33)$$

so that

$$n_1 = \lambda p_1 \sigma_1, \; n_2 = \lambda p_2 \sigma_2, \; \ldots, \; n_k = \lambda p_k \sigma_k$$

i.e.

$$n_i = \lambda p_i \sigma_i, \; i = 1, 2, \ldots, k \tag{7.34}$$

Hence

$$n = n_1 + n_2 + \ldots + n_k$$

$$= \lambda \sum_{i=1}^{k} p_i \sigma_i$$

so that the constant λ is given by

$$\lambda = \frac{n}{\sum_{i=1}^{k} p_i \sigma_i}$$

Substituting the value of λ in the expression of n_i given by (7.34), we get

$$n_i = \frac{p_i \sigma_i}{\Sigma p_i \sigma_i} n, \; i = 1, 2, \ldots, k \tag{7.35}$$

which gives the sample size to be taken from stratum i for which variance $(\bar{x}_{st})$ will be minimum for given total sample size $n = \sum_{i=1}^{k} n_i$. Since $Np_i = N_i$ we can write (7.35) also as

$$n_i = \frac{n}{\Sigma N_i \sigma_i} N_i \sigma_i = (\text{constant}) \times N_i \sigma_i \tag{7.35a}$$

since $\Sigma N_i \sigma_i$ is constant for a given population from which sample is drawn. In other words

$$n_i \propto N_i \sigma_i \tag{7.35b}$$

Substituting the value of n_i given by (7.35) in (7.8) we get

$$\text{var } (\bar{x}_{st}) \text{ (under optimum allocation)}$$

$$\equiv \text{var } (\bar{x}_{st})_{opt}$$

$$= \frac{(\Sigma p_i \sigma_i)^2}{n} \tag{7.36}$$

This allocation is known as *Neyman allocation.*

Particular case. Suppose that $\sigma_1 = \sigma_2 = \ldots = \sigma_k$, then from (7.35) we get

$$n_i = \frac{p_i}{\Sigma p_i} n = p_i \; n \quad \text{since } \Sigma \, p_i = 1$$

$$= \frac{N_i}{N} n, \; i = 1, 2, \ldots, k$$

which reduces to n_i given under proportional allocation (see (7.11)), i.e. with constant sampling fraction. From (7.13), we get

$$\text{var } (\bar{x}_{st}) \text{ (under proportional allocation)}$$

$$\equiv var \; (\bar{x}_{st})_{prop}$$

$$= \frac{\Sigma p_i \, \sigma_i^2}{n}$$

Remarks

It can be shown that

$$\text{var } (\bar{x}_{st})_{opt} < \text{var } (\bar{x}_{ran}) = \frac{\sigma^2}{n}$$

when

(i) stratum means μ_i differ from each other, or

(ii) stratum s.d. σ_i differ from each other, or

(iii) both (i) *and* (ii) occur.

Again, it can be shown that

$$\text{var } (\bar{x}_{st})_{prop} < \text{var } (\bar{x}_{ran}) = \frac{\sigma^2}{n}$$

when

(i) stratum means μ_i differ from each other. As σ_i are equal for all i, (ii) and (iii) do not hold in this case.

(b) Optimum Allocation under fixed cost

We consider now the allocation of sample size between strata such that the error is minimised given the total cost of the survey.

The cost of a survey consists of an overhead cost a of the organisation conducting the survey as well as of covering the units to be sampled. Suppose that the cost of covering each of the units in stratum i is c_i, then the cost of sampling n_i units in stratum i is $n_i \, c_i$. Thus the total cost of the stratified sampling design is given by the cost function

$$C = a + \sum_{i=1}^{k} c_i \, n_i \tag{7.37}$$

Suppose that N is large compared to n so that var $(\bar{x}_{st})$ is given by (7.8). Thus the problem of optimum allocation is reduced to the problem of finding n_i such that

$$\text{var}\,(\bar{x}_{st}) = \sum_{i=1}^{k} \left(\frac{N_i}{N}\right)^2 \frac{\sigma_i^2}{n_i}$$

is minimised subject to the constraint of total cost given by (7.37).

The problem can be solved by using the method of Lagrange's multiplier for finding the optimum. This is beyond the scope of this book. We simply state the result below.

The optimum sample size n_i is given by

$$n_i = \frac{n\,N_i\,\sigma_i/\sqrt{c_i}}{\sum_{i=1}^{k} (N_i\,\sigma_i/\sqrt{c_i})}, \quad i = 1, 2, \ldots, k \tag{7.38}$$

This allocation will minimise var $(\bar{x}_{st})$ subject to a given fixed cost. This gives an optimum allocation of n_i.

Substituting the value of n_i from (7.38) in (7.37) we get

$$C - a = \sum_{i=1}^{k} \left[\frac{nN_i\,\sigma_i\,\sqrt{c_i}}{\Sigma(N_i\,\sigma_i\,/\,\sqrt{c_i})}\right]$$

so that

$$n = \frac{(C - a)\ \Sigma(N_i\,\sigma_i/\sqrt{c_i})}{\Sigma N_i\,\sigma_i\sqrt{c_i}} \tag{7.39}$$

which gives the total sample size.

In particular, when $c_1 = c_2 = \ldots = c_k$, i.e. cost per unit is the same whatever the stratum to which the sample belongs, then from (7.38) we get

$$n_i = \frac{nN_i\,\sigma_i}{\Sigma N_i\,\sigma_i}$$

which is the allocation for Neyman allocation (as given by (7.35a)).

Thus when the cost per unit is the same for all units of the sample, then the optimum allocation for fixed cost reduces to optimum allocation for fixed sample size n.

(c) Optimum Allocation with given error at minimum cost

One can consider the other optimum allocation problem : minimisation of total cost for given variance v_0. In other words, one has to find the minimum

sample size n required for estimation of the population mean from a stratified sample with fixed variance var $(\bar{x}_{st}) = v_0$. The value of n is given in terms of N, N_i, σ_i, c_i and v_0.

In particular, suppose that $c_1 = c_2 = \ldots = c_k$. (Neyman allocation).

The expression for the optimum (minimum) sample size for estimation of population mean with fixed variance v_0 under Neyman allocation is given by

$$n = \frac{\left(\sum_{i=1}^{k} N_i \sigma_i\right)^2}{N^2 v_0 + \sum_{i=1}^{k} N_i \sigma_i^2}$$

14.7.5. Comparison of Stratified Random Sampling with Simple Random Sampling

We have considered stratified sampling design and indicated the sampling error specially with reference to the estimation of the population mean Likewise methods are available for estimation of other characteristics such as population total and population proportions.

Stratification is possible only when additional information about the population is available. Unless we have supplementary information according to which the population can be divided into distinct strata, we cannot go for stratified sampling and simple random sampling design will have to be adopted. This should be obvious from the fact that the error formulas involve N_i or $p_i = N_i / N$. Moreover, the process of stratification of the population may be difficult, costly and time consuming, and so may not be worthwhile.

Allocation of sample sizes between different strata is an important point for consideration. Any kind of allocation will not do. For example, allocation of the whole sample to any particular stratum will not serve any useful purp se.

Several methods of allocation of sample sizes are available. The methods generally lead to smaller sampling error than that of the simple random sampling without stratification. It can be shown that

$$\text{var}\,(\bar{x}_{ran}) \geq \text{var}\,(\bar{x}_{st})_{prop} \geq \text{var}\,(\bar{x}_{st})_{Neyman}$$

so that the sampling error associated with the estimation of the population mean is larger for simple random sampling than for stratified random sampling under proportional allocation as well as under Neyman allocation. Neyman allocation is thus the best so far as the precision of the estimate is concerned. However, it may be noted that Neyman allocation

$$n_i = \frac{p_i \sigma_i}{\Sigma p_i \sigma_i} n, i = 1, 2, \ldots, k$$

involves σ_i, the variance of the mean of each of the strata, $i = 1, 2, \ldots, k$. These may sometimes be available from a previous inquiry. If these are not available, the question of estimating these will arise. Proportional allocation, however, requires only the values N_i and the total sample size n. It gives more precise result (estimate of population mean) than simple random sampling but less precise than under Neyman allocation.

14.8. SYSTEMATIC SAMPLING

Systematic sampling is a convenient method when a sampling frame i.e. complete list of the sampling units is readily available or can be easily prepared. Here every kth unit is selected, where k is determined from the population size N and the sample size n; k is taken as the integer nearest to $\frac{N}{n}$ and is called sampling interval. The population can be divided into n groups, each of the first $(n-1)$ groups consisting of exactly k units and the nth (or the last group) consisting of the remaining units from the population. The first member of the sample is selected *at random from the first group of k units* and thereafter every kth unit is selected. In other words, select a number at random between 1 and k; suppose that j is the number selected at random $(1 \le j \le k)$: the jth unit is selected at first and then $(j + k)$ th, and then $(j + 2k)$ units and so on. The method is known as *linear systematic sampling.*

For example, suppose that $N = 103$ and $n = 10$, then $\frac{N}{n}$ is 10.3, and the nearest integer is 10 so that $k = 10$. Suppose that the unit selected at random from the first group of 10 units corresponds to number 7 (say): then the sample consists of units corresponding to the numbers 7, 17, 27, 37, 47, 57, 67, 77, 87, 97. Here the last group consists of 13 units instead of 10 as in other groups.

There is another method, known as *circular systematic sampling* (due to Lahiri) in which the first unit is selected *at random from the whole population*, i.e. one has a completely random start. Let k be the integer nearest to N/n when $N \ne nk$, and let m be the number selected at random between 1 and N $(1 \le m \le N)$. Then select every kth unit thereafter $m + k, m + 2k, \ldots, m + jk$ so long as $m + jk < N$ but $m + (j + 1)k > N$. Then select the unit corresponding to $m + (j + 1)k - N$ and every kth unit thereafter till a total of n units is selected. For example, let $N = 103$ and $n = 10$; then $k = 10$; let the number selected at random between 1 and 103 be 46 $(= m)$. The sample consists of units corresponding to the numbers 46, 56, 66, 76, 86, 96 and 3, 13, 23, 33; these can be renumbered serially as 3, 13, 23, 33, 46, 56, 66, 76, 86, 96.

It may be noted that when $N = nk$, then the two methods of systematic sampling, linear and circular, become identical.

For estimation of the population mean, the sampling error (i.e. the s.e. of the estimator under systematic sampling) can be calculated and compared with the sampling error of the estimators of other sampling methods. This would give an idea of the precision of systematic sampling design as compared to other sampling designs. Without going into algebraic details we may give some idea.

A systematic sample leads to more precision compared to a simple random sample without replacement if the units selected within the sample are heterogeneous. If, on the other hand, there exists some periodicity in the list leading to selection of more or less homogeneous units, then one would get completely biased results.

Systematic sampling is a case of mixed sampling, partly probabilistic and partly non-probabilistic. Here at first a selection is made at random (probabilistic) while subsequent selections are non-probabilistic. It has some resemblance to stratified random sampling with each stratum comprising of k (or so) units. Then in stratified random sampling, selection is made on random sampling basis from each stratum. Stratified random sampling presents a case of probabilistic sampling. Comparison of precision of systematic sampling design with that of stratified random sampling design would depend on some more properties of the population sampled. This is not discussed here.

Systematic sampling plans have certain advantages. These are indicated below.

(1) The method of selecting the sample is very simple and straight forward. The sample can be selected by an enumerator even in the field itself. What he would need is a sheet of random number table. It is inexpensive compared with other methods.

(2) Because of the simplicity, organisation of field work is easy. For example, for households to be sampled in a village, one needs only to know the total number N of units on the basis of which the systematic sampling of the plan can be designed; one can then start from one end of the village sampling the units according to the plan.

(3) The sample is more or less evenly distributed over the whole population.

(4) Systematic sampling gives more precise results than simple random sampling when the units sampled are heterogeneous.

The disadvantages are indicated below:

(1) If there is some periodicity in the list then biased results may occur when the sampling interval is equal to or is a multiple of the period. For example, consider the question of sampling trees in a forest with neatly planted trees with k in a row and suppose that a big drain with pollutants (or a big wall obstructing sunlight) runs along the last column. If the first unit selected is by the side of the drain (or wall) that is, the last one in the row, then every other unit selected will be the last one in every other row when k is the sampling interval. Thus the systematic sample will give biased result about the mean growth (or other such characteristics) of trees in the plantation.

(2) The sample mean is the estimator of the population mean. The variance of the estimator cannot be estimated from the sample. This makes the computation of the precision of the estimator difficult.

14.9. OTHER METHODS

The simple random sampling, the stratified random sampling and the systematic sampling are some of the methods of sampling more commonly employed. There are other methods, as mentioned in section 14.5.1. We describe them briefly.

Multistage sampling. Here the population is at first divided into first stage sampling units. A random sample is taken from the first stage units; usually selection is done with probabilities of selection proportional to their size.

Further division is then made of the first-stage sampling units selected, and a random sample is then taken from the second-stage sampling units. The process can be continued for a number of stages. For example, a state can be divided into districts as first stage units, from where a number of districts are selected. Then the selected districts can be divided into villages; a number of villages can be selected from these second stage units (villages). The villages can be divided into households as third stage units and then a number of households can be selected. These will be the ultimate sampling units in this three-stage sampling plan. If the sampling is confined to two stages only, it is called *two-stage sampling* or *sub-sampling*.

One of the advantages of multi-stage sampling is its flexibility. One can utilise the available information of existing divisions and subdivisions of the population as sampling units at different stages. Consider the example cited above of a three-stage sampling plan. The first stage units are districts, from which a number of districts are selected. From the selected districts a number of villages are selected as second stage units. Here information about divisions into districts and villages are readily available; this presents little difficulty. Then a sampling frame is to be made of the households in the selected villages, from which a number of households are selected as third and final stage sampling units. The task is very much easier. Imagine the difficulty of selecting households from a state through one-stage sampling. Thus it is simpler to carry out a multi-stage sampling than one-stage (or single-stage sampling). However, multistage sampling is, in general, less precise than some one-stage sampling, when suitably chosen.

Multiphase sampling. Here at first a sample is taken and then in a second phase a sample is taken from the original sample. The sampling units in a two phase sampling are the same in the two phases, unlike in two-stage sampling where the sampling units are different in the two stages. When sampling is done in several phases it is called multiphase sampling.

Quota sampling. This is a purposive sampling usually adopted when stratification is difficult, the objective being to derive the benefits of stratification without going for probabilistic sampling. Quotas may be fixed for each stratum. This method is usually employed for opinion polls and in some cases for socio-economic investigation. For example, for an opinion poll about the popularity of a leader (or about the appropriateness of a certain scheme or measure), quotas may be fixed from people divided into strata such as office workers, factory workers, agriculture labour and so on.

While it is simple in practice this method suffers the disadvantage that sampling theory cannot be applied as it is not based on probabilistic sampling.

14.10 SAMPLE SURVEYS IN PRACTICE IN INDIA

India is a vast country. It is difficult to collect information and data by taking complete enumeration even in a state or even of a community. Sample surveys therefore play an important role in collection of data on diverse aspects and needed for various purposes. Sample surveys are conducted in large number with specific objectives by different organisations. Apart from such adhoc surveys, some sample surveys are conducted on continuous basis at national and subnational levels.

The Sample Registration System (SRS) which was initiated in 1964-65 by the Government of India, is a dual record system with the main objective of providing reliable estimates of vital rates. The sample unit in the rural area is a village (or a segment of a village) and in the urban area is a census block. At present it covers over 6000 sample units. Apart from continuous enumeration of births and deaths by a resident enumerator, an independent survey is conducted every six months with a view to matching the data.

SRS results are available in the official journal *Sample Registration Bulletin*, Registrar General, India.

Basic structures of the SRS : main components are :

1. Base-line survey of the sample unit to obtain usual resident population of the same area.
2. Continuous enumeration of vital events pertaining to usual population by the enumerator.
3. An independent half-yearly survey by the supervisor : recording of births and deaths during the half-year also updating house list and household schedule.
4. Matching of events recorded under 2 and 3 above.
5. Verification in the field of unmatched and partially matched events.

The National Sample Surveys (NSS) organisation was set up in 1950 by the Government of India with the objective of providing socio-economic data. The NSSO conducts continuing surveys in what is known as rounds, which are distinct survey periods varying from 3 months to one year. Data on different aspects are collected in different rounds : these cover topics such as national income, acreage and yield rates of major crops, land holdings, small-scale industries, family budgets, demography and several other socio-economic issues. The general sample design is a stratified two-stage plan with villages as first-stage units and with households as second-stage units for socio economic investigations and with clusters of plots as second stage units for enquiries relating to crops. The overall governance is left to a Governing Council, an autonomous apex body, to maintain credibility of the NSSO as an independent fact finding agency.

Survey results are available in the NSSO journal *Sarvekshana*. Some of the current surveys conducted by the NSSO are as follows.

A. Socio-economic Surveys

(a) **44 th Round (July 1988-June 1989)**
Information collected relate to living conditions of the tribals, housing conditions of general population, current building construction activity etc.

(b) **45 th Round (July 1989-June 1990)**
In this round, a survey of enterprises engaged in manufacturing and repairing activities was carried to fill the vital gap in information relating to unorganised segments of the non-agricultural sectors.

(c) 46 th Round (July 1990-June 1991)
The survey was directed for collecting information on small traders (of unorganised sectors) as well as consumer expenditure of households.

(d) 47 th Round (July 1991-June 1992)
The survey just launched would collect information, for the first time, on socio-cultural (literacy etc.) aspects, demographic aspects (on handicapped ètc.) besides other economic aspects, usually studied.

B. Among other surveys mention may be made of

(e) Annual Survey of Industries,
(f) Enterprise Survey, 1988-89,
(g) Survey under the scheme for improvement of Crop Statistics, 1989-90.

The wealth of data provided by the NSSO proves very valuable for policy making and plan formulation as well as for research investigation.

EXERCISES-14

Sections 14.1-14.5

1. What is a sample survey? Explain the basic idea behind sampling.
2. Enumerate the advantages of a sample survey over complete enumeration.
3. Distinguish between sample survey and complete enumeration. Mention the circumstances under which a complete enumeration can be recommended in preference to a sample survey.
4. What are the various types of error that may creep up in an inquiry through (i) a sample survey, (ii) a complete enumeration? Discuss.
5. What is meant by sampling error? When and how can sampling errors occur? Can these be reduced? If so, explain how.
6. How is sampling error measured? Explain.
7. Explain why sampling error need be computed.
8. What are non-sampling errors? Explain how do these occur.
9. Enumerate and explain the various stages of a sample survey.
10. Indicate the problems involved in the various stages of a sample survey. How would you try to tackle them?
11. (i) Explain clearly the following with suitable examples
    (a) Population
    (b) Sample
    (c) Sampling Units
    (d) Sampling Frame

    (ii) Write notes on
    (e) Method of data collection
    (f) Schedules and Questionnaires
    (g) Pilot Survey
12. Mention the aspects that are to be taken into account while selecting an appropriate sampling design.
13. A sample survey is to be conducted to determine the reading habits (such as hours devoted to study at home, at library as well as hours devoted to study for the course enroled, hours devoted to newspaper and other reading and so on) of students of your college. Indicate a sample design and prepare a draft questionnaire for the survey.
    Give an outline of the tables you would prepare.
14. A sample survey is to be organised for assessing the extent of unemployment and under-employment of students of your college, who graduated two years back.
    Indicate a sample design and prepare a draft questionnaire for the survey.
15. A family-budget survey is to be conducted in a village of your district. Give an outline of the schedule you would convass.
16. Enumerate the types and methods of sampling. Discuss their merits and demerits.

Section 14.6

17. Explain what is meant by simple random sampling. Under what circumstances, will simple random sampling with replacement be more or less similar to simple random sampling without replacement?
18. Mention the various methods of drawing a simple random sample. Explain the use of random number tables in this context.
19. Discuss the points to be taken into consideration while determining the sample size of a simple random sample.
20. Describe the method of determining the sample size of a simple random sample for estimation of the population mean from the sample mean with a specified margin of error and confidence limit when the population to be sampled is (i) infinite, (ii) finite. State the assumptions you make and the important results that you use.
21. Find the sample size (of a simple random sample) needed to estimate the population mean from the sample mean with a margin of error 5 per cent and confidence level 5 per cent taking the population to be infinite with variance 125.
22. A population consists of 100 sampling units and has s.d. 8. Find the sample size required for estimation of the population mean from the sample mean, margin of error permissible being 10% and confidence level 1 per cent.

Section 14.7

23. Define stratified sampling. Stratified sampling is said to combine the advantages of both random selection and purposive selection. Explain.
24. When would you recommend use of stratified sampling design in preference to simple random sampling design? Explain with the help of some suitable illustration.
25. Explain the purpose of stratification in sample surveys. When is stratification possible?
26. A population is divided into k strata with N_i units in stratum i, $\sum_{i=1}^{k} N_i = N$. Random samples of size i are drawn from the strata i, $i = 1, 2, \ldots, k$, and $\bar{x}_i$ is the sample mean of ith stratum. Find the mean and variance of the estimator

$$\bar{x}_{st} = \sum_{i=1}^{k} \left(\frac{N_i}{N}\right) \bar{x}_i$$

Find the variance when N_i is large compared to n_i for each i.

27. Compare the variance of the estimated mean of stratified random sample with that of the estimated mean in case of a simple random sample. Does the stratified random sampling always give smaller variance ? Discuss.
28. What is meant by allocation of sample size between strata in case of a stratified random sampling? Why is allocation considered important? What are the methods of allocation? Discuss.
29. Define proportional allocation. Find the variance of the estimated mean in proportional allocation. Compare the precision of proportional allocation with that in case of simple random sample.
30. What is meant by optimum allocation? Indicate the criteria to be considered for optimality.
31. Discuss Neyman allocation. Find the variance of the estimated mean under Neyman allocation. When is Neyman allocation identical with proportional allocation?
32. Discuss optimum allocation under fixed cost function $C = a + \Sigma c_i n_i$ when N is large compared to n. What happens when $c_1 = c_2 = \ldots = c_k$ i.e. cost per unit is the same for all units in the sample?
33. Explain the various methods of allocation of sample size between strata in a stratified random sampling. When does stratification produce large gains in precision? Discuss.

Section 14.8

34. Explain the methods of systematic sampling. When does systematic sampling lead to more precision compared to simple random sampling? Discuss.
35. Explain Lahiri's method of circular systematic sampling. When is this method adopted? When do the two methods of systematic sampling, linear and circular, become identical?
36. Enumerate the merits and demerits of systematic sampling in relation to other methods of sampling.

Sections 14.9-14.10

37. **Describe multistage and multiphase sampling methods. Mention some situations where you would recommend them.**
38. **What is quota sampling? When is this method suitable ?**
39. **Describe the Sample Registration Scheme of India. Discuss about the nature and importance of (SRS) data.**
40. **Explain the role and importance of National Sample Survey Organisation in providing data on Indian economy.**

Answers to Exercises

Exercises 6

1. (a) (i) $\frac{1}{4}$ (ii) $\frac{5}{16}$ (iii) $\frac{1}{16}$
2. (a)(i) WW, WB, WR, BW, BB, BR, RW, RB, RR : 9 points
   (ii) WB, WR, BW, BR, RB, RW : 6 points
3. (a) $\frac{1}{9}$ (b) $\frac{2}{9}$ (c) $\frac{5}{9}$
4. (a) 0 (b) $\frac{1}{3}$ (c) $\frac{2}{3}$
5. (i) $\frac{1}{6}$ (ii) $\frac{1}{4}$ (iii) $\frac{1}{6}$ (iv) $\frac{1}{36}$
6. (i) $\frac{1}{2}$ (ii) $\frac{1}{2}$ (iii) $\frac{1}{12}$ (iv) $\frac{1}{12}$
7. (i) $\frac{1}{52}$ (ii) $\frac{1}{4}$ (iii) $\frac{1}{13}$
8. (i) $\left(\frac{1}{13}\right)^2$ (ii) $\left(\frac{1}{52}\right)^2$ (iii) $4 \cdot \left(\frac{1}{52}\right)^2$
9. $\frac{1}{13} \cdot \frac{3}{51}$
10. (a) yes (b) yes (c) no (d) no.
11. $\frac{1}{8}$
13. $\frac{8}{13}$
14. $\frac{4}{13}$
15. $\frac{3}{4}$; yes
16. $\frac{1}{3}$; no
17. $\frac{11}{36}$
18. $\frac{1}{36}, \frac{5}{18}$
21. $\frac{2}{5} \cdot \frac{3}{5}$

22. $\frac{2}{5} \cdot \frac{3}{4}$

23. $\left(\frac{3}{8}\right)^2 + \left(\frac{5}{8}\right)^2$

24. $\frac{5}{8} \cdot \frac{4}{7} + \frac{3}{8} \cdot \frac{2}{7}$

25. (i) 6 (ii) 0 (iii) $\frac{1}{2}$ (iv) 0

Exercises 7

1. If X represents the number of heads, then

$$P(X=k) = \frac{\binom{4}{k}}{2^4}, \; k = 0, 1, 2, 3, 4. \qquad E(X) = 2$$

2. If X represents the number of heads, then
$P(X = 0) = 0.064$, $P(X = 1) = 0.288$, $P(X = 2) = 0.432$, $P(X = 3) = 0.216$.
$E(X) = 1.8$
3. $E(X)=5.5$, $E(X^2) = 38.5$, $\text{var}(X) = 8.25$
4. $E(X) = 7$
5. $E(Y) = 16$, $\text{var}(Y) = 33$
6. $\text{var}(Z) = 1.458$
7. $E(X) = -0.3$, $\text{var}(X) = 0.61$
$E(Z) = -0.767$, $\text{var}(Z) = 0.0678$
8. $E(X) = 5/6$
9. If X is the number of spade cards, then

(i) $P(X = 0) = \left(\frac{39}{52}\right)^2$, $P(X = 1) = \frac{2 \times 39 \times 13}{(52)^2}$, $P(X = 2) = \left(\frac{1}{4}\right)^2$

(ii) $P(X = 0) = \frac{39 \times 38}{52 \times 51}$, $P(X = 1) = \frac{2 \times 39 \times 13}{52 \times 51}$, $P(X = 2) = \frac{13 \times 12}{52 \times 51}$

10. Let X be the number of white balls:

X	:	0	1	2	3
Prob	: (i)	0.216	0.432	0.288	0.064
	: (ii)	$\frac{1}{6}$	$\frac{1}{2}$	$\frac{3}{10}$	$\frac{1}{30}$

12. $E(Z) = 0$, $\text{var}(Z) = \frac{35}{6}$

13. (i) $\frac{1}{2}$ (ii) $\frac{1}{2}$ (iii) 1 (iv) $\frac{1}{4}$

14. (i) $\frac{1}{100}$ (ii) 1 (iii) 1 (iv) $\frac{1}{4}$

15. (i) $\frac{3}{4}$ (ii) 1 (iii) 1

17. $E(X) = 0.1$, var $(X) = 0.09$

18. (a)

X :	0	1	2	3	4
Prob :	$\left(\frac{2}{3}\right)^4$	$4 . \frac{1}{3} . \left(\frac{2}{3}\right)^3$	$6\left(\frac{1}{3}\right)^2\left(\frac{2}{3}\right)^2$	$4\left(\frac{1}{3}\right)^3\left(\frac{2}{3}\right)$	$\left(\frac{1}{3}\right)^4$

(b) Rs. 480 gain

19. (i) $\left(\frac{1}{2}\right)^6$, (ii) $\left(\frac{1}{2}\right)^6$, (iii) $\frac{21}{32}$, (iv) $\frac{21}{32}$

20. (i) 0.0486 (ii) 0.0037 (iii) 0.9963
mean = 0.4, variance = 0.36

21. (i) $\frac{5}{16}$ (ii) $\frac{1}{32}$ (iii) $\frac{3}{16}$

22. (i) 1, (ii) $\frac{13}{256}$, (iii) $\frac{81}{256}$

23. (i) No. (ii) Yes; $p = \frac{1}{9}$, $n = 9$

24. Parameters $n = 6, p = \frac{1}{3}; \gamma_1 = \frac{2\sqrt{3}}{9}, \gamma_2 = -\frac{1}{4}$

25. Let X be the number of black balls

X :	0	1	2	3
Prob :	$\frac{1}{30}$	$\frac{3}{10}$	$\frac{1}{2}$	$\frac{1}{6}$

mean = $\frac{9}{5}$, variance = 0.56

26. (i) $\binom{5}{0}\binom{95}{3} / \binom{100}{3}$ (ii) $\binom{5}{2}\binom{95}{1} / \binom{100}{3}$

(iii) $\left[\binom{5}{0}\binom{95}{3} + \binom{5}{1}\binom{95}{1}\right] / \binom{100}{3}$

27. (i) $\frac{1}{5}$ (ii) $\frac{4}{5}$ (iii) 0 ; Mean = 1

28. (i) $\frac{16}{125}$ (ii) $\frac{61}{125}$

29. Prob. = $(104)(99)^4/(100)^5$; Expected number = 0.05

30. (i) $\dfrac{\binom{20}{2}\binom{5}{2}\binom{4}{1}\binom{1}{0}}{\binom{30}{5}}$

(ii) $1 - \dfrac{\binom{20}{4}\binom{10}{1} + \binom{20}{5}\binom{10}{0}}{\binom{30}{5}}$

(iii) $1 - \dfrac{\binom{4}{0}\binom{26}{5}}{\binom{30}{5}}$

31. (i) $(0.9)^3\ (0.1)$, (ii) $0.1 + 0.9 \times 0.1 + (0.9)^2 \times 0.1$

32. Prob. = $\frac{4}{27}$, Exp. cost = Rs. 200

33. Prob. = $(0.8)^4(0.2)$, Exp. cost = Rs. 5

34. (a) (i) $\frac{1}{4}$, (ii) $\frac{1}{5}$

(b) $(0.8)^3\ (0.2)$; mean = 5; var = 20

35. (i) 0.05, (ii) 0.024 (iii) 0.926, (iv) 0.05

36. (i) 3 (ii) 3, (iii) $1 - e^{-3}$ (iv) $(8.5)e^{-3}$ (v) $e^{-3}3^5/5\,!$

37. (i) e^{-2}, (ii) $3e^{-2}$

38. (i) e^{-5}, (ii) $1 - \sum_{k=0} \dfrac{e^{-5}\ 5^k}{k!} = 0.551$

39. (i) e^{-10} (ii) $e^{-10}10^3/3\,!$

40. $1 - \sum_{k=0}^{3} \dfrac{e^{-1.5}\ (1.5)^k}{k\,!} = 0.0662$

41. (i) 2 (ii) 0.27

42. (i) e^{-1} (ii) $\frac{3}{2}e^{-1}$ (iii) 1

43. (i) 7;13 (ii) 4;16 (iii) 1;19 (iv) 0.9938 (v) 0.2514

44. (i) 0.8413 (ii) 0.0228 (iii) 0.9759

45. (i) 0.0002 (ii) 0.7745 (iii) 0.0228

Number failed = 228; Number getting distinction = 2

46. (i) 0.5 (ii) 0.5 (iii) 0.683

47. Mean = 115; s.d. = 10

49. (i) 0.1587 (ii) 0.0668, (iii) 0.1974

Exercises 8

2. Mean ($\bar{x}$) = 4; var ($\bar{x}$) = 1.67
3. Mean = 4; var = 5/9
4. $E(T)$ = 2, var (T) = 1
5. E ($\bar{x}$) = 1.5, var ($\bar{x}$) = 0.375
8. E ($\bar{x}$) = 0.5, var ($\bar{x}$) = 0.0625
9. E (x) = 1/6, var ($\bar{x}$) = 5/144
10. (i) $P(x > 35) = 0.5$ (ii) $P(\bar{x} < 35) = 0.5$
    (iii) $P(x > 34 = 0.9992$, (iv) $(\bar{x} < 34) = 0.0008$
    (v) $P(34 < \bar{x} < 36) = 0.9984$
11. (i) $P (\bar{x} < 2.5) = 0.5$
    (ii) $P(\bar{x} > 2.5) = 0.5$
    (iii) $P(2.4 < \bar{x} < 2.6) = 0.9548$
14. (i) $P(\bar{x} < 400) = 0.5$ (ii) $P(\bar{x} > 400) = 0.5$
    (iii) $P(380 < \bar{x} < 420) = 1$
15. Mean = 50; s.d. = 1.2247
16. (i) Mean = 100; var = 10
    (ii) Mean = 100; var = 9.009
17. (i) $P (\hat{p} > 0.7) = 0.5$
    (ii) $P (\hat{p} < 0.7) = 0.5$
    (iii) $P (0.65 < \hat{p} < 0.75) = 1$
18. (i) $P(\hat{p} > 0.01) = 0.5$, $P (\hat{p} < 0.01) = 5$
    (ii) $P(0.015 < \hat{p} < 0.02) = 0.1498$
19. (i) $P(\hat{p} > 1/2) = 0.4364$ (ii) $P(\hat{p} < 1/2) = 0.5636$
20. Mean = 0.05, s.d. = 0.108. Not normal

Exercises 9

8. $r = 0.7$
9. $r = -0.35$
10. $r = 0.935$
11. $r = -0.278$
12. $r = 0.506$
13. $r = 0.949$

14. $r = 0.964$
15. $r = 0.95$
21. $y = 146.05 - 0.35\, x$, $x = 141.5 - 0.35\, y$
22. $y = 0.2163 + 1.0758\, x$, $x = 7.046 + 8129\, y$
23. $y = 45.48 - 0.1032\, x$, $x = 172.5 - 0.75\, y$
24. $y = 1.2392 + 0.6268\, x$, $x = 3.955 + 0.409\, y$
25. $y = -17.265 + 1.99\, x$, $x = 9.753 + 0.466\, y$
26. (a) (i) $\hat{y} = 30$ (ii) $\hat{y} = 24.84$
    (b) (i) $\hat{x} = 142.5$ (ii) $\hat{x} = 135$
27. $r = 0.9165$
28. $r = 0.6257$
29. $y = -\frac{2}{3}x + 2$, $x = -\frac{7}{5}y + \frac{12}{5}$
30. (a) (i) $r = -0.966$ (ii) $\bar{x} = -6$, $\bar{y} = 6$
    (b) No. $\frac{\sigma_y}{\sigma_x} = 0.69$
31. 36 %
36. $r = 0.438$
37. $y = -42.49 + 0.709\, x$; $x = 144.05 + 0.271\, y$
    $\hat{y} = 83.003$, $\hat{x} = 161.665$
38. $r = 0.024$
39. $y \doteq -1.984 + 0.00435\, x$; $x = -19.523 + 0.1336\, y$
40. $r = 0.719$; $y = -126.948 + 0.9524\, x$; $x = -47.158 + 0.542\, y$

Exercises 10

4. (i) 5.83 (ii) 9.85 (iii) 8.23
5. (i) 0.2689 (ii) 0.2096 (iii) 0.1465
6. (i) 0.5378, 0.8067 (ii) 0.4192, 0.6289 (iii) 0.2931, 0.4396
8. $n = 97$
9. 4 times, 16 times
10. (i) 100, 400 (ii) 144, 576 (iii) 400, 1600
13. (i) $(\bar{x} - 3.92, \bar{x} + 3.92)$, (ii) $(\bar{x} - 5.16, \bar{x} + 5.16)$
14. 21.213
15. (i) 10.954, (ii) 16
17. $\hat{p} = 0.03$ (i) (0, 0.063), (0, 0.074)
18. $\hat{p} = 0.75$

19. $n = 73$
20. (0.1047, 0.2553)
21. (i) (0.4864, 0.7136), (ii) (0.4642, 0.7358), (iii) (0.4212, 0.7788)
26. (i) –2.34 and 2.34 (ii) –1.44 and 1.44
28. Z = –2.58, Rejected at 5% level :
29. Z = –15, Rejected
30. Z = 8.486; Rejected
31. Z = 5.2; Rejected
32. Z = 0.3246; Accepted
33. Z = 0.67; Accepted
34. Z = – 4; Rejected
35. Z = 1.698; Accepted
36. Acceptance region (– 0.2098, 0.2098) at 5%; Rejected at 5% level
37. Acceptance region (– 0.0486, 0.0486) at 5%; Rejected at 5% level
38. Acceptance region (– 7.01, 7.01) at 5%; Rejected at 5% level
39. Acceptance region (– 0.093, 0.093) at 5% level; Accepted at 5% level
40. Acceptane region (– 0.054, 0.054) at 5% level; Accepted at 5% level

Exercises 11

6. $t = 0.415$ with 11 d.f.
7. $t = 1.5$ with 15 d.f.
9. $t = 1.904$ with 18 d.f.
10. $t = 1.44$ with 12 d.f.
    As paired t, $t = 1.39$ with 6 d.f.
14. $F = 1.13$ with 14 and 9 d.f.
15. $F = 1.48$ with 6 and 6 d.f.
21. $\chi^2 = 10.125$ with 9 d.f.
22. $\chi^2 = 11.85$ with 15 d.f.
23. $\chi^2 = 13.65$ with 11 d.f.
25. $\chi^2 = 0.640$ with 3 d.f. (last 3 classes pooled)
26. $\chi^2 = 0.156$ with 3 d.f. (last 3 classes pooled)
27. $\chi^2 = 1.165$ with 4 d.f. (last 3 classes pooled)
28. $\chi^2 = 26.465$ with 1 d.f.
29. $\chi^2 = 60.155$ with 1 d.f.
30. $\chi^2 = 16.355$ with 1 d.f.

Exercises 12

8.

Years	Trend Values	Years	Trend Vlaues
72-73	617.1	78-79	692.1
73-74	647.2	79-80	693.3
74-75	667.3	80-81	694.2
75-76	677.5	81-82	713.1
76-77	670.0	82-83	734.9
77-78	687.9	83-84	753.3
		84-85	778.7

9. Year and Trend Values : (71 : 15.83, 72 : 15.77, 73 : 15.63, 74 : 15.30, 75 : 15.13, 76 : 15.20, 77 : 14.63, 78 : 13.90)

10. Year and Trend Values : (70-71 : 2.95, 71-72 : 2.72, 72-73 : 2.51, 73-74 : 2.28, 74-75 : 2.125, 75-76 : 2.02, 76-77 : 1.92, 77-78 : 1.84)

16. The trend equations are
 $y = 28.786 + 2.643\ x$
 $z = 1.875 + 0.375\ x$

17. The trend equation is : $y = 188.12 - 4.86\ x$
 Year and Trend values (81-82 : 188.12, 82-83 : 183.26, 83-84 : 178.4, 84-85 : 173.54, 84-85 : 168.68)

18. The trend equation and (trend values) are
 India : $y = 6.2 - 0.45\ x$ (6.2, 5.75, 5.3, 4.85, 4.4)
 Bangladesh : $z = 6.94 - 0.12\ x$ (6.94, 6.82, 6.7, 6.58, 6.46)

38. The seasonal indices are : From April to March
 (98.507, 99.883, 100.380, 101.319, 102.284, 101.590, 101.331, 100.100, 99.128, 98.840, 98.426, 98.212)

41. The seasonal indices are (Monday to Sunday)
 (110.42, 104.80, 96.38, 91.60, 93.45, 103.35;

42. The seasonal indices are (Jan-Mar to Oct-Dec)
 (96.69, 101.71, 103.28, 98.36)

Bibliography

1. Allen, R.G.D. (1966). *Statistics for Economists,* Hutchinson's University Library, London.

2. Bhattacharyya, G.K. and Johnson R.A. (1977). *Statistical Concepts and Methods,* Wiley, New York.

3. Croxton, Frederick E., Cowden, Dudley J. and Klein Sidney (1969) *Applied General Statistics.* (3rd edition). Prenctice-Hall of India Ltd., New Delhi.

4. Draper, N.R. and Smith, H. (1965). *Applied Regression Analysis,* Wiley & Sons, New York.

5. Fox, K.A. (1968) *Intermediate Economic Statistics,* 4th Edition, Wiley, New York.

7. Karmel, P.H. and Polasek, M. (1970). *Applied Statistics for Economists.* (3rd edition) Pitman Publishing, London.

8. Keeping, E.S. (1963). *Intrdouction to Statistical Inference,* Van Nostrand, New York.

9. Moore, P.G. (1969) *Principles of Statistical Techniques,* 2nd Edition, Cambridge University Press, London.

10. Mudgett, B.D. (1951). *Index Numbers.* Wiley, New York.

11. Pearson, E.S. and Hartly, H.O., Editors (1954) *Statistical Tables for Statisticians,* Vol. I, Cambridge University Press, London.

12. Simpson, G. and Kafka, F. (1957) *Basic Statistics,* W.W. Norton & Co. Inc., New York.

13. Yamane, Toro (1969). *Statistics : An Introductory Analysis.* Harper & Row, New York.

14. Yule G.U. and Kendall, M.G. (1950). *An Introduction to the Theory of Statistics,* Griffin, London.

References

CHAPTER-1

T.A. Bancroft : The American Statistical Association : A Single Scientific and educational community. *Jour. Am. Stat. Ass.* 66(1971) 7-12

A.F. Goodman : The Interface of Computer Science and Statistics. *Nov. Res. Log. Qrly. 18(1971)* 215-230.

J. Neyman : Indeterminism in Science and New demands for Statisticians. *J. Am. Slat. Ass. 55* (1960) 625-639.

J. Tanur (Ed) : *Statistics : A Guide to the Unkniown.* Holden Day Inc., (1972).

CHAPTER-5

Kali S. Banerjee, (1975). *Cost of Living Index Numbers : Practice, Precision, and Theory.* Marcel Dekker Inc., New York

Ministry of Finance, Govt. of India-Economic Survey, 1986-87, 1987-88, 1988-89, 1989-90, 1990-91, 1999-2000.

CHAPTER-9

D.A. Kenny. *Correlation and Causality*, Wiley, New York (1988).

CHAPTER-12

M.G. Kendall and A. Stuart : *The Advanced Theory of Statistics,* Vol III. Griffin, London (1963). P.A. Samuelson : *Economics,* 10th Edition, McGraw-Hill Kogakusha Ltd. Tokyo (1976). 'P.Kennedy : *The Rise and Fall of the Great Powers,* Random House New York (1987). [For an analysis of dynamic for change driven by economic and technological developments]

CHAPTER-13

Registrar General, Govt. of India—Census? *Reports,* New Delhi.

ESCAP : *Country Population Monographs No. 10. INDIA.* Bangkok (1982).

ESCAP : 1986 Population Data Sheet, Bangkok (1987).

U.S. Bureau of the Census, *Country Demographic Profiles No. 16 :* INDIA

G.W. Barclay, (1958). *Techniques of Population Analysis,* Wiley, New York.

DJ. Bogue (1969). *Principles of Demography,* Wiley, New York.

R.Ramakumar, *Technical Demography,* Wiley Eastern. New Delhi (1986).

UNFPA : *The State of World Population Report,* 1990 & 1992, United Nations, New York.

CHAPTER-14

W.G. Cochran, (1963). *Sampling Techniques,* Wiley, New York

D. Singh and F.S. Chaudhary, (1986). *Sample Survey Designs,* Wiley Eastern, New Delhi.

APPENDICES

APPENDIX A

TABLES

Table 1. The Normal Probability Function
Table 2. Percentage Points of the chi-square (χ^2) distribution
Table 3. Percentage Points of the *t*-distribution
Table 4. (a) & 4 (b). Percentage Points of the *F*-distribution
Table 5. Individual Terms of Poisson distribution

Table 1

The Normal probability function. the integral $P(X)$ and ordinate $Z(X)$ in terms of the standardized deviate X.

X	$P(X)$	$Z(X)$	X	$P(X)$	$Z(X)$
.00	.5000000	.3989423	.50	.6914625	.3520653
.01	.5039894	.3989223	.51	.6949743	.3502919
.02	.5079783	.3988525	.52	.6984682	.3484925
.03	.5119665	.3987628	.53	.7019440	.3466677
.04	.5159534	.3986233	.54	.7054015	.3448180
.05	.5199388	.3984439	.55	.7088403	.3429439
.06	.5239222	.3982248	.56	.7122603	.3410458
.07	.5279032	.3982248	.57	.7156612	.3391243
.08	.5318814	.3976677	.58	.7190427	.3371799
.09	.5358564	.3973298	.59	.7224047	.3352132
.10	.5398278	.3969525	.60	.7257469	.3332246
.11	.5437953	.3965360	.61	.7290691	.3312147
.12	.5477584	.3960802	.62	.7323711	.3291840
.13	.5517168	.3955854	.63	.7356527	.3271330
.14	.5556700	.3950517	.64	.7389137	.3250623
.15	.5596177	.3944793	.65	.7421539	.3229724
.16	.5635595	.3938684	.66	.7453731	.3208638
.17	.5674949	.3932190	.67	.7485711	.3187371
.18	.5714237	.3925315	.68	.7517478	.3165929
.19	.5753454	.3918060	.69	.7549029	.3144317
.20	.5792597	.3910427	.70	.7580363	.3122539
.21	.5831662	.3902419	.71	.7611479	.3100603
.22	.5870644	.3894038	.72	.7642375	.3078513
.23	.5909541	.3884286	.73	.7673049	.3056274
.24	.5948349	.3876166	.74	.7703500	.3033893
.25	.5987063	.3866681	.75	.7733726	.3011374
.26	.6025681	.3856834	.76	.7763727	.2988724
.27	.6064199	.3846627	.77	.7793501	.2965948
.28	.6102612	.3836063	.78	.7823046	.2943050
.29	.6140919	.3825146	.79	.7852361	.2920038
.30	.6179114	.3813878	.80	.7881446	.2896916
.31	.6217195	.3802264	.81	.7910299	.2873689
.32	.6255158	.3790305	.82	.7938919	.2850364
.33	.6293000	.3778007	.83	.7967306	.2826945
.34	.6330717	.3765372	.84	.7995458	.2803438
.35	.6368307	.3752403	.85	.8023375	.2779849
.36	.6405764	.3739106	.86	.8051055	.2756182
.37	.6443088	.3725483	.87	.8078498	.2732444
.38	.6480273	.3711539	.88	.8105703	.2708640
.39	.6517317	.3697277	.89	.8132671	.2684774
.40	.6554217	.3682701	.90	.8159399	.2660852
.41	.6590970	.3667817	.91	.8185887	.2636880
.42	.6627573	.3652627	.92	.8212136	.2612863
.43	.6664022	.3637136	.93	.8238145	.2588805
.44	.6700314	.3621349	.94	.8263912	.2564713
.45	.6736448	.3605270	.95	.8289439	.2540591
.46	.6772419	.3588903	.96	.8314724	.2516443
.47	.6808225	.3572253	.97	.8339768	.2492277
.48	.6843863	.3555325	.98	.8364569	.2468095
.49	.6879331	.3538124	.99	.8389129	.2443904
.50	.6914625	.3520653	1.00	.8413447	.2419707

$$Z(X) = e^{-\frac{1}{2}X^2} \Big/ \sqrt{(2\pi)}, \quad P(X) = 1 - Q(X) = \int_{-\infty}^{X} Z(u)\, du$$

Table 1 *(Contd.)*

X	$P(X)$	$Z(X)$	X	$P(X)$	$Z(X)$
1.00	.8413447	.2419707	1.50	.9331928	.1295176
1.01	.8437524	.2395511	1.51	.9344783	.1275830
1.02	.8461358	.2371320	1.52	.9357445	.1256646
1.03	.8484950	.2347138	1.53	.9369916	.1227628
1.04	.8508300	.2322970	1.54	.9382198	.1218775
1.05	.8531409	.2298821	1.55	.9394292	.1200090
1.06	.8554277	.2274696	1.56	.9406201	.1181573
1.07	.8576903	.2250599	1.57	.9417924	.1163225
1.08	.8599289	.2226355	1.58	.9429466	.1145048
1.09	.8621834	.2202508	1.59	.9440826	.1127042
1.10	.8643339	.2178522	1.60	.9452007	.1109208
1.11	.8665005	.2134582	1.61	.9463011	.1091548
1.12	.8686431	.2130691	1.62	.9473839	.1074061
1.13	.8707619	.2106856	1.63	.9484493	.1056748
1.14	.8728568	.2083078	1.64	.9494974	.1039611
1.15	.8749281	.2059363	1.65	.9505285	.1022649
1.16	.8769756	.2035714	1.66	.9515428	.1005864
1.17	.8789995	.2012135	1.67	.9525403	.0989255
1.18	.8809999	.1988631	1.68	.9535213	.0972823
1.19	.8829768	.1965205	1.69	.9544860	.0956568
1.20	.8849303	.1941861	1.70	.9554345	.0940491
1.21	.8868606	.1918602	1.71	.9563671	.0924591
1.22	.8887676	.1895432	1.72	.9572838	.0908870
1.23	.8906514	.1872354	1.73	.9581849	.0893326
1.24	.8925123	.1849373	1.74	.9590705	.0877961
1.25	.8943502	.1826491	1.75	.9599408	.0862773
1.26	.8961653	.1803712	1.76	.9607961	.0847764
1.27	.8979577	.1781038	1.77	.9616364	.0832932
1.28	.8997274	.1758474	1.78	.9624620	.0818278
1.29	.9014747	.1736022	1.79	.9632730	.0803801
1.30	.9031995	.1713686	1.80	.9640697	.0789502
1.31	.9049021	.1691468	1.81	.9648521	.0775379
1.32	.9065825	.1669370	1.82	.9656205	.0761433
1.33	.9082409	.1647397	1.83	.9663750	.0747663
1.34	.9098773	.1625551	1.84	.9671159	.0734068
1.35	.9114920	.1603833	1.85	.9678432	.0720649
1.36	.9130850	.1582248	1.86	.9685572	.0707404
1.37	.9146565	.1560797	1.87	.9692581	.0694333
1.38	.9162067	.1539483	1.88	.9699460	.0681436
1.39	.9177356	.1518308	1.89	.9706210	.0668711
1.40	.9192433	.1497275	1.90	.9712834	.0656158
1.41	.9207302	.1476385	1.91	.9719334	.0643777
1.42	.9221962	.1455641	1.92	.9725711	.0631566
1.43	.9236415	.1435046	1.93	.9731966	.0619524
1.44	.9250663	.1414600	1.94	.9738102	.0607652
1.45	.9264707	.1394306	1.95	.9744119	.0595947
1.46	.9278550	.1374165	1.96	.9750021	.0584409
1.47	.9292191	.1354181	1.97	.9755808	.0573038
1.48	.9305634	.1334353	1.98	.9761482	.0561831
1.49	.9318879	.1314684	1.99	.9767045	.0550789
1.50	.9331928	.1295176	2.00	.9772499	.0539910

$$Z(X) = e^{-\frac{1}{2}X^2} \Big/ \sqrt{(2\pi)}\,, \quad P(X) = 1 - Q(X) = \int_{-\infty}^{X} Z(u)\,du$$

Table 1 *(Contd.)*

X	$P(X)$	$Z(X)$	X	$P(X)$	$Z(X)$
2.00	.9772499	.0539910	2.50	.9937903	.0175283
2.01	.9777844	.0529192	2.51	.9939634	.0170947
2.02	.9783083	.0518636	2.52	.9941323	.0166701
2.03	.9788217	.0508239	2.53	.9942969	.0162545
2.04	.9793248	.0498001	2.54	.9944574	.0158476
2.05	.9798178	.0487920	2.55	.9946139	.0154493
2.06	.9803007	.0477996	2.56	.9947664	.0150596
2.07	.9807738	.0468226	2.57	.9949159	.0146782
2.08	.9812372	.0458611	2.58	.9950600	.0143051
2.09	.9816911	.0449148	2.59	.9952012	.0139401
2.10	.9821356	.0439836	2.60	.9953388	.0135830
2.11	.9825708	.0430674	2.61	.9954729	.0132337
2.12	.9829970	.0421661	2.62	.9956035	.0128921
2.13	.9834142	.0412795	2.63	.9957308	.0125581
2.14	.9838226	.0404076	2.64	.9958547	.0122315
2.15	.9842224	.0395500	2.65	.9959754	.0119122
2.16	.9846137	.0387069	2.66	.9960930	.0116001
2.17	.9849966	.0378779	2.67	.9962074	.0112951
2.18	.9853713	.0370629	2.68	.9963189	.0109969
2.19	.9857379	.0362619	2.69	.9964274	.0107056
2.20	.9860966	.0354746	2.70	.9965330	.0104209
2.21	.9864474	.0347009	2.71	.9966358	.0101428
2.22	.9867906	.0339408	2.72	.9967359	.0098712
2.23	.9871263	.0331939	2.73	.9968333	.0096058
2.24	.9874545	.0324603	2.74	.9969280	.0093466
2.25	.9877755	.0317397	2.75	.9970202	.0090936
2.26	.9880894	.0310319	2.76	.9971099	.0088465
2.27	.9883962	.0303370	2.77	.9971972	.0086052
2.28	.9886962	.0296546	2.78	.9972821	.0083697
2.29	.9889893	.0289847	2.79	.9973646	.0081398
2.30	.9892759	.0283270	2.80	.9974449	.0079155
2.31	.9895559	.0276816	2.81	.9975229	.0076965
2.32	.9898296	.0270481	2.82	.9975988	.0074829
2.33	.9900969	.0264065	2.83	.9976726	.0072744
2.34	.9903581	.0258166	2.84	.9977443	.0070711
2.35	.9906133	.0250182	2.85	.9978140	.0068728
2.36	.9908625	.0246313	2.86	.9978818	.0066793
2.37	.9911060	.0240556	2.87	.9979476	.0064907
2.38	.9913437	.0234910	2.88	.9980116	.0063067
2.39	.9915758	.0229374	2.89	.9980738	.0061274
2.40	.9918025	.0223945	2.90	.9981342	.0059525
2.41	.9920237	.0218624	2.91	.9981929	.0057821
2.42	.9922397	.0213407	2.92	.9982498	.0056160
2.43	.9924506	.0208294	2.93	.9983052	.0054541
2.44	.9926564	.0203284	2.94	.9983589	.0052963
2.45	.9928572	.0198374	2.95	.9984111	.0051426
2.46	.9930531	.0193563	2.96	.9984618	.0049929
2.47	.9932443	.0188550	2.97	.9985110	.0048470
2.48	.9934309	.0184233	2.98	.9985110	.0048470
2.49	.9936128	.0179711	2.99	.9986051	.0045666
2.50	.9937903	.0175283	3.00	.9986501	.0044318

$$Z(X) = e^{-\frac{1}{2}X^2} \Big/ \sqrt{(2\pi)}\,, \quad P(X) = 1 - Q(X) = \int_{-\infty}^{X} Z(u)\,du$$

Table 1. *(Contd)*

X	P(X)	Z(X)	X	P(X)	Z(X)
3.00	.9986501	.0044318	3.50	.9997674	.0008727
3.01	.9986938	.0043007	3.51	.9997759	.0008426
3.02	.9987361	.0041729	3.52	.9997842	.0008135
3.03	.9987772	.0040486	3.53	.9997922	.0007853
3.04	.9988171	.0039276	3.54	.9997999	.0007581
3.05	.9988558	.0038098	3.55	.9998074	.0007317
3.06	.9988933	.0036951	3.56	.9998146	.0007061
3.07	.9989297	.0035836	3.57	.9998215	.0006814
3.08	.9989650	.0034751	3.58	.9998282	.0006575
3.09	.9989992	.0033695	3.59	.9998347	.0006343
3.10	.9990324	.0032668	3.60	.9998409	.0006119
3.11	.9990646	.0031669	3.61	.9998469	.0005902
3.12	.9990957	.0030698	3.62	.9998527	.0005693
3.13	.9991260	.0029754	3.63	.9998583	.0005490
3.14	.9991553	.0028835	3.64	.9998637	.0005294
3.15	.9991836	.0027943	3.65	.9998689	.0005105
3.16	.9992112	.0027075	3.66	.9998739	.0004921
3.17	.9992378	.0026231	3.67	.9998787	.0004744
3.18	.9992636	.0025412	3.68	.9998834	.0004573
3.19	.9992886	.0024615	3.69	.9998879	.0004408
3.20	.9993129	.0023841	3.70	.9998922	.0004248
3.21	.9993363	.0023089	3.71	.9998964	.0004093
3.22	.9993590	.0022358	3.72	.9999004	.0003944
3.23	.9993810	.0021649	3.73	.9999043	.0003800
3.24	.9994024	.0020960	3.74	.9999080	.0003661
3.25	.9994230	.0020290	3.75	.9999116	.0003526
3.26	.9994429	.0019641	3.76	.9999150	.0003396
3.27	.9994623	.0019010	3.77	.9999184	.0003271
3.28	.9994810	.0018397	3.78	.9999216	.0003149
3.29	.9994991	.0017803	3.79	.9999247	.0003032
3.30	.9995166	.0017226	3.80	.9999277	.0002919
3.31	.9995335	.0016666	3.81	.9999305	.0002810
3.32	.9995499	.0016122	3.82	.9999333	.0002705
3.33	.9995658	.0015595	3.83	.9999359	.0002604
3.34	.9995811	.0015084	3.84	.9999385	.0002506
3.35	.9995959	.0014587	3.85	.9999409	.0002411
3.36	.9996103	.0014106	3.86	.9999433	.0002320
3.37	.9996242	.0013639	3.87	.9999456	.0002232
3.38	.9996376	.0013187	3.88	.9999478	.0002147
3.39	.9996505	.0012748	3.89	.9999499	.0002065
3.40	.9996631	.0012322	3.90	.9999519	.0001987
3.41	.9996752	.0011910	3.91	.9999539	.0001910
3.42	.9996869	.0011510	3.92	.9999557	.0001837
3.43	.9996982	.0011122	3.93	.9999575	.0001766
3.44	.9997091	.0010747	3.94	.9999593	.0001698
3.45	.9997197	.0010383	3.95	.9999609	.0001633
3.46	.9997299	.0010030	3.96	.9999625	.0001569
3.47	.9997398	.0009689	3.97	.9999641	.0001508
3.48	.9997493	.0009358	3.98	.9999655	.0001449
3.49	.9997585	.0009037	3.99	.9999670	.0001393
3.50	.9997674	.0008727	4.00	.9999683	.0001338

$$Z(X) = e^{-\frac{1}{2}X^2} \Big/ \sqrt{(2\pi)}, \quad P(X) = 1 - Q(X) = \int_{-\infty}^{X} Z(u)\, du$$

* **Abridged from Table 1 of the *Biometrika Tables for Statisticians*, Vol. I, with the kind permission of the Biometrika Trustees.**

Table 2.
Percentage points of the χ^2 distribution

n/Q	0.995	0.950	0.050	0.025	0.010	0.005	0.001
1	392704.10^{-10}	393214.10^{-8}	3.84146	5.02389	6.63490	7.87944	10.828
2	0.0100251	0.102587	5.99147	7.37776	9.21034	10.5966	13.816
3	0.0717212	0.351846	7.81473	9.34840	11.3449	12.8381	16.266
4	0.206990	0.710721	9.48773	11.1433	13.2767	14.8602	18.467
5	0.411740	1.145476	11.0705	12.8325	15.0863	16.7496	20.515
6	0.675727	1.63539	12.5916	14.4494	16.8119	18.5476	22.458
7	0.989265	2.16735	14.0671	16.0128	18.4753	20.2777	24.322
8	1.344419	2.73264	15.5073	17.5346	20.0902	21.9550	26.125
9	1.734926	3.32511	16.9190	19.0228	21.6660	23.5893	27.877
10	2.15585	3.94030	18.3070	20.4831	23.2093	25.1882	29.588
11	2.60321	4.57481	19.6751	21.9200	24.7250	26.7569	31.264
12	3.07382	5.22603	21.0261	23.3367	26.2170	28.2995	32.909
13	3.56503	5.89186	22.3621	24.7356	27.6883	29.8194	34.528
14	4.07468	6.57063	23.6848	26.1190	29.1413	31.3193	36.123
15	4.60094	7.26094	24.9958	27.4884	30.5779	32.8013	37.697
16	5.14224	7.96164	26.2962	28.8454	31.9999	34.2672	39.252
17	5.69724	8.67176	27.5871	30.1910	33.4087	35.7185	40.790
18	6.26481	9.39046	28.8693	31.5264	34.8053	37.1564	42.312
19	6.84398	10.1170	30.1435	32.8523	36.1908	38.5822	43.820
20	7.43386	10.8508	31.4104	34.1696	37.5662	39.9968	45.315
21	8.03366	11.5913	32.6705	35.4789	38.9321	41.4010	46.797
22	8.64272	12.3380	33.9244	36.7807	40.2894	42.7956	48.268
23	9.26042	13.0905	35.1725	38.0757	41.6384	44.1813	49.728
24	9.88623	13.8484	36.4151	39.3641	42.9798	45.5585	51.179
25	10.5197	14.6114	37.6525	40.6465	44.3141	46.9278	52.620
26	11.1603	15.3791	38.8852	41.9232	45.6417	48.2899	54.052
27	11.8076	16.1513	40.1133	43.1944	46.9630	49.6449	55.476
28	12.4613	16.9279	41.3372	44.4607	48.2782	50.9933	56.892
29	13.1211	17.7083	42.5569	45.7222	49.5879	52.3356	58.302
30	13.7867	18.4926	43.7729	46.9792	50.8922	53.6720	50.703
40	20.7065	26.5093	55.7585	59.3417	63.6907	66.7659	73.402
50	27.9907	34.7642	67.5048	71.4202	76.1539	79.4900	86.661
60	35.5346	43.1879	79.0819	83.2976	88.3794	91.9517	99.607
70	43.2752	51.7393	90.5312	95.0231	100.425	104.215	112.317
80	51.1720	60.3915	101.879	106.629	112.329	116.321	124.839
90	59.1963	69.1260	113.145	118.136	124.116	128.299	137.208
100	67.3276	77.9295	124.342	129561	135.807	140.169	149.449

$$Q = Q(\chi^2 \mid n) = 1 - P(\chi^2 \mid n) = 2^{-\frac{1}{2}n}\left\{\Gamma\left(\frac{1}{2}n\right)\right\}^{-1}\int_{\chi^2}^{\infty} e^{-\frac{1}{2}x}\, x^{\frac{1}{2}n-1}\, dx.$$

For large value of n, the quantity $\sqrt{(2\chi^2)} - \sqrt{(2n-1)}$ may be taken as a standard normal variable.

*** Abridged from Table 8 of *Biometrika Tables for Statisticians*, I with the kind permission of the Biometrika Trustees.**

Table 3.
Percentage points of the t-distribution

n	Q = 0.4	0.25	0.1	0.05	0.025	0.01	0.005	0.0025	0.001	0.0005
	$2Q$ = 0.8	0.5	0.2	0.1	0.05	0.02	0.01	0.005	0.002	0.001
1	0.325	1.000	3.078	6.314	12.706	31.821	63.657	127.32	318.31	636.62
2	.289	0.816	1.866	2.920	4.303	6.965	9.925	14.089	22.326	31.598
3	.277	.765	1.638	2.353	3.182	4.541	5.841	7.453	10.213	12.924
4	.271	.741	1.533	2.132	2.776	3.747	4.604	5.598	7.173	8.610
5	0.267	0.727	1.476	2.015	2.571	3.365	4.032	4.773	5.893	6.869
6	.265	.718	1.440	1.943	2.447	3.143	3.707	4.317	5.208	5.959
7	.263	.711	1.415	1.895	2.365	2.998	3.499	4.029	4.785	5.408
8	.262	.706	1.397	1.860	2.306	2.896	3.355	3.833	4.501	5.041
9	.261	.703	1.383	1.833	2.262	2.821	3.250	3.690	4.297	4.781
10	0.260	0.700	1.372	1.812	2.228	2.764	3.169	3.581	4.144	4.587
11	.260	.697	1.363	1.796	2.201	2.718	3.106	3.497	4.025	4.437
12	.259	.695	1.356	1.782	2.179	2.681	3.055	3.428	3.930	4.318
13	.259	.694	1.350	1.771	2.160	2.650	3.012	3.372	3.852	4.221
14	.258	.692	1.353	1.761	2.145	2.624	2.977	3.326	3.787	4.140
15	0.258	0.691	1.341	1.753	2.131	2.602	2.947	3.286	3.733	4.073
16	.258	.690	1.337	1.746	2.120	2.583	2.921	3.252	3.686	4.015
17	.257	.689	1.333	1.740	2.110	2.567	2.898	3.222	3.646	3.965
18	.257	.688	1.330	1.734	2.101	2.552	2.878	3.197	3.610	3.922
19	.257	.688	1.328	1.729	2.093	2.539	2.861	3.174	3.579	3.883
20	0.257	0.687	1.325	1.725	2.086	2.528	2.845	3.153	3.552	3.850
21	.257	.686	1.323	1.721	2.080	2.518	2.831	3.135	3.527	3.819
22	.256	.686	1.321	1.717	2.074	2.508	2.819	3.119	3.505	3.792
23	.256	.685	1.319	1.714	2.069	2.500	2.807	3.104	3.485	3.767
24	.256	.685	1.318	1.711	2.064	2.492	2.797	3.091	3.467	3.745

Table 3. *(Contd.)*

n	Q = 0.4 $2Q$ = 0.8	0.25 0.5	0.1 0.2	0.05 0.1	0.025 0.05	0.01 0.02	0.005 0.01	0.0025 0.005	0.001 0.002	0.0005 0.001
25	0.256	0.684	1.316	1.708	2.060	2.485	2.787	3.078	3.450	3.725
26	.256	.684	1.315	1.706	2.056	2.479	2.779	3.067	3.435	3.707
27	.256	.684	1.314	1.703	2.052	2.473	2.771	3.057	3.421	3.690
28	.256	.683	1.313	1.701	2.048	2.467	2.763	3.047	3.408	3.674
29	.256	.683	1.311	1.699	2.045	2.462	2.756	3.038	3.396	3.659
30	0.256	0.683	1.310	1.697	2.042	2.457	2.750	3.030	3.385	3.646
40	.255	.681	1.303	1.684	2.021	2.423	2.704	2.971	3.307	3.551
60	.254	.679	1.296	1.671	2.000	2.390	2.660	2.951	3.232	3.460
120	.254	.677	1.289	1.658	1.980	2.358	2.617	2.860	3.160	3.373
∞	.253	.674	1.282	1.645	1.960	2.326	2.576	2.807	3.000	3.291

Abridged from the *Biometrike Tables for Statisticians,* Vol. I Table 12, with the kind permission of the Biometrika Trustees.

Table 4 (a)
Percentage points of the F-distribution
Upper 5% points

$n_2 \backslash n_1$	1	2	3	4	5	6	7	8	9	10	12	15	20	24	30	40	60	120	∞
1	161.4	199.5	215.7	224.6	230.2	234.0	236.8	238.9	240.5	241.9	243.9	245.9	248.0	249.1	250.1	251.1	252.2	253.3	254.3
2	18.51	19.00	19.16	19.25	19.30	19.33	19.35	19.37	19.38	19.40	19.41	19.43	19.45	19.45	19.46	19.47	19.48	19.49	19.50
3	10.13	9.55	9.28	9.12	9.01	8.94	8.89	8.85	8.81	8.79	8.74	8.70	8.66	8.64	8.62	8.59	8.57	8.55	8.53
4	7.71	6.94	6.59	6.39	6.26	6.16	6.09	6.04	6.00	5.96	5.91	5.86	5.80	5.77	5.75	5.72	5.69	5.66	5.63
5	6.61	5.79	5.41	5.19	5.05	4.95	4.88	4.82	4.77	4.74	4.68	4.62	4.56	4.53	4.50	4.46	4.43	4.40	4.36
6	5.99	5.14	4.76	4.53	4.39	4.28	4.21	4.15	4.10	4.06	4.00	3.94	3.87	3.84	3.81	3.77	3.74	3.70	3.67
7	5.59	4.74	4.35	4.12	3.97	3.87	3.79	3.73	3.68	3.64	3.57	3.51	3.44	3.41	3.38	3.34	3.30	3.27	3.23
8	5.32	4.46	4.07	3.84	3.69	3.58	3.50	3.44	3.39	3.35	3.28	3.22	3.15	3.12	3.08	3.04	3.01	2.97	2.93
9	5.12	4.26	3.86	3.63	3.48	3.37	3.29	3.23	3.18	3.14	3.07	3.01	2.94	2.90	2.86	2.83	2.79	2.75	2.71
10	4.96	4.10	3.71	3.48	3.33	3.22	3.14	3.07	3.02	2.98	2.91	2.85	2.77	2.74	2.70	2.66	2.62	2.58	2.54
11	4.84	3.98	3.59	3.36	3.20	3.09	3.01	2.95	2.90	2.85	2.79	2.72	2.65	2.61	2.57	2.53	2.49	2.45	2.40
12	4.75	3.89	3.49	3.26	3.11	3.00	2.91	2.85	2.80	2.75	2.69	2.62	2.54	2.51	2.47	2.43	2.38	2.34	2.30
13	4.67	3.81	3.41	3.18	3.03	2.92	2.83	2.77	2.71	2.67	2.60	2.53	2.46	2.42	2.38	2.34	2.30	2.25	2.21
14	4.60	3.74	3.34	3.11	2.96	2.85	2.76	2.70	2.65	2.60	2.53	2.46	2.39	2.35	2.31	2.27	2.22	2.18	2.13
15	4.54	3.68Q	3.29	3.06	2.90	2.79	2.71	2.64	2.59	2.54	2.48	2.40	2.33	2.29	2.25	2.20	2.16	2.11	2.07
16	4.49	3.63	3.24	3.01	2.85	2.74	2.66	2.59	2.54	2.49	2.42	2.35	2.28	2.24	2.19	2.15	2.11	2.06	1.01
17	4.45	3.59	3.20	2.96	2.81	2.70	2.61	2.55	2.49	2.45	2.38	2.31	2.23	2.19	2.15	2.10	2.06	2.01	1.96
18	4.41	3.55	3.16	2.93	2.77	2.66	2.58	2.51	2.46	2.41	2.34	2.27	2.19	2.15	2.11	2.06	2.02	1.97	1.92
19	4.38	3.52	3.13	2.90	2.74	2.63	2.54	2.48	2.42	2.38	2.31	2.23	2.16	2.11	2.07	2.03	1.98	1.93	1.88

Table 4 (a) *F*-distribution *(Contd.)*

$n_2 \backslash n_1$	1	2	3	4	5	6	7	8	9	10	11	12	15	20	24	30	40	60	∞
20	4.35	3.49	3.10	2.87	2.71	2.60	2.51	2.45	2.39	2.35	2.28	2.20	2.12	2.08	2.04	1.99	1.95	1.90	1.84
21	4.32	3.47	3.07	2.84	2.68	2.57	2.49	2.42	2.37	2.32	2.25	2.18	2.10	2.05	2.01	1.96	1.92	1.87	1.81
22	4.30	3.44	3.05	2.82	2.66	2.55	2.46	2.40	2.34	2.30	2.23	2.15	2.07	2.03	1.98	1.94	1.89	1.84	1.78
23	4.28	3.42	3.03	2.80	2.64	2.53	2.44	2.37	2.32	2.27	2.20	2.13	2.05	2.01	1.96	1.91	1.86	1.81	1.76
24	4.26	3.40	3.01	2.78	2.62	2.51	2.42	2.36	2.30	2.25	2.18	2.11	2.03	1.98	1.94	1.89	1.84	1.79	1.73
25	4.24	3.39	2.99	2.76	2.60	2.49	2.40	2.34	2.28	2.24	2.16	2.09	2.01	1.96	1.92	1.87	1.82	1.77	1.71
26	4.23	3.37	2.98	2.74	2.59	2.47	2.39	2.32	2.27	2.22	2.15	2.07	1.99	1.95	1.90	1.85	1.80	1.75	1.69
27	4.21	3.35	2.96	2.73	2.57	2.46	2.37	2.31	2.25	2.20	2.13	2.06	1.97	1.93	1.88	1.84	1.79	1.73	1.67
28	4.20	3.34	2.95	2.71	2.56	2.45	2.36	2.29	2.24	2.19	2.12	2.04	1.96	1.91	1.87	1.82	1.77	1.71	1.65
29	4.18	3.33	2.93	2.70	2.55	2.43	2.35	2.28	2.22	2.18	2.10	2.03	1.94	1.90	1.85	1.81	1.75	1.70	1.64
30	4.17	3.32	2.92	2.69	2.53	2.42	2.33	2.27	2.21	2.16	2.09	2.01	1.93	1.89	1.84	1.79	1.74	1.68	1.62
40	4.08	3.23	2.84	2.61	2.45	2.34	2.25	2.18	2.12	2.08	2.00	1.92	1.84	1.79	1.74	1.69	1.64	1.58	1.51
60	4.00	3.15	2.76	2.53	2.37	2.25	2.17	2.10	2.04	1.99	1.92	1.84	1.75	1.70	1.65	1.59	1.53	1.47	1.39
120	3.92	3.07	2.68	2.45	2.29	2.17	2.09	2.02	1.96	1.91	1.83	1.75	1.66	1.61	1.55	1.50	1.43	1.35	1.25
∞	3.84	3.00	2.60	2.37	2.21	2.10	2.01	1.94	1.88	1.83	1.75	1.67	1.57	1.52	1.46	1.39	1.32	1.22	1.00

* Abridged from Table 18 of the *Biometrika Tables for Statisticians*, Vol. I with the kind permission of the Biometrika Trustees.

Table 4 (b)
Percentage points of the F-distribution
Upper 1% points

$n_2 \backslash n_1$	1	2	3	4	5	6	7	8	9	10	12	15	20	24	30	40	60	120	∞
1	4052	4999.5	5403	5625	5764	5859	5928	5982	6022	6056	6106	6157	6209	6235	6261	6287	6313	6339	6366
2	98.50	99.00	99.17	99.25	99.30	99.33	99.36	99.37	99.39	99.40	99.42	99.43	99.45	99.46	99.47	99.47	99.48	99.49	99.50
3	34.42	30.82	29.46	28.71	28.24	27.91	27.67	27.49	27.35	27.23	27.05	26.87	26.69	26.60	26.50	26.41	26.32	26.22	26.13
4	21.20	18.00	16.69	15.98	15.52	15.21	14.98	14.80	14.66	14.55	14.37	14.20	14.02	13.93	13.84	13.75	13.65	13.56	13.46
5	16.26	13.27	12.06	11.39	10.97	10.67	10.46	10.29	10.16	10.05	9.89	9.72	9.55	9.47	9.38	9.29	9.20	9.11	9.02
6	13.75	10.92	9.78	9.15	8.75	8.47	8.26	8.10	7.98	7.87	7.72	7.56	7.40	7.31	7.23	7.14	7.06	6.97	6.88
7	12.25	9.55	8.45	7.85	7.46	7.19	6.99	6.84	6.72	6.62	6.47	6.31	6.16	6.07	5.99	5.91	5.82	5.74	5.65
8	11.26	8.65	7.59	7.01	6.63	6.37	6.18	6.03	5.91	5.81	5.67	5.52	5.36	5.28	5.20	5.12	5.03	4.95	4.86
9	10.56	8.02	6.99	6.42	6.06	5.80	5.61	5.47	5.35	5.26	5.11	4.96	4.81	4.73	4.65	4.57	4.48	4.40	4.31
10	10.04	7.56	6.55	5.99	5.64	5.39	5.20	5.06	4.94	4.85	4.71	4.56	4.41	4.33	4.25	4.17	4.08	4.00	3.91
11	9.65	7.21	6.22	5.67	5.32	5.07	4.89	4.74	4.63	4.54	4.40	4.25	4.10	4.02	3.94	3.86	3.78	3.69	3.60
12	9.33	6.93	5.95	5.41	5.06	4.82	4.64	4.50	4.39	4.30	4.16	4.01	3.86	3.78	3.70	3.62	3.54	3.45	3.36
13	9.07	6.70	5.74	5.21	4.86	4.62	4.44	4.30	4.19	4.10	3.96	3.82	3.66	3.59	3.51	3.43	3.34	3.25	3.17
14	8.86	6.51	5.56	5.04	4.69	4.46	4.28	4.14	4.03	3.94	3.80	3.66	3.51	3.43	3.35	3.27	3.18	3.09	3.00
15	8.68	6.36	5.42	4.89	4.56	4.32	4.14	4.00	3.89	3.80	3.67	3.52	3.37	3.29	3.21	3.13	3.05	2.96	2.87
16	8.53	6.23	5.29	4.77	4.44	4.20	4.03	3.89	3.78	3.69	3.55	3.41	3.26	3.18	3.10	3.02	2.93	2.84	2.75
17	8.40	6.11	5.18	4.67	4.34	4.10	3.93	3.79	3.68	3.59	3.46	3.31	3.16	3.08	3.00	2.92	2.83	2.75	2.65
18	8.29	6.01	5.09	4.58	4.25	4.01	3.84	3.71	3.60	3.51	3.37	3.23	3.08	3.00	2.92	2.84	2.75	2.66	2.57
19	8.18	5.93	5.01	4.50	4.17	3.94	3.77	3.63	3.52	3.43	3.30	3.15	3.00	2.92	2.84	2.76	2.67	2.58	2.49

Table 4(b). F-distribution *(Contd.)*

$n_2 \backslash n_1$	1	2	3	4	5	6	7	8	9	10	12	15	20	24	30	40	60	120	∞
20	8.10	5.85	4.94	4.43	4.10	3.87	3.70	3.56	3.46	3.37	3.23	3.09	2.94	2.86	2.78	2.69	2.61	2.52	2.42
21	8.02	5.78	4.87	4.37	4.04	3.81	3.64	3.51	3.40	3.31	3.17	3.03	2.88	2.80	2.72	2.64	2.55	2.46	2.36
22	7.95	5.72	4.82	4.31	3.99	3.76	3.59	3.45	3.35	3.26	3.12	2.98	2.83	2.75	2.67	2.58	2.50	2.40	2.31
23	7.88	5.66	4.76	4.26	3.94	3.71	3.54	3.41	3.30	3.21	3.07	2.93	2.78	2.70	2.62	2.54	2.45	2.35	2.26
24	7.82	5.61	4.72	4.22	3.90	3.67	3.50	3.36	3.26	3.17	3.03	2.89	2.74	2.66	2.58	2.49	2.40	2.31	2.21
25	7.77	5.57	4.68	4.18	3.85	3.63	3.46	3.32	3.22	3.13	2.99	2.85	2.70	2.62	2.54	2.45	2.36	2.27	2.17
26	7.72	5.53	4.64	4.14	3.82	3.59	3.42	3.29	3.18	3.09	2.96	2.81	2.66	2.58	2.50	2.42	2.33	2.23	2.13
27	7.68	5.49	4.60	4.11	3.78	3.56	3.39	3.26	3.15	3.06	2.93	2.78	2.63	2.55	2.47	2.38	2.29	2.20	2.10
28	7.64	5.45	4.57	4.07	3.75	3.53	3.36	3.23	3.12	3.03	2.90	2.75	2.60	2.52	2.44	2.35	2.26	2.17	2.06
29	7.60	5.42	4.54	4.04	3.73	3.50	3.33	3.20	3.09	3.00	2.87	2.73	2.57	2.49	2.41	2.33	2.23	2.14	2.03
30	7.56	5.39	4.51	4.02	3.70	3.47	3.30	3.17	3.07	2.98	2.84	2.70	2.55	2.47	2.39	2.30	2.21	2.11	2.01
40	7.31	5.18	4.31	3.83	3.51	3.29	3.12	2.99	2.89	2.80	2.66	2.52	2.37	2.29	2.20	2.11	2.02	1.92	1.80
60	7.08	4.98	4.13	3.65	3.34	3.12	2.95	2.82	2.72	2.63	2.50	2.35	2.20	2.12	2.03	1.94	1.84	1.73	1.60
120	6.85	4.79	3.95	2.48	2.17	2.96	2.79	2.66	2.56	2.47	2.34	2.19	2.03	1.95	1.86	1.76	1.66	1.53	1.38
∞	6.63	4.61	3.78	3.32	3.02	2.80	2.64	2.51	2.41	2.32	2.18	2.04	1.88	1.79	1.70	1.59	1.47	1.32	1.00

Abridged from Table 18 of the *Biometrika Tables for Statisticians*, Vol. I with the kind permission of the Biometrika Trustees.

Table 5.

Individual terms $\frac{e^{-m} m^i}{i!}$ of Poisson distribution

i	m = 0.1	0.2	0.3	0.4	0.5	0.6	0.7	0.8	0.9	1.0	i
0	.904837	.818731	.740818	.670320	.606531	.548812	.496585	.449329	.406570	.367879	0
1	.090484	.163746	.222245	.268128	.303265	.329287	.347610	.359463	.365913	.367879	1
2	.004524	.016375	.033337	.053626	.075816	.098786	.121663	.143785	.164661	.183940	2
3	.000151	.001092	.003334	.007150	.012636	.019575	.028388	.038343	.048398	.061313	3
4	.000004	.000055	.000250	.000715	.001580	.002964	.004968	.007667	.011115	.015328	4
5	—	.000002	.000015	.000057	.000158	.000356	.000696	.001227	.002001	.003066	5
6	—	—	.000001	.000004	.000013	.000036	.000081	.000164	.000300	.000511	6
7	—	—	—	—	.000001	.000003	.000008	.000019	.000039	.000073	7
8	—	—	—	—	—	—	.000001	.000002	.000004	.000009	8
9	—	—	—	—	—	—	—	—	—	.000001	9

i	m = 1.1	1.2	1.3	1.4	1.5	1.6	1.7	1.8	1.9	2.0	i
0	.332871	.301194	.272532	.246597	.223130	.201897	.182684	.165299	.149569	.135335	0
1	.366158	.361433	.354291	.345236	.334695	.323034	.310562	.297538	.284180	.270671	1
2	.201387	.216860	.230289	.241665	.251021	.258428	.263978	.267784	.269971	.270671	2
3	.073842	.086744	.099792	.112777	.125510	.137828	.149587	.160671	.170982	.180447	3
4	.020307	.026023	.032432	.039472	.047067	.055131	.063575	.072302	.081216	.090221	4
5	.004467	.006246	.008432	.011052	.014120	.017642	.021615	.26029	.030862	.036089	5
6	.000819	.001249	.001827	.002589	.003530	.004705	.006124	.007809	.009773	.012030	6
7	.000129	.000214	.000339	.000516	.000756	.001075	.001487	.002008	.002653	.003437	7
8	.000018	.000032	.000055	.000090	.000142	.000215	.000316	.000452	.000630	.000859	8
9	.000002	.000004	.000008	.000014	.000024	.000038	.000060	.000090	.000133	.000191	9
10	—	.000001	.000001	.000002	.000004	.000006	.000010	.000016	.000025	.000038	10
11	—	—	—	—	—	.000001	.000002	.000003	.000004	.000007	11
12	—	—	—	—	—	—	—	—	.000001	.000001	12

Table 5. *(Contd.)*

	m										
i	2.1	2.2	2.3	2.4	2.5	2.6	2.7	2.8	2.9	3.0	*i*
0	.122456	.110803	.100259	.090718	.082085	.074274	.067206	.060180	.055023	.049787	0
1	.257159	.243767	.230595	.217723	.205212	.193111	.181455	.170268	.159567	.149361	1
2	.270016	.268144	.265185	.261268	.256516	.251045	.244964	.238375	.231373	.224042	2
3	.189012	.196639	.203308	.209014	.213763	.217572	.220468	.222484	.223660	.224042	3
4	.099231	.108151	.116902	.125409	.133602	.141422	.148816	.155739	.162154	.168031	4
5	.041677	.047587	.053775	.060196	.066801	.073539	.080360	.087214	.094049	.100819	5
6	.014587	.017448	.020614	.024078	.027834	.031867	.036162	.040700	.045457	.050409	6
7	.004376	.005484	.006773	.008255	.009941	.011836	.013948	.016280	.018832	.021604	7
8	.001149	.001508	.001947	.002477	.003106	.003847	.004708	.005698	.006827	.008102	8
9	.000268	.000369	.000498	.000660	.000863	.001111	.000412	.001773	.002200	.002701	9
10	.000056	.000081	.000114	.000158	.000216	.000289	.000381	.000496	.000638	.000810	10
11	.000011	.000016	.000024	.000035	.000049	.000068	.000094	.000126	.000168	.000221	11
12	.000002	.000003	.000005	.000007	.000010	.000015	.000021	.000029	.000041	.000055	12
13	—	.000001	.000001	.000001	.000002	.000003	.000004	.000006	.000009	.000013	13
14	—	—	—	—	—	.000001	.000001	.000001	.000002	.000003	14
15	—	—	—	—	—	—	—	—	—	.000001	15

Table 5. *(Contd.)*

					m						
i	3.1	3.2	3.3	3.4	3.5	3.6	3.7	3.8	3.9	4.0	*i*
0	.045049	.040762	.036883	.033873	.030197	.027324	.024724	.022371	.020242	.018316	0
1	.139653	.130439	.121714	.113469	.105691	.098365	.091477	.085009	.078943	.073263	1
2	.216461	.208702	.200829	.192898	.184959	.177058	.169233	.161517	.153940	.146525	2
3	.223677	.222616	.220912	.218617	.215785	.212469	.208720	.204588	.200122	.195367	3
4	.173350	.178093	.182252	.185825	.188812	.191222	.193066	.194359	.195119	.195367	4
5	.107477	.113979	.120286	.126361	.132169	.137680	.142869	.147713	.152193	.156293	5
6	.055530	.060789	.066158	.071604	.077098	.082608	.088102	.093551	.098925	.104196	6
7	.024592	.027789	.031189	.034779	.038549	.042484	.046568	.050785	.055115	.059540	7
8	.009529	.011116	.012865	.014781	.016865	.019118	.021538	.024123	.026869	.029770	8
9	.003282	.003952	.004717	.005584	.006559	.007647	.008854	.010185	.011643	.013231	9
10	.001018	.001265	.001557	.001899	.002296	.002756	.003276	.003870	.004541	.005292	10
11	.000287	.000368	.000167	.000587	.000730	.000901	.001102	.001137	.001610	.001925	11
12	.000074	.000098	.000128	.000166	.000213	.000270	.000340	.000423	.000523	.000642	12
13	.000018	.000024	.000033	.000043	.000057	.000075	.000097	.000124	.000157	.000197	13
14	.000004	.000006	.000008	.000011	.000014	.000019	.000026	.000034	.000044	.000056	14
15	.000001	.000001	.000002	.000002	.000003	.000005	.000006	.000009	.000011	.000015	15
16	—	—	—	.000001	.000001	.000001	.000001	.000002	.000003	.000004	16
17	—	—	—	—	—	—	—	—	.000001	.000001	17

Table 5. *(Contd.)*

i	*m* = 4.1	4.2	4.3	4.4	4.5	4.6	4.7	4.8	4.9	5.0	*i*
0	.016573	.014996	.013569	.012277	.011109	.010052	.009095	.008230	.007447	.006738	0
1	.067948	.062981	.058345	.054020	.049990	.046238	.042748	.039503	.036488	.033690	1
2	.139293	.132261	.125441	.118845	.112479	.106348	.100457	.091807	.089396	.084224	2
3	.190368	.185165	.179799	.174305	.168718	.163068	.157383	.151691	.146014	.140374	3
4	.195127	.194424	.193284	.191736	.198808	.187528	.184925	.182029	.178867	.175467	4
5	.160004	.163316	.166224	.168728	.170827	.172525	.173830	.174748	.175290	.175467	5
6	.109336	.114321	.119127	.123734	.128120	.132270	.136167	.139798	.143153	.146223	6
7	.064040	.068593	.073178	.077775	.082363	.086920	.091426	.095862	.100207	.104445	7
8	.032820	.036011	.039333	.042776	.046329	.049979	.053713	.057517	.061377	.065278	8
9	.014951	.016806	.018793	.020913	.023165	.025545	.028050	.030676	.033416	.036266	9
10	.006130	.007058	.008081	.009202	.010424	.011751	.013184	.014724	.016374	.018133	10
11	.002285	.002695	.003159	.003681	.004264	.004914	.005633	.006425	.007294	.008242	11
12	.000781	.000943	.001132	.001350	.001599	.001884	.002206	.002570	.002978	.003434	12
13	.000246	.000305	.000374	.000457	.000554	.000667	.000798	.000949	.001123	.001321	13
14	.000072	.000091	.000115	.000144	.000178	.000219	.000268	.000325	.000393	.000472	14
15	.000020	.000026	.000033	.000042	.000053	.000067	.000084	.000104	.000128	.000157	15
16	.000005	.000007	.000009	.000012	.000015	.000019	.000025	.000031	.000039	.000049	16
17	.000001	.000902	.000002	.000003	.000004	.000005	.000007	.000009	.000011	.000014	17
18	—	—	.000001	.000001	.000001	.000001	.000002	.000002	.000003	.000004	18
19	—	—	—	—	—	—	—	.000001	.000001	.000001	19

* Abridged from Table 39 of the *Biometrika Tables for Statisticians*, Vol. I with the kind permission of the Biometrika Trustees.

APPENDIX B

COMPUTER PROGRAMS FORTRAN 77

CP 1. Program to Compute Mean, Variance and STD for Ungrouped Data
CP 2. Program to Compute Mean, Variance and STD for Grouped Data
CP 3. Program to Compute Correlation Coefficient and to Fit Regression Line of Y on X (Ungrouped Data)
CP 4. Program to Compute Correlation Coefficient (Ungrouped data): With Arbitrary Origin
CP 5. Program to Fit Linear Trend to Time-Series Data
CP 6. Program to Fit Quadratic Trend to Time-Series Data
CP 7. Program to Fit Exponential Curve of the Type $Y = AB^X$ (Y = A*B**X) to Time Series Data
CP 8. Program to Compute Correlation Coefficient and to Fit Regression Lines (Grouped Data)

THE AUTHOR ACKNOWLEDGES THE EXPERTISE OF AND HELP RENDERED BY S. K. SINHA WITH DEEP APPRECIATION.

N.B.The Programs have been written (in FORTRAN 77) in accordance with the procedure used in the text; these have been illustrated in some cases.

Suitable alternative Programs can be easily written (in FORTRAN as well as in other languages).

CP 1

```
C      PROGRAM TO COMPUTE MEAN, VARIANCE AND STD. DEVIATION
C      FOR UNGROUPED DATA
C      NC = NO. OF CLASSES
C      X (I) = MID POINT OF I TH CLASS
C      F (I) = I TH FREQUENCY
C      C = CLASS WIDTH
C      A = ORIGIN TAKEN AT MID-POINT
       INTEGER I, NC
       REAL X (10), F (10), D, U. FU. FU2. A, C, SUMF, SUMFU, SUMFU2
       DATA SUMF, SUMFU, SUMFU2 /3*0.0/
       WRITE (*, *) 'ENTER A, C, NC'
       READ (*, *) A, C, NC
       WRITE (*, *) 'ENTER X-VALUES'
       READ (*, *) (X (I), I = 1, NC)
       WRITE (*, *) 'ENTER F-VALUES'
       READ (*, *) (F (I), I = 1, NC)
       WRITE (*, 33)
       DO 10 I = 1, NC
          D = X (I) - A
          U = D / C
          FU = F (I) * U
          FU2 = FU * U
          SUMF = SUMF + F (I)
          SUMFU = SUMFU + FU
          SUMFU2 = SUMFU2 + FU2
          WRITE (*, 11) X (I), D, U, F (I), FU, FU2

10     CONTINUE
       UBAR = SUMFU / SUMF
       XBAR = A + C * UBAR
       VAR = C * C * (SUMFU2 / SUMF - (SUMFU / SUMF) ** 2)
       STD = SQRT (VAR)
       WRITE (*, 44)
       WRITE (*, 22) SUMF, SUMFU, SUMFU2, UBAR, XBAR, VAR, STD

11     FORMAT (2X, 6F 10.3)
22     FORMAT (2X, 7F 10.3)
33     FORMAT (/ / 6X, 'X', 9X, 'D', 9X, 'U', 9X, 'F', 9X, 'FU', 8X, 'FU2' / /)
44     FORMAT (/ /7X, 'SUMF', 6X, 'SUMFU', 5X, 'SUMFU2', 4X, 'UBAR', 6X,
     * 'XBAR', 6X, 'XBAR', 6X, 'STD')
       STOP
       END
```

CP 2

```
C      PROGRAM TO COMPUTE MEAN, VARIANCE & STD. DEVIATION
C      FOR GROUPED DATA
C      NC = NO. OF CLASSES
C      X (I) = MID POINT OF I TH CLASS
C      F (I) = I TH FREQUENCY
C      C = CLASS WIDTH
C      A = ORIGIN TAKEN AT ONE OF TWO MIDDLE CLASSES
C      LL (I) = LOWER LIMIT UL (I) = UPPER LIMIT OF I TH CLASS
       INTEGER I, NC, MAXI
       REAL X, F (10), D, U, FU, FU2, A, C, SUMF, SUMFU, SUMFU2, LL (20), UL (20)
       DATA SUMF, SUMFU, SUMFU2 /3* 0.0/
       WRITE (*, *) 'ENTER A, C, NC'
       READ (*, *) A, C, NC
       WRITE (*, *) 'ENTER LL, UL, F FOR NC CLASSES'
       READ (*, *) (LL (I), UL (I), F (I), I = 1, NC)
       WRITE (*, 33)
       DO 20 I = 1, NC
            X = (LL (I) + UL (I)) /2.0
            D = X - A
            U = D / C
            FU = F (I) * U
            FU2 = FU * U
            SUMF = SUMF + F (I)
            SUMFU = SUMFU + FU
            SUMFU2 = SUMFU2 + FU2
            WRITE (*, 11) LL (I), UL(I), X, D, U, F (I), FU, FU2
20     CONTINUE
       UBAR = SUMFU / SUMF
       XBAR = A + C * UBAR
       VAR = C * C * (SUMFU2 / SUMF - (SUMFU / SUMF) ** 2)
       STD = SQRT (VAR)
       WRITE (*, 44)
       WRITE (*, 22) SUMF, SUMFU, SUMFU2, UBAR, XBAR, VAR, STD
11     FORMAT (2X, F6.1, '--', F6.1, 6F 10.3)
22     FORMAT (2X, 7F 10.3)
33     FORMAT (/ /7X, 'L', 6X, 'U', 7X, 'X', 9X, 'D', 9X, 'U', 9X, 'F', 9X, 'FU',
     * 8X, 'FU2' / /)
44     FORMAT (/ /7X, 'SUMF', 6X, 'SUMFU', 5X, 'SUMFU2', 4X, 'UBAR', 6X, 'XBAR'
     * 6X, 'VAR', 6X, 'STD')
       STOP
       END
```

CP 3

```
C     PROGRAM TO COMPUTE CORRELATION COEFFICIENT AND TO FIT
C     REGRESSION LINE OF Y ON X
C     N = NUMBER OF PAIRS OF OBSERVATIONS
      REAL X (10), Y (10), SUMX, SUMY, XBAR, YBAR, SUMDXY, SUMDX2, SUMDY2
      REAL R, BCAP, ACAP
      INTEGER I, N
      WRITE (*, *) 'ENTER NO. OF OBS. (N)'
      READ (*, *) N
      WRITE (*, *) 'ENTER X - VALUES'
      READ (*, *) (X (I), I = 1, N)
      WRITE (*, *) 'ENTER Y - VALUES'
      READ (*, *) (Y (I), I = 1, N)
      DO 10 I = 1, N
         SUMX = SUMX + X (I)
         SUMY = SUMY + Y (I)
10    CONTINUE
      XBAR = SUMX/FLOAT (N)
      YBAR = SUMY/FLOAT (N)
      DO 20 I = 1, N
         DX = X (I) - XBAR
         DY = Y (I) - YBAR
         SUMDXY = SUMDXY + DX * DY
         SUMDX2 = SUMDX2 + DX * DX
         SUMDY2 = SUMDY2 + DY * DY
20    CONTINUE
      R = (SUMDXY/SQRT (SUMDX2 * SUMDY2)
      BCAP = SUMDXY / SUMDX2
      ACAP = YBAR - BCAP * XBAR
      WRITE (*, 22) R
      WRITE (*, 33) ACAP, BCAP
22    FORMAT (/ /1X, 'R = ', F 10.3)
33    FORMAT (1X, 'EQUATION IS Y = ', F 10.3,' + ', F 10.3,' * X')
      STOP
      END
```

CP 4

```
C       PROGRAM TO COMPUTE CORRELATION COEFFICIENT FOR
C       UNGROUPED DATA
C       WITH ARBITRARY ORIGIN (OX, OY)
C       N = NUMBER OF PAIRS OF OBSERVATIONS
        REAL X (10), Y (10), DX, DY, SUMDX, SUMDY, SUMDXY, SUMDX2. SUMDY2
        REAL R, NUM, DEN, OX, OY
        INTEGER I, N
        WRITE (*, *) 'ENTER NO. OF PAIRS OF OBSERVATION'
        READ (*, *) N
        WRITE (*, *) 'ENTER THE ORIGIN (OX, OY)'
        READ (*, *) OX, OY
        WRITE (*, *) 'ENTER THE VALUES OF X'
        READ (*, *) (X (I), I = 1, N)
        WRITE (*, *) 'ENTER THE VALUES OF Y'
        READ (*, *) (Y (I), I = 1, N)
        DO 20 I = 1, N
            DX = X (I) - OX
            DY = Y (I) - OY
            SUMDX = SUMDX + DX
            SUMDY = SUMDY + DY
            SUMDXY = SUMDXY + DX * DY
            SUMDX2 = SUMDX2 + DX * DX
            SUMDY2 = SUMDY2 + DY * DY
20      CONTINUE
        NUM = N * SUMDXY - SUMDX * SUMDY
        DEN = SQRT ((N * SUMDX2 - SUMDX * SUMDX) * (N * SUMDY2 - SUMDY *
     1 SUMDY))
        R = NUM / DEN
        WRITE (*, 22) R
22      FORMAT (1X, 'R = ', F 10.3)
        STOP
        END
```

```
        ILLUSTRATION : EXAMPLE 3(b), CHAPTER 9
ENTER NO. OF PAIRS OF OBSERVATION
10
ENTER THE ORIGIN (OX, OY)
165, 165
ENTER THE VALUES OF X
170 167 162 163 167 166 169 171 164 165
ENTER THE VALUES OF Y
168 167 166 166 168 165 168 170 165 168
R = 0.760
Stop - Program terminated.
```

CP 5

```
C      PROGRAM TO FIT LINEAR TREND TO TIME SERIES DATA
       INTEGER N, I, MID
       REAL Y (20), X (20), X2, XY, SUMY, SUMXY, SUMX2, ACAP, BCAP, YCAP
       DATA SUMY, SUMXY, SUMX2 /3* 0.0/
       WRITE (*, *) 'ENTER NO OF OBSERVATIONS'
       READ (*, *) N
       WRITE (*, *) 'ENTER VALUES OF Y'
       READ (*, *) (Y (I), I = 1, N)
       IF (MOD N, 2). EQ. O) THEN
C           EVEN NO. OF OBSERVATIONS X UNITS 1/2 YEAR
            DO 10 I = 1, N
                X (I) = 2 * I - N - 1
10          CONTINUE
       ELSE
C           ODD NO OF OBSERVATIONS X UNITS 1 YEAR
            MID = (N + 1) / 2
            DO 20 I = 1, N
                X (I) = I - MID
20          CONTINUE
       ENDIF
       WRITE (*, *)'        X          Y          X2          XY'
       DO 30 I = 1, N
            X2 = X (I) * X (I)
            XY = X (I) * Y (I)
            SUMY = SUMY + Y (I)
            SUMXY = SUMXY + XY
            SUMX2 = SUMX2 + X2
            WRITE (*, 33) X (I), Y (I), X2, XY
30     CONTINUE
       ACAP = SUMY / FLOAT (N)
       BCAP = SUMXY / SUMX2
       WRITE (*, 11) ACAP, BCAP
       WRITE (*, *) 'TREND VALUES ARE'
       DO 40 I = 1, N
            YCAP = ACAP + BCAP * X (I)
            WRITE (*, 22) YCAP
40     CONTINUE
11     FORMAT (/ / 1X, 'TREND EQUATION IS YCAP = ', F10.3,' + ', F10.3,' * X')
22     FORMAT (5X, F 10.3)
33     FORMAT (1X, 4F 10.3)
       STOP
       END
```

CP 6

```
C       PROGRAM TO FIT QUADRATIC TREND TO TIME SERIES DATA
        INTEGER N, I, MID
        REAL Y (20), X (20), X2, XY, SUMY, SUMXY, SUMX2, SUMX2Y, SUMX4
        REAL ACAP, BCAP, YCAP, DEN
        DATA SUMY, SUMXY, SUMX2, SUMX2Y, SUMX4 / 5 *0.0/
        WRITE (*, *) 'ENTER NO OF OBSERVATIONS'
        READ (*, *) N
        WRITE (*, *) 'ENTER VALUES OF Y'
        READ (*, *) (Y (I), I = 1, N)
        IF (MOD (N, 2). EQ. 0) THEN
C           EVEN NO. OF OBSERVATIONS X UNITS 1/2 YEAR
            DO 10  I = 1, N
                X (I) = 2 * I - N - 1
10          CONTINUE
        ELSE
C           ODD NO OF OBSERVATIONS X UNITS 1 YEAR
            MID = (N + 1) / 2
            DO 20  I = 1, N
                X (I) = I - MID
20          CONTINUE
        ENDIF
        WRITE (*, *)'        X         Y         X2        XY        X2Y        X4'
        DO 30  I = 1, N
            X2 = X (I) * X (I)
            XY = X (I) * Y (I)
            X2Y = X2 * Y (I)
            X4 = X2 * X2
            SUMY = SUMY + Y (I)
            SUMXY = SUMXY + XY
            SUMX2 = SUMX2 + X2
            SUMX2Y = SUMX2Y + X2Y
            SUMX4 = SUMX4 + X4
            WRITE (*, 33) X (I), Y (I), X2, XY, X2Y, X4
30      CONTINUE
        DEN = N * SUMX4 - SUMX2 * SUMX2
        ACAP = (SUMX4 * SUMY - SUMX2 * SUMX2Y) / DEN
        BCAP = SUMXY / SUMX2
        CCAP = (SUMX2Y * N - SUMX2 * SUMY) / DEN
        WRITE (*, 11) ACAP, BCAP, CCAP
        WRITE (*, *) 'TREND VALUES ARE'
        DO 40  I = 1, N
            YCAP = ACAP + (BCAP + CCAP * X (I)) * X (I)
            WRITE (*, 22) YCAP
40      CONTINUE
11      FORMAT (/ / 1X, 'TREND EQUATION IS YCAP = ', F10.4,' + ', F10.4,
     1  * X + ', F 10.4,' * X2')
22      FORMAT (5X, F 10.3)
33      FORMAT (1X, 6F 10.3)
        STOP
        END
```

ILLUSTRATION :

EXAMPLE 5(e), CHAPTER 12

ENTER NO OF OBSERVATIONS

7

ENTER VALUES OF Y

177.2 185.0 224.9 254.0 304.9 359.9 438.8

X	Y	X2	XY	X2Y	X4
−3.000	177.200	9.000	−531.600	1594.800	81.000
−2.000	185.000	4.000	−370.000	740.000	16.000
−1.000	224.900	1.000	−224.900	224.900	1.000
.000	254.000	.000	.000	.000	.000
1.000	304.900	1.000	304.900	304.900	1.000
2.000	359.900	4.000	719.800	1439.600	16.000
3.000	438.800	9.000	1316.400	3949.200	81.000

(WITH ORIGIN AT 1981-82, X UNITS AS 1 YEAR)

TREND EQUATION IS YCAP = 255.2143 + 43.3786 * X + 5.6500 * X2

TREND VALUES ARE

175.929
191.057
217.486
255.214
304.243
364.571
436.200

Stop - Program terminated.

EXAMPLE 5(f), CHAPTER 12

ENTER NO OF OBSERVATIONS

6

ENTER VALUES OF Y

1.05 1.34 1.26 1.98 2.24 2.28

X	Y	X2	XY	X2Y	X4
−5.000	1.050	25.000	−5.250	26.250	625.000
−3.000	1.340	9.000	−4.020	12.060	81.000
−1.000	1.260	1.000	−1.260	1.260	1.000
1.000	1.980	1.000	1.980	1.980	1.000
3.000	2.240	9.000	6.720	20.160	81.000
5.000	2.280	25.000	11.400	57.000	625.000

(WITH ORIGIN MIDWAY BETWEEN 1941-51 and 1951 - 61, X UNITS AS 1/2 YEAR)

TREND EQUATION IS YCAP = 1.6859 + 0.1367 * X + 0.0005 * X2

TREND VALUES ARE

1.015

CP 7

```
C     PROGRAM TO FIT EXPONENTIAL TREND CURVE
C     OF TYPE Y = A * B * * X TO TIME-SERIES DATA
C     YL = LOG 10 (Y) YLCAP TREND VALUE (LOG)
      INTEGER N, I, MID
      REAL Y (20), X (20), X2, YL, XYL, SUMY, SUMYL, SUMXYL
      REAL ACAP, BCAP, YLCAP, A, B
      DATA SUMY, SUMX2, SUMYL, SUMXYL / 4 * 0.0/
      WRITE (*, *) 'ENTER NO OF OBSERVATIONS'
      READ (*, *) N
      WRITE (*, *) 'ENTER VALUES OF Y'
      READ (*, *) (Y (I), I = 1, N)
      IF (MOD (N, 2). EQ. 0) THEN
C        EVEN NO. OF OBSERVATIONS X UNITS 1/2 YEAR
         DO 10  I = 1, N
            X (I) = 2 * I - N - 1
10       CONTINUE
      ELSE
C        ODD NO OF OBSERVATIONS X UNITS 1 YEAR
         MID = (N + 1) / 2
         DO 20  I = 1, N
            X (I) = I - MID
20       CONTINUE
      ENDIF
      WRITE (*, *)'          X          Y          X2          LOGY          XLOGY'
      DO 30  I = 1, N
         X2 = X (I) * X (I)
         YL = ALOG 10 (Y (I))
         XYL = X (I) * YL
         SUMY = SUMY + Y (I)
         SUMX2 = SUMX2 + X2
         SUMYL = SUMYL + YL
         SUMXYL = SUMXYL + XYL
         WRITE (*, 33) X (I), Y (I), X2, YL, XYL
30    CONTINUE
      ACAP = SUMYL / FLOAT (N)
      BCAP = SUMXYL / SUMX2
      A = 10 * * ACAP
      B = 10 * * BCAP
      WRITE (*, 11) ACAP, BCAP
      WRITE (*, 44) A, B
      WRITE (*, *) 'TREND VALUES ARE          YCAP          YCAL'
      DO 40  I = 1, N
      YLCAP = ACAP + BCAP * X (I)
      AYCAP = 10 * * YLCAP
      WRITE (*, 22) YLCAP, AYCAP
40    CONTINUE
11    FORMAT (/ / 1X, 'TREND EQUATION (LINE) IS LOG 10Y = ', F10.4,' + ',
    1 F10.4, ' * LOG 10 X')
22    FORMAT (20X, 2F 10.3)
33    FORMAT (1X, 5F 10.3)
44    FORMAT (/ / 1X,' TREND EQUATION (EXPO) IS Y = ', F10.3, ' * ' , F10.3,
    1 * * X' / /)
      STOP
      END
```

CP 8

```
C     PROGRAM TO COMPUTE CORRELATION COEFFICIENT AND
C     TO FIT REGRESSION LINE FROM GROUPED DATA
C     K = NO. OF X CLASSES, M = NO. OF Y CLASSES
C     C = C. I. OF X, D = C.I. OF Y
C     Xi, Yi = MID VALAUE OF TH X, Y CLASS
      INTEGER I, J, KP1, MP1, K, M, C, D, MIDX, MIDY
      INTEGER F (20, 20), U (20), V (20)
      REAL X (20), Y (20), SUMHV, SUMHV2, SUMGU, SUMGU2, SUMFUV
      REAL XBAR, YBAR, T1, T2
      DATA SUMHV, SUMHV2, SUMGU, SUMGU2, SUMFUV/5 * 0.0./
      WRITE (*, *) 'ENTER NO OF X, Y CLASSES, CI OF X, Y (K, M, C, D)'
      READ (*, *) K, M, C, D
      WRITE (*, *) 'ENTER THE MID-VALUES OF X CLASSES'
      READ (*, *) (X (I), I = 1, K)
      WRITE (*, *) 'ENTER THE MID-VALUES OF Y CLASSES'
      READ (*, *) (Y (I), I = 1, M)
      KP1 = K + 1
      MP1 = M + 1
      MIDX = X (KP1/2)
      MIDY = Y (MP 1/2)
      WRITE (*, *) 'ENTER FREQUENCIES F'
      DO 10 I = 1, M
          DO 10 J = 1, K
              READ (*, *) F (I, J)
              F (I, KP1) = F (I, KP1) + F (I, J)
              F (MP1, J) = F (MP1, J) + F (I, J)
              N = N + F (I, J)
10    CONTINUE
      DO 20 J = 1, K
          U (J) = (X (J) - MIDX)/C
          GU = F (MP1, J) * U (J)
          SUMGU = SUMGU + GU
          SUMGU2 = SUMGU2 + GU * U (J)
20    CONTINUE
      DO 30 I = 1, M
          V (I) = (Y (I) - MIDY) /D
          HV = F (I, KP1) * V (I)
          SUMHV = SUMHV + HV
          SUMHV2 = SUMHV2 + HV * V (I)
30    CONTINUE
      DO 40 I = 1, K
      DO 40 J = 1, M
              SUMFUV = SUMFUV + F (J, I) * U (I) * V (J)
40    CONTINUE
      XBAR = MIDX + C * SUMGU / FLOAT (N)
      YBAR = MIDY + D * SUMHV/FLOAT (N)
      RNUM = FLOAT (N) * SUMFUV - SUMHV * SUMGU
      T1 = FLOAT (N) * SUMHV2 - SUMHV * SUMHV
      T2 = FLOAT (N) * SUMGU2 - SUMGU * SUMGU
      R = RNUM/SQRT (ABS (T1 * T2)
      BCAP = RNUM/T1
      DCAP = RNUM / T2
      WRITE (*, 11) R
      WRITE (*, 22) YBAR, BCAP, XBAR
      WRITE (*, 33) XBAR, DCAP, YBAR
```

```
11      FORMAT (//1X, 'CORRELATION COEFFICIENT R = ', F 12.6 //)
22      FORMAT (//1X, 'REGRESSION LINE OF Y ON X IS' /10X, ' (Y - ', F 10.3,')
     * = ', F 10.5, ' (X - ', F 10.3, ')' //)
33      FORMAT (//1X, 'REGRESSION LINE OF X ON Y IS' /10X,' (X - ', F 10.3, ')
     * = ', F 10.5, ' (Y - ', F 10.3, ')' //)
        STOP
        END
```

ILLUSTRATION : EXAMPLE 8(a), CHAPTER 9

```
ENTER NO OF X, Y CLASSES, CI OF X, Y (K, M, C, D)
6, 7, 5, 3
ENTER THE MID-VALUES OF X CLASSES
152, 157, 162, 167, 172, 177
ENTER THE MID-VALUES OF Y CLASSES
52, 55, 58, 61, 64, 67, 70
ENTER FREQUENCIES F

 1   1   0   0   0   0
 1   2   1   0   0   0
 0   2   2   5   0   0
 0   1  15  23   1   0
 0   0   6  18   1   0
 0   0   1   3   7   1
 0   0   0   1   3   4


CORRELATION COEFFICIENT R = 0.736893

REGRESSION LINE OF Y ON X IS

(Y - 62.500) = 0.56135 (X - 165.950)

REGRESSION LINE OF X ON Y IS

(X - 165.950) = 0.96733 (Y - 62.500)

Stop - Program terminated.
```

Index